Embedded Systems Design Using the Rabbit 3000 Microprocessor

Embedded Systems Design Using the Rabbit 3000 Microprocessor

Interfacing, Networking and Application Development

By

Kamal Hyder
Bob Perrin

AMSTERDAM • BOSTON • HEIDELBERG • LONDON
NEW YORK • OXFORD • PARIS • SAN DIEGO
SAN FRANCISCO • SINGAPORE • SYDNEY • TOKYO

Newnes is an imprint of Elsevier

Newnes is an imprint of Elsevier
30 Corporate Drive, Suite 400, Burlington, MA 01803, USA
Linacre House, Jordan Hill, Oxford OX2 8DP, UK

Copyright © 2005, Elsevier Inc. All rights reserved.

No part of this publication may be reproduced, stored in a retrieval system, or transmitted in any form or by any means, electronic, mechanical, photo-copying, recording, or otherwise, without the prior written permission of the publisher.

Permissions may be sought directly from Elsevier's Science & Technology Rights Department in Oxford, UK: phone: (+44) 1865 843830, fax: (+44) 1865 853333, e-mail: permissions@elsevier.com.uk. You may also complete your request on-line via the Elsevier homepage (http://elsevier.com), by selecting "Customer Support" and then "Obtaining Permissions."

 Recognizing the importance of preserving what has been written, Elsevier prints its books on acid-free paper whenever possible.

Library of Congress Cataloging-in-Publication Data

(Application submitted.)

British Library Cataloguing-in-Publication Data
A catalogue record for this book is available from the British Library.

ISBN: 0-7506-7872-0

For information on all Newnes publications
visit our Web site at www.books.elsevier.com

04 05 06 07 08 09   10 9 8 7 6 5 4 3 2 1

Printed in the United States of America

Internet Explorer, HyperTerminal, and Windows are trademarks of Microsoft Corporation. Dynamic C and RCM34xx are trademarks of Rabbit Semiconductor. Softools is the trademark of Softools, Inc. All other trademarks are the property of their respective owners. Readers should contact the appropriate companies for more complete information regarding trademarks and registration.

Dedications

Kamal Hyder

To my parents Rasheed and Najma, who supported my vision of coming to America.

To my wife Mariam, whose support and patience got me through the long nights of work.

To my friend and mentor Anugrah, who showed me through his life that anything is achievable through sincerity and perseverance.

Bob Perrin

For my mom, who started it all.

For my wife, who never lets up.

Contents

Preface xi
   Organization xi
   Example Projects xii
Acknowledgments xiii
Chapter 1: Introduction 1
   1.1 Embedded Systems and Embedded Controllers 1
   1.2 Embedded Systems—Case Studies 3
   1.3 Available Off-the-Shelf Solutions 10
   1.4 Software Development Tools 14
   1.5 Design Trade-offs 16
   1.6 Migration to Higher Volume Production 17
Chapter 2: The Basics 19
   2.1 Evaluating Controllers 19
   2.2 Defining the Problem 28
   2.3 A Survey of Solutions 29
   2.4 A Rabbit's Roots 36
   2.5 Rabbit in Detail 38
   2.6 In Summary 64
Chapter 3: Starting Out 65
   3.1 Introduction to the RCM3200 Rabbit Core 65
   3.2 Introduction to the Dynamic C Development Environment 66
   3.3 Brief Introduction to Dynamic C Libraries 67
   3.4 Memory Spaces in Dynamic C 68
   3.5 How Code is Compiled and Run 76
   3.6 Setting Up a PC as an RCM3200 Development System 79
   3.7 Time to Start Writing Code! 79
   3.8 What's Next? 91
Chapter 4: Debugging 92
   4.1 The Zen of Embedded Systems Development and Troubleshooting 92

Contents

 4.2 Avoid Debugging Altogether—Code Smart 97
 4.3 Common Problems 98
 4.4 Dynamic C Debugging Tools 101
 4.5 Isolating the Problem.. 105
 4.6 Run-Time Errors... 109
 4.7 Miscellaneous Advanced Techniques 111
 4.8 Final Thoughts.. 115

Chapter 5: Interfacing to the External World 116
 5.1 Introduction 116
 5.2 Digital Interfacing 116
 5.3 High Current Outputs 130
 5.4 CPLDs and FPGAs... 141
 5.5 Analog Interfacing—An Overview 143
 5.6 Conclusion.. 156

Chapter 6: Introduction to Rabbit Assembly Language 157
 6.1 Introduction to the Rabbit 3000 Instruction Set 158
 6.2 Some Unique Rabbit Instructions... 178
 6.3 Starting to Code Assembly with Dynamic C.. 180
 6.4 Passing Parameters Between C and Assembly 189
 6.5 Project 1: Creating a Delay Routine 196
 6.6 Project 2: Blinking an LED 200
 6.7 Project 3: Debouncing a Switch 205
 6.8 Project 4: Driving a Multiplexed LED Display.. 211
 6.9 Project 5: Setting Up a Real-time Clock 221

Chapter 7: Interrupts Overview 225
 7.1 Interrupt Details.. 228
 7.2 Writing an Interrupt Service Routine 236
 7.3 Project 1: Polled vs. Interrupt-Driven Serial Communication 244
 7.4 Project 2: Using Timer Interrupts... 254
 7.5 Project 3: Using the Watchdog Timer.. 270
 7.6 Project 4: Setting Up a Real-time Clock 280

Chapter 8: Multitasking Overview 288
 8.1 Why Use Multitasking?.. 288
 8.2 Some More Definitions.. 295
 8.3 Cooperative Multitasking 297
 8.4 Preemptive Multitasking 298
 8.5 What to Be Careful About in Multitasking.. 301
 8.6 Beginning to Multitask with Dynamic C 305
 8.7 Dynamic C's Implementation of Cooperative Multitasking 306
 8.8 Dynamic C's Implementation of Preemptive Multitasking....................................... 310
 8.9 Project 2: Flashing LEDs with Multitasking 312

8.10 Project 3: Using Linux to Display Real Time Data 321
8.11 Project 4: Designing an Analog Sensor Task 326
8.12 Back to the State Machine from Project 1 331
8.13 Final Thought 333

Chapter 9: Networking 334
9.1 Dynamic C Support for Networking Protocols 335
9.2 Typical Network Setup 338
9.3 Setting up a Core Module's Network Configuration 340
9.4 Project 1: Bringing up a Rabbit Core Module for Networking 344
9.5 The Client Server Paradigm 348
9.6 The Berkeley Sockets Interface 349
9.7 Using TCP vs. UDP in an Embedded Application 352
9.8 Important Dynamic C Library Functions for Socket Programming...................... 353
9.9 Project 2: Implementing a Rabbit TCP/IP Server 355
9.10 Project 3: Implementing a Rabbit TCP/IP Client.. 361
9.11 Project 4: Implementing a Rabbit UDP Server 369
9.12 Project 5: Web Enabling the Sensor Routine... 374
9.13 Project 6: Building an Ethernet-Connected Sprinkler Controller 384
9.14 Some Useful (and Free!) Networking Utilities 406
9.15 Final Thought 409

Chapter 10: Softools—The Third Party Tool 410
10.1 Who is Softools? 410
10.2 The Rabbit WinIDE 411
10.3 SCRabbit Optimizer 417
10.4 SCRabbit Segments 419
10.5 SCRabbit #pragmas 419
10.6 Near and Far Functions 421
10.7 Inline Assembly 423
10.8 Library Support 423
10.9 WinIDE's SLINK Linker... 424
10.10 Debugging in the WinIDE 426
10.11 Memory Layout 430
10.12 Real Time Operating Systems 435
10.13 Ethernet and TCP/IP 436
10.14 WinIDE and the Book's Example Programs 436
10.15 Conclusion 437

Appendix A: Rabbit 3000A—Extending the Rabbit 3000's Architecture 438
About the Authors 449
Index 451

Preface

Welcome! Are you new to Rabbit Semiconductor's products? Are you new to networking? Are you new to embedded controller design? Then this book is for you.

This book is written by embedded developers for their peers. The authors asked each other "if we were starting to design with a new microprocessor today, what would we want to know about it? How would a book help us achieve an efficient design quickly?" and developed the book accordingly. A number of concepts presented here are not just specific to the Rabbit 3000 microprocessor; they are equally applicable to any microprocessor.

Organization

The book starts simple and brings readers along quickly to a level where they can assemble hardware, wiggle bits, and blink lights. Then the real fun begins—web-enabling embedded controllers.

The first two chapters introduce the key concepts needed for embedded system design. Next, the reader is given an architectural overview of the Rabbit 3000 microprocessor and introduced to an easy-to-use development environment—Dynamic C. Simple and advanced debugging techniques are covered with examples.

Chapter 5 explains common hardware interfacing issues. Since we believe that hardware and software in embedded systems are inexorably woven together, we use a mix of freely available Linux-based tools to show how analog data can be recovered from an embedded controller and analyzed.

Chapters 6, 7 and 8 take the reader on a succinct tour of Rabbit assembly language, interrupts and multitasking. This is where readers familiar with embedded system design but not with the Rabbit will probably want to start.

Chapter 9 is a comprehensive treatment of how to bring the web to an embedded system. We have done projects ranging from simple UDP and TCP clients and servers on both PCs and Rabbit Core Modules to a data-acquisition system and an automated sprinkler controller with a web browser interface. This chapter also introduces RabbitWeb, a novel and powerful scripting language that makes it easy to create powerful web interfaces.

Chapter 10 introduces a very powerful and professional development environment for Rabbit 3000 code—the Softools ANSI C compiler. Softools brought almost two decades of

Preface

experience with optimizing compilers, assemblers and very clever linkers together to create an easy-to-use development environment. We'll discuss it at length.

The book closes with Appendix A which covers the enhancements made to the Rabbit 3000 with the release of the Rabbit 3000A. Both processors are fully compatible with all of the code and examples used throughout this book.

Example Projects

The authors firmly believe that people learn by example. This philosophy pervades the book. Each concept, once introduced and discussed, is used in a project. The projects range from simple assembly code that blinks an LED to a web-enabled sprinkler controller.

Today, there are many software tools available to engineers. Recognizing that not all engineers are proficient with all tool sets, we have included examples in languages ranging from assembly language to Java, Perl, C/C++/C#, and Bash. Target environments include Windows and Linux and of course the Rabbit Semiconductor hardware. The idea is that every reader will likely take away tricks for tools that here-to-fore have not been in their toolbox.

The authors have deliberately chosen simple examples, so that the reader would focus on the key concepts instead of getting bogged down with implementation details of complex projects. For example, the objective of a number of examples is just to flash an LED. While this may seem overly simple, flashing an LED requires the right I/O ports to be set up and the right logic and timing to be in place. These are key elements in most embedded systems designs.

Our intention in writing this book is to bring to the reader our sense of excitement for embedded systems design as well as embarking on an adventure with the Rabbit 3000 as our faithful companion. We hope our enjoyment and adventurism will rub off on you. May you enjoy reading this book as much as we have enjoyed writing it.

Sincerely,

Kamal Hyder and Bob Perrin
September 2004

Acknowledgments

This book has benefited from contributions great and small from a long list of professionals. Here we take a moment to recognize and thank the following individuals for their contributions.

Norm Rogers and Carrie Maha (Rabbit Semiconductor)
Thank you for supporting and encouraging the authors thoughout the development of this book and for arranging for quick and in-depth technical support when the authors had questions.

Bill Auerbach (Softools Inc)
Thank you for your time to review our work and offer suggestions on improving it. But most of all, thank you for writing the bulk of Chapter 10—Softools. When you want a job done right, go to an expert.

Kelly Hall (Saint Bernard Software)
Thank you for contributing the Linux based data acquisition projects (code, data and prose) found in Chapter 5 and Chapter 8.

Greg Young (Progressive Solutions)
Thank you for helping the authors maintain a good perspective on this project.

Pedram Abolghasem (Rabbit Semiconductor)
Thank you for being unafraid to tackle any issue at Rabbit Semiconductor for us. You were man on point for us at Rabbit. Thank you.

Brian Murtha (Rabbit Semiconductor)
Thank you for contributing most of the technical information about debugging in Dynamic C found in Chapter 4. Especially, thank you for writing the FASTSERMACS.LIB for this book. Readers will be using these macros for years! Nice work, Brian.

Larry Cicchinelli (Rabbit Semiconductor)
Thank you for reviewing the first seven chapters of this book. Sometimes you turned your edits around in just a day or so. You answered a lot of questions for us, and this book would not be what it is without your input. Thank you.

Eugene Fodor, Joel Baumert, Owen Magee and Steve Hardy (Rabbit Semiconductor)
The Rabbit Semiconductor engineers really stepped up to the plate as reviewers for the last half of the book. But mostly, your advice, support, assistance and dedication helped us work though a difficult piece of work—Chapter 9. Thank you.

Acknowledgments

Qingyi H. Perrin
For photography in Chapters 1, 5 and 9.

Raymond D. Payne (Ebara Technologies)
For all the hours you have spent researching, understanding and teaching AC snubbing techniques. Chapter 5 and Chapter 9 have benefited from your experience and wisdom. Thank you for being generous with your time.

Scott Henion (SH Designs)
For helping the authors understand the Rabbit 3000 Timer B, and for graciously allowing us to include your ST-timerb.zip and timerb.ZIP in our CD. These libraries and examples will be useful to many readers.

Carol Lewis and Tiffany Gasbarrini (Elsevier Science and Technology Books)
For midwifing this book.

Michael Caisse
For his careful review of Chapter 9 and for moral support throughout the development of the book.

Karan Bajaj
For helping us verify the networking code. His background with Microsoft Corporation and .net development really helped.

Prameela Mukkavilli
For helping us verify the networking code. Her Java background was very helpful.

Baktha Muralidharan (Cisco Systems)
For reviewing Chapter 9. He is a software engineer with Cisco Systems and has contributed towards the popular Ethereal tool.

Aamer Akhter (Cisco Systems)
For reviewing Chapter 9. He is a technical leader with Cisco Systems and has contributed towards the popular Ethereal tool.

Marco Molteni (Cisco Systems)
For reviewing Chapter 9. He is a Software Engineer with Cisco Systems and works on future platforms and IPv6. He tries to contribute back to Open Source tools like Ethereal.

Elias Kesh (Pioneer Electronics)
For reviewing Chapter 8. He has designed and implemented a number of RTOSes and is active with embedded Linux development.

CHAPTER 1

Introduction

This book focuses on methods and practices for embedded system design, and it takes a top-down approach in presenting the material. The discussion will progress from questions of what an embedded system is, to detailed examples of how to solve specific design problems with Rabbit Semiconductor's technology.

This chapter examines broad issues surrounding embedded development and discusses common solutions. Chapter 1 is not concerned with specific technology but rather over-arching issues of embedded system development. The chapter narrows the scope of embedded system problems to those with which the remainder of the book is concerned.

1.1 Embedded Systems and Embedded Controllers

In the 1960s, mini-computers found their way into dedicated control applications. Someone coined the phrase "OEM computers" to differentiate these machines from "business computers." As time passed and technology shrunk, the expression "dedicated controller" came into vogue and was promptly supplanted by "embedded controller."

The phrase "embedded system" was coined to describe systems that contain an embedded controller. Nowadays, all but the simplest electronic devices have some sort of microprocessor in them. Hence today, the phrase "embedded system" describes almost any electronic product.

The embedded community has adapted various industry "computer" standards for use in industrial control systems. For example, the PCI bus has four common form factors. The first is the "standard" PCI form factor found in desktop PCs. The second is the stackable PC104+ form factor. The third is the rack mountable CPCI (Compact PCI). The fourth is the PCI Industrial Computer Manufacturers Group (PICMG) adaptation.

PC104+, CPCI and PICMG are all attempts to adapt a "computer" technology for industrial use. The electrical specifications are almost identical, but the form factors are significantly altered.

Until recently, embedded PCs have been specialized PCs. For example, Ampro invented the PC104 form factor to allow PCs to be squeezed into a physical envelope more conducive to embedded applications than a full-sized desktop PC motherboard. These embedded PCs have been easily identified as "embedded controllers" and are quite distinct in form from desktop computers.

In 2002, the Mini-ITX form-factor x86 motherboards hit shelves everywhere. Mini-ITX was developed ostensibly to provide a smaller footprint for desktop PCs. Embedded systems

Chapter 1

designers seized on the low cost, high volume miniaturized full-up PC motherboards for control applications. Companies began packaging the Mini-ITX motherboards with power supplies and I/O mixes suitable for embedded applications. What was initially designed as a "desktop computer" is now serving as an embedded controller. Let's take a closer look at the evolution from desktop to embedded PC.

Figure 1.1 shows a PC motherboard. To make a complete system, several PCI or ISA cards must be added to provide video and I/O. Of course, a power supply is required. A hard drive is required to store an operating system and application software. Enclosures are available to rack-mount this type of system, but most cases are designed for consumer use.

Figure 1.1: A desktop PC motherboard is big, and not very functional without additional cards, power supply and hard disk.

Figure 1.2 shows a PC104 stack. The system is shown with a power supply card, processor card and five expansion cards that provide video, storage, parallel I/O and numerous serial channels. The super-rugged enclosure is made of a thick aluminum extrusion and uses dense rubbery rails to isolate the electronics from vibration and shock. The PC104 stack is an example of how PC technology was adapted to an embedded form factor.

Figure 1.2: A PC104 stack is more compact and rugged than a desktop PC.

Figure 1.3 shows a JK Microsystems Mini-ITX based embedded PC. This system's footprint is little larger than a compact disc. The system has a power supply board that takes 7–30 volts DC as an input.

Storage is provided by up to two Compact Flash cards. These rugged, solid-state devices are more consumer technology suitable for use in embedded systems. They are lightweight and rugged. If more storage is required, Type II Compact Flash mechanical hard drives can be used.

A watchdog timer is also provided on the same PCB that contains the power supply and Compact Flash

Figure 1.3: The Mini-ITX was adopted directly into the embedded systems market.

connectors. The watchdog timer is a device that many embedded systems contain to improve reliability. In the event of a software crash, the watchdog timer will reset the system.

A compact disc and hard drive can also be added. Figure 1.3 shows these devices installed. Both devices were designed for laptop computers. They are lightweight and tolerate environmental stresses gracefully. The popularity of laptops has pushed the prices of these components almost as low as desktop PC components.

The processor on the motherboard is a fanless low power device. This too is coincidentally well-suited to the embedded market.

JK Microsystems sells the embedded PC shown in Figure 1.3 with an aluminum enclosure. I/O is limited to keyboard, mouse, video, Ethernet, USB, serial and parallel ports, but can be expanded using the PCI slot shown. The overall system is less expensive than a similar PC104 system shown in Figure 1.2.

In just a few years, the embedded PC market has gone from having ill-suited desktop PC technology to having expensive but rugged PC104 technology and currently to having inexpensive compact rugged PCs. With the pressures and economic realities of consumer markets, PC technologies can be expected to become less expensive, smaller and more rugged. The embedded systems sector will certainly adopt these technologies directly.

Technology's perpetual march is oblivious to the delicacies of human semantics. Today, the distinction between embedded controller and computer is rapidly blurring. We can expect to see more and more "computer" technologies adopted directly into embedded control applications.

An engineer faced with automating a process or building an instrument has access to a wide variety of products. There are embedded controllers available as printed circuit boards (PCBs), as packaged controllers, or as hundreds of flavors of embedded PCs. There are microprocessors and microcontrollers that range from 8-pin devices costing less than a dollar to many hundred-pin ball grid array (BGA) packaged devices costing hundreds of dollars. If none of these devices suit a particular application, perhaps the developer might fancy a half a million-gate field programmable gate array (FPGA).

The term "embedded system" is applied to everything from coffee makers to communications satellites. All of these systems have some form of microprocessor lurking behind the scenes orchestrating behavior.

This book focuses mainly on small- to medium-scale embedded systems and associated instrumentation. These type of systems share many common attributes. There are sensors. There are actuators. There are desired behaviors. There are human interfaces—Man Machine interfaces or MMIs. Above all, there is an embedded controller operating behind the scenes, tying everything together.

1.2 Embedded Systems—Case Studies

To help the reader understand the scope of embedded applications with which this book is concerned, three systems are detailed here. The systems as presented here have been simplified from their actual implementations both to protect proprietary intellectual property and for brevity.

Chapter 1

The applications range from an underwater torque tool to a 30 megawatt generator. Each application presented its designers with unique problems. The three projects share common threads. Each project shows different methods for addressing specific control problems. These projects give the reader a flavor of the diversity in the embedded control industry.

1.2.1 Underwater Torque Tool

Four hundred meters below the North Atlantic is most inhospitable. It's also home to sizable petroleum reserves. In an increasingly audacious quest for oil, humans have run pipelines and placed wellheads deep under the ocean.

One task that must be performed is the simple act of rotating mechanical valves on the sea floor. One technique used is to mount a hydraulic torque tool on the end of a remotely operated vehicle's (ROV's) manipulator. Fly the ROV and tool down to the site. Insert the torque tool in the valve manifold. Turn the valve.

The combination of high-cost equipment and environmental impact makes turning underwater valves a more considered task than turning on a garden hose. If the wrong valve is turned, or rotated the wrong direction, or moved the wrong amount, or is over-tightened, or is stuck or is broken, the environmental impact can be disastrous and the economic costs staggering.

The system described here was designed to rotate 28-inch ball valves with up to a quarter-million foot-pounds of torque.

The torque tool is a robot and contains sensors to monitor pressures, strain, speed and temperatures. The on-board computer communicates through an umbilical to the surface ship. Electrically controlled hydraulic valves control the tool's actions.

Figure 1.4 shows a block diagram of the tool. The front of the tool has the coupler and latches. Undersea valves have a port designed to capture torque tools. This port is referred to as a "bucket." The latches engage the bucket to secure the tool. This also provides a stationary anchor for the tool to press against as it produces torque on the valve.

To ensure that the correct valve is turned, each valve has a different orifice geometry. Much like common screws require a slotted or crossed screwdriver to turn the head, underwater valves require different couplers. The coupler can be changed to accommodate different valves.

The ROV pilot flies the tool down to the valve. Using a video camera, the manipulator arm is used to position the torque tool's nose in the bucket. Next, the tool operator sends a command to the tool engaging the latches securing the tool in the bucket. If conditions are right, the valve rotation can commence.

An embedded controller inside the torque tool monitors conditions in the tool, provides a communication link with the ship and directs the tool's behavior. This particular tool uses an industrial PC with stackable I/O cards, a PC104 stack.

The ROV contains a hydraulic pressure unit (HPU), and the pressurized fluid is delivered to the tool through a high-pressure hose. Hydraulic pressure data is acquired through sensors and monitored by the PC104 stack.

Hall-effect sensors on the gearbox monitor the motor and coupler speed. Strain gauges monitor the actual torque applied to the valve.

Introduction

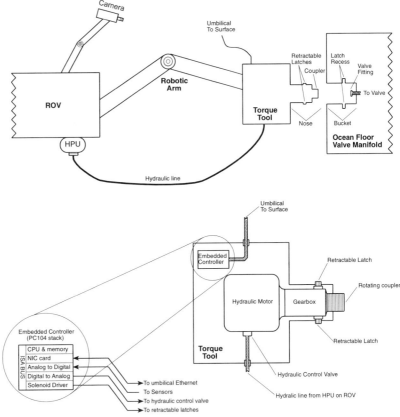

Figure 1.4: Underwater torque tool designed to turn valves.

The PC104 stack constantly watches the sensors. If any problem is detected, the tool is shut down and the operator alerted.

The inclusion of an embedded controller in the tool allows the implementation of algorithms to gradually ramp up torque and speed on the valve. It also allows for an operator to command the valve movement precisely. If a valve is stuck, the operator can command the tool to apply specific torques for specific times or angular distances.

The PC104 stack ties together numerous sensors and actuators, allowing high forces to be applied to undersea valves with safety and precision.

A PC was chosen as the embedded platform. The developers desired a multi-tasking Linux operating system. This allowed code development to occur on inexpensive desktop PCs with a low initial investment in software tools. Another big factor was the Ethernet connectivity supported by Linux.

The tool can be operated through an IP (internet protocol) network from a geographically remote location. The tool operator does not have to be aboard the ship to control the tool. TCP/IP packets are easily routed over the Internet.

In fact, during initial debugging sessions, the tool was deployed from a ship at sea while the torque tool operator was located in a cubicle 3,000 miles away. This configuration allowed the R & D team to interrogate the tool, operate it and update software on the tool, all over the Internet. This arrangement was less expensive than flying the R & D team and their lab to a ship.

A derivative design of this controller used a serial port to communicate between the tool and the ship. Point-to-point protocol (PPP) was used to route TCP/IP packets over the serial connection, while still retaining the diverse network features of the design.

1.2.2 Industrial Vacuum Pump for Semiconductor Processing Equipment

The process of turning silicon wafers into silicon chips has many steps. Some of these operations are carried out in low-pressure environments. There are vacuum pumps designed specifically to create low-pressure environments for use in wafer processing.

A silicon wafer can yield hundreds of individual dice (or chips). Depending on complexity, each die may be worth a considerable sum of money. Each wafer is worth several hundred times the price of a die.

During processing, wafers are ganged together in carriers or caddies. Each caddy carries tens of wafers. A caddy of silicon wafers increases in value as it moves through the fabrication process. By the time the caddy is halfway through a process run, it is not unusual for the value of the caddy's contents to be several hundred thousand dollars.

Silicon fabrication plant operators consider it "bad form" on the part of process tool vendors to allow a process tool, such as a vacuum pump, to ruin a batch of wafers. Considerable care goes into the design of vacuum pumps destined for silicon fabrication plants.

Figure 1.5 shows a block diagram of an industrial vacuum unit. In the unit, two pumps are cascaded to develop the low pressures required by the wafer fabrication process. Each pump is driven by an eight-horsepower three-phase electric motor.

The tool chamber is filled with highly toxic vapors released by the process chemicals. Nitrogen gas is mixed with the exhaust gasses to dilute the toxic gases to safer levels.

The pumps require water-cooling to maintain acceptable operating temperatures. If the pump housing is too hot or is unevenly heated, the mechanical parts fail to stay in tolerance and leaks develop—such leaks can reduce vacuum in the process chambers, which can in turn upset the process chemistry and diminish wafer yield.

Flow sensors monitor the nitrogen and water supplied to the unit. A disruption in the flow of nitrogen is a safety concern. A disruption in the flow of water affects pump efficiency and mechanical wear.

Thermocouples monitor pump housing and bearing temperature. Elevated temperatures indicate excessive friction, or reduced water circulation. Slightly elevated temperatures require a maintenance engineer to review the system, while severely elevated temperatures require the pump to be shut down.

The man-machine interface (MMI) shown in Figure 1.5 consists of an LCD mounted on the front of the equipment cabinet. The MMI has no front panel controls. The inclusion of front panel controls would only open the possibility of a human manually changing the pump behavior and ruining a caddy of wafers. The process is fully automated. The silicon-wafer process-tool controls the vacuum pump through the "control interface."

Introduction

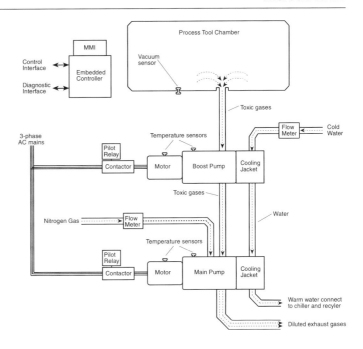

Figure 1.5: Industrial vacuum pump used on silicon fab lines.

The control interface is configured to meet the requirements of the customer. This interface may be a simple digital interface, or a set of dry-contacts, or one of three serial interfaces—RS-485, RS-232 or Echelon's LonTalk™.

The maintenance port provides access to a pump's performance logs and to real-time operating conditions. A service technician uses a laptop with an RS-232 connection to access the pump's maintenance port.

This system is an example of an embedded application that requires limited control functions but numerous monitoring and logging features. The embedded controller can turn the pump motors on and off but has no ability to control the speed of the motors. The temperatures may be monitored and logged but there is no valve affording electronic control over coolant. The performance of this type of system is fixed by the mechanical design. The embedded controller can monitor conditions and shut down the system if a critical fault occurs. System performance can be measured and recorded for later inspection by engineers, but control algorithms aren't needed as part of the embedded controller.

When it came to selecting/designing the pump's embedded controller, the engineers were faced with a build vs. buy decision. The decision factored in cost, size, availability and computational requirements. In the end, the engineers opted to design a custom controller from the ground up.

The embedded controller was tailored to the application. Extra digital and analog channels were included to support future expansion. The controller served the application well for many years.

Chapter 1

1.2.3 Controlling a 30-Megawatt Generator

An embedded controller may only be a discrete subsystem in a larger electronic control project. As an example, Figure 1.6 shows a 30-megawatt natural gas powered turbine-driven generator. These systems were built to provide electricity in remote locations in Mexico, Central America and parts of South America.

Variants of this system replaced the generator with large compressors or other mechanical equipment necessary for industrial facilities to operate in remote locales. However, all variants of this system used the same natural gas feeder valve and valve-controller.

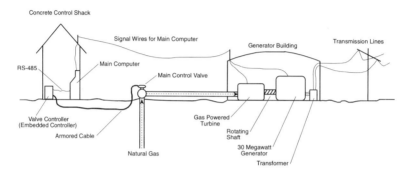

Figure 1.6: An automated natural gas powered electrical generating station.

The overall system required the monitoring and control of a variety of devices associated with the gas turbine and the accompanying generator or compressor. A main computer was responsible for all the high-level data acquisition and control tasks. However, in the case of the natural gas feeder valve, a separate valve-controller was developed to offload some of the low level monitoring and control functions.

The valve-controller allowed the main computer to issue commands such as "open feeder valve to x%," and "emergency off." The details of controlling the valve's stepper motor and monitoring the position feedback were left to the valve-controller.

Figure 1.7 shows a detailed view of the embedded system designed to control the natural gas feeder valve. The gas valve was moved with a large stepper motor and gearbox. There was feedback indicating the position of the valve.

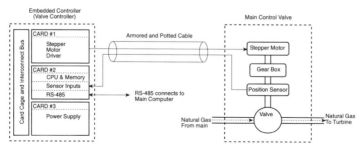

Figure 1.7: Valve-controller block diagram.

Safety requirements dictated that the valve-controller be located in the concrete control room—see Figure 1.6. The armored and potted cable that ran out to the automated valve protected the wires from mechanical damage and prevented natural gas from seeping up the cable and into the control shack.

The system was designed in the late 1980s. The valve-controller consisted of three rack-mounted cards. The first card contained the power-supply. The second card contained a custom controller design based on the 8051 microcontroller, and the third card was a commercially available stepper motor driver.

Embedded systems integrated into larger projects are no less challenging to design than stand alone designs. The engineer is always faced with build vs. buy decisions. As was done with this project, a portion of the design may be bought (the stepper motor controller) while the rest of the system can be built. The process of putting together the pieces is called "system integration."

Some systems can be pieced together entirely by a systems integrator. An example is a desktop PC built up by an electronics enthusiast. The motherboard, PCI cards, case, fans, power supplies, disk drives, keyboards, mouse and monitor are all purchased components. The assembly is fairly mechanical.

Other types of systems integration projects can take a lot of effort. For instance, putting together devices that don't communicate using same protocols or don't have well thought out and well-designed interfaces may consume lots of time. Moreover, testing of systems or components and deploying pilot projects can be nontrivial tasks.

System integration can reduce development time, and in small volume production it can save money. In the case of the valve-controller, buying the stepper motor controller saved design time.

1.2.4 Case Study Summary

Each of these case studies appears on the surface to be quite unique. However, they all share common design challenges. Sensors must be monitored. Communication with other devices is required. Actuators must be energized. Control decisions must be made locally while some decisions are made remotely by humans or probably by another control system.

The sensory tasks for these applications are as demanding as many instrumentation projects. Signal conditioning for a pressure transducer in a hydraulic torque-tool is similar to the circuitry for monitoring pressure (or lack thereof) in a vacuum pump.

Communicating with a valve controller via a command-line interface over RS-485 is very similar to RS-485 communication in a silicon wafer-processing tool. The commands and communications protocols may be different but the underlying physical interface is the same.

The underlying technology required for these applications is common to many other projects. An engineer that understands the issues and trade-offs involved with these case studies can design a wide variety of embedded systems.

Chapter 1

1.3 Available Off-the-Shelf Solutions

Commercially available embedded controllers can be divided into four categories—packaged controllers, board-level controllers, core modules, and chip level devices. Each has advantages and disadvantages.

1.3.1 Packaged Controllers

Commercially available packaged controllers have an enclosure. Packaged control solutions often have I/O targeted at industrial applications. The controller industry offers a range of enclosures from vented plastic to explosion-proof cases. Some packaged controllers have generic MMIs—touch-screens, LCDs and keypads are the most common. Figure 1.8 shows an example of a Z-World packaged controller with a generic MMI.

Figure 1.8: The OP7200 is a full-featured controller with built-in LCD and keypad.

Packaged controllers are the most expensive of the four embedded controller classes. In applications that are not too cost-sensitive, these devices can also provide the quickest solution to automation problems. The systems integrator needs only to mount the controller's enclosure, hook up wires and write some code.

Of the four classes of commercially available controllers, the packaged controllers are the most expensive, but are also the most complete and easy to use solutions. Their cost often causes them to be most economically acceptable in low-volume applications where accelerated time-to-market and reduced development costs can be traded off against higher per-product costs and reduced assembly operations.

1.3.2 Board-Level Controllers

Board-level controllers are the most diverse class of controllers. They range from $2500 PICMG (PCI Industrial Computer Manufacturers Group) PCs to low cost postage-stamp sized controllers. Board-level controllers require the systems integrator to provide physical protection for the electronics. An engineer considering a board-level solution has a vast selection of I/O mixes, processor types, memory capacities and I/O mixes from which to pick. Figure 1.9 shows an example of a feature-rich Z-World board-level controller.

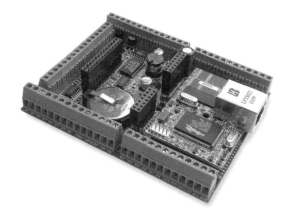

Figure 1.9: The BL2100 offers numerous I/O points and Ethernet.

Introduction

Board-level controllers require mounting, some form of physical protection, and sometimes cooling. Many times an equipment enclosure, suitable for mounting the embedded controller, already exists in an industrial system. Other applications may require the developer to design or purchase a separate enclosure. Some board-level controller manufacturers offer optional enclosures for their board level products.

Connecting wires to board level controllers requires careful attention. Screw terminals are a popular choice for wire termination. Figures 1.10a and 1.10b show examples of two types of screw terminals. These are easy to prototype with but can make production wiring harnesses difficult or time consuming to attach to the controller.

Screw terminal connectors that are fixed to the controller board, as shown in Figure 1.9 and Figure 1.10b, require the production staff to mount the controller first and then to screw wires into the controller. This means that people making the wiring harness can't complete the wire termination until the harness is installed with the controller. Depending on the product and manufacturing process, this may not be desirable.

Screw terminal connectors that are fixed to the controller board can also pose challenges to field service technicians. A technician may want to swap out a controller with a known good unit. Having to unscrew a large number of screw terminals, making sure that a large number of wires are marked for proper reattachment, and then reattaching all the wires to a new controller is time consuming and potentially error prone.

An attractive alternative to screw terminals that are fixed to a controller board is a class of connectors called "pluggable" screw terminals. Figure 1.10a shows a Z-World packaged controller (PK2500) with pluggable screw terminals. This system allows the wire harness to be terminated with a screw terminal socket that plugs into a fixed header on the controller board. This arrangement allows for rapid assembly of the embedded systems as well as swift and error-free field replacement of the embedded controller. Of course the trade-off for this convenience is a small increased cost.

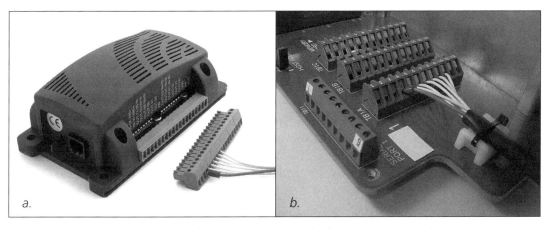

Figure 1.10: Pluggable screw terminals are more expensive but are more flexible than fixed screw terminals.

Systems in which the cable-harnesses will be assembled separately and integrated into a system will benefit from the use of pluggable screw terminals. Completed cable-harnesses can be easily tested. Final assembly will go much smoother if the complete and tested cable-harness is simply plugged into the PCB mounted header.

One ramification of using fixed screw terminals is that as part of final assembly, individual wires from the prefabbed cable-harness must be terminated to the PCB. This can be an error-prone task. Furthermore, testing of the cable-harness is quite difficult with a bunch of loose wires flying around. This implies that the first time the cable harness is tested is in the completed system. Errors will be found, and depending on the wiring error, serious and costly damage can result.

Field service technicians overwhelmingly prefer pluggable screw terminals to fixed screw terminals. When a few wires must be changed, say to replace a sensor, both types of screw terminals are equally well suited. However, when a controller board must be changed, the time required to carefully remove, label and reinstall a lot of individual wires is much longer than simply removing a pluggable screw terminal.

Another popular connector choice for controller manufacturers is the D-Subminiature or D-sub connector. These have limited current carrying capacity, and can be expensive to integrate into production wiring harnesses. On the positive side, shielded D-sub back-shells are available to help minimize EMI.

A very inexpensive connector found on board-level controllers is the pin-header. Depending on the pin size and pitch, reasonable voltage isolation and current carrying capacity can be realized. Crimp pins, insulation displacement connectors (IDC), and mass-termination ribbon cables are available for production wire harnesses. At first glance, pin-headers seem a bit dubious, but they have found widespread industry acceptance.

Board level-controllers in the form of edge cards, epoxy-encapsulated modules, DIP, SIP or SIMM boards require a "carrier-board." The carrier-board will have to be designed to accommodate the purchased controller, while providing a pragmatic method of pinning out the controller's I/O to the devices and sensors that make up the control application.

1.3.3 Core Modules

The third product class is the core module. A core module is a physically small controller consisting of a central processing unit (CPU), memory, glue logic and simplistic input/output (I/O). A core module is built on a printed circuit board (PCB) and is similar to many of the small board-level controllers. The primary feature differentiating a core module from a board level controller is intellectual property (IP).

Board-level products are proprietary controllers. The manufacturer of these devices will not freely license the controller designs. Any attempt to copy a board-level controller infringes on the manufacturer's IP rights.

Core modules are designed to be copied. A core model is in many respects a practical reference design. An engineer can purchase a low-cost core module and develop a complete

Introduction

embedded application. Depending on the economics of the project and where it is in its life cycle, the designer can freely copy the core module design and migrate to a lower cost, higher volume chip level solution.

All microprocessor manufacturers have reference designs available. Many of these are available on demonstration or evaluation boards. These reference designs are designed for use on an engineer's desktop. By contrast, core modules are designed for integration into production systems. Core modules are controllers first and reference designs second.

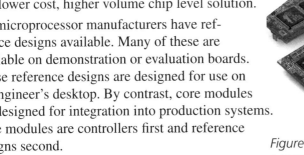

Figure 1.11: Rabbit Semiconductor has a core module for every occasion.

Figure 1.11 shows a sample of the array of available core modules available from Rabbit Semiconductor. As of the time of this writing, Rabbit offers ten core modules, and has more in development.

These are physically small, inexpensive and easily plugged into an application-specific carrier board. Rabbit core modules are great for fast product development. The large assortment of available products ensures there is a core module with features for almost any application.

1.3.4 Chip Solutions

Chip level solutions comprise the last class of controllers. The complexity available varies from very simple to extremely complex. All chip level solutions require a PCB to be designed. Most chips don't offer rugged I/O. The design must protect the delicate ICs against heat, cold, humidity, Electro Static Discharge (ESD), over-voltage, back Electro Magnetic Force (EMF), brown-outs, Electro Magnetic Interference (EMI) and other forms of abuse.

Traditionally, engineers have classified computational chips as being either microprocessors or microcontrollers. This classification isn't particularly useful, and as technology has progressed, has become outmoded.

The classic differentiation between microcontrollers and microprocessors has been on-chip ROM. Microcontrollers have onboard ROM, microprocessors don't. Over time, integrated I/O has also come to be associated with microcontrollers.

The Intel 8051 is possibly the most widely recognized microcontroller architecture. In addition to a modest amount of ROM and RAM, this Harvard architecture sports built-in serial ports, counter/timers and a number of digital I/O pins.

The 8051 has been so successful that companies besides Intel have now released 8051 cores surrounded by on-chip peripherals such as analog-to-digital converters (ADCs), control area network (CAN) interfaces, universal serial bus (USB) ports and others. Some of the 8051 variants permit or even require external memory. The lack of internal ROM in some of these variants technically makes them microprocessors.

Chapter 1

Today, microprocessors routinely include peripherals that have long been identified with microcontrollers. Consider the Rabbit 3000 processor, with six serial ports, real-time clock, internal watchdog, built-in pulse width modulator (PWM), quadrature decoding and 56 digital I/O pins. The requirement for external memory makes it a microprocessor. The Rabbit's I/O feature set makes it more identifiable with control applications than as a numeric processor.

The technology that further blurs the old microprocessor/microcontroller classification scheme is the field programmable gate array (FPGA). Popular CPU designs are available as FPGA-based soft cores. Memory can be provided on-chip or off. Some FPGA families include SRAM cells that can be ganged and initialized at boot to provide program storage While FPGAs are neither microprocessors nor microcontrollers, they can be configured as either.

Originally, the concept of a microcontroller was intended to convey the idea of a "single chip solution." This looked great in marketing brochures, but seldom worked out in copper. The delicate nature of the I/O pins on most microcontrollers required additional interface ICs, as did power conditioning and reset management.

All but the simplest chip-level applications required multi-IC designs.

The trend in chip level design is to make parts as small as possible. This, combined with the demand for high I/O count, and therefore pin count, has pushed IC manufacturers to offer processors in surface mount (SMT) packages. Figure 1.12 shows three Rabbit Semiconductor microprocessors. These are all SMT packages. The Rabbit 2000 is the largest package and is the least powerful part. The Rabbit 3000 is offered in two packages. The smallest package is a ball grid array (BGA) and is shown on the left of Figure 1.12.

Figure 1.12: The Rabbit 2000 and Rabbit 3000 processors have enough I/O that only SMT packaging is offered.

1.4 Software Development Tools

Just as silicon has advanced, so have software development techniques. The old days of writing code on punch cards, toggling in binary bootstrap loaders or keying in hexadecimal opcodes are long gone. The tried, true and tiresome technique of "burn and learn" is still with us, but in a greatly reduced capacity. Most applications are developed using assemblers, compilers, linkers, loaders, simulators, emulators, EPROM programmers and debuggers.

Selecting software development tools suited to a particular project is important and complex. Bad tool choices can greatly extend development times. Tools can cost thousands of dollars per developer, but the payoff can be justifiable because of increased productivity. On the other hand, initial tool choice can adversely affect the product's maintainability years down the road.

For example, deciding to use JAVA to develop code for a PIC® microcontroller in a coffee maker is a poor choice. While there are tools available to do this, and programmers willing to do this, code maintenance is likely to be an issue. Once the JAVA-wizard programmer moves on to developing code for websites, it may be difficult to find another JAVA-enabled programmer willing to sustain embedded code for a coffee maker. Equally silly would be to use an assembler to write a full-up GUI (graphical user interface)-based MMI.

A quick trip to the Embedded Systems Conference will reveal a wide array of development tools. Many of these are ill suited for embedded development, if not for reasons of scale or cost, then for reasons of code maintainability or tool stability.

The two time-tested industry-approved solutions for embedded development are assembly and C. Forth, BASIC, JAVA, PLM, Pascal, UML, XML and a plethora of other obscure languages have been used to produce functioning systems. However, for low-level fast code, such as Interrupt Service Routines (ISRs), assembly is the only real option. For high-level coding, C is the best choice due to the availability of software engineers that know the language and the wide variety of available libraries.

Selecting a tool vendor is almost as important as selecting a language. Selecting a tool vendor without a proven track record is a risk. If the tool proves problematic, good tech-support will be required.

This in no way implies that a small shop isn't capable of producing an excellent tool and providing responsive support. For example, Softools Inc. is a firm with relatively low headcount and an outstanding record of fast and comprehensive technical support. Their ANSI compliant C compiler for the Rabbit microprocessor is a solid tool worthy of consideration for any Rabbit-based project.

Sometimes, tool vendors are little more than new college grads trying to pawn off warmed-over senior projects as development tools. While the tools may look good, dealing with such an embryonic company is a risk.

Public domain tools have uncertain histories and no guarantee of support. The idea behind open source tools is that if support is needed, the user can tweak the tool's code-base to force the tool to behave as desired. For some engineers, this is a fine state of affairs. On the other hand, many embedded software engineers may not know, or even desire to know, how to tweak, for example, a backend code generator on a compiler.

Rabbit Semiconductor and Z-World offer a unique solution to the tool dilemma facing embedded systems designers. Rabbit Semiconductor designs ICs and core modules. Z-World designs board-level and packaged controllers based on Rabbit chips. Both companies share the development and maintenance of Dynamic C™.

Dynamic C offers the developer an integrated development environment (IDE) where C and assembly can be written and blended. Once an application is coded, Dynamic C will download the executable image to the target system over a serial cable. Debugging tools such as single stepping, break points, and watch-windows are provided within the IDE, without the need for an expensive In-Circuit Emulator (ICE).

Between Z-World and Rabbit Semiconductor, all four classes of controllers are available as well as a complete set of highly integrated development tools. Libraries support a file

Chapter 1

system, Compact Flash interfaces, TCP/IP, IrDA, SDLC/HDLC, SPI, I²C, AES, FFTs, and the uCOS/II RTOS.

On of the most attractive features of Dynamic C is that the TCP/IP stack is royalty free. This is unusual in the embedded industry, where companies are charging thousands of dollars for TCP/IP support. If TCP/IP is required for an application, the absence of royalties makes Dynamic C a very attractive tool.

1.5 Design Trade-offs

The stock in trade of engineers is making trade-offs. An embedded systems designer has many options to weigh. Striking a balance for a given project depends primarily on the anticipated production volume of a project.

A product that is going to be manufactured by the millions will have an entirely different set of trade-offs than a product that is going to be manufactured in the thousands. Low volume production has yet a different set of trade-offs.

Something that will have high production volumes like a cell phone will be optimized for the lowest possible manufacturing costs. Up front tooling for packaging will be more than paid for by the high volume sales. Custom liquid crystal displays (LCDs) will be made. Application-specific integrated circuits (ASICs) may be considered.

A product produced in medium quantities, say the tens-of-thousands, will almost certainly be designed using surface mount technology (SMT). Designers will probably stay away from using off-the-shelf controllers that come prepackaged or as a printed circuit board (PCB). Spending the time up front to develop or license a controller design will pay off over the course of product's lifetime in manufacturing. ASICs probably won't be cost effective because the required minimum quantities won't be reached. This type of product will benefit from a chip-level solution.

A product produced in low quantities (up to hundreds of units) has a number of other challenges. Many surface mount parts are available only on reels of thousands of parts. Many small, low volume products benefit from a ready-made embedded controller.

This book deals with projects that will be produced in low to medium quantities.

Many products that aspire to medium quantity production are forced by market conditions to start out being produced as low quantity products. Over the years this has been addressed primarily in two ways, neither of which has been entirely satisfactory.

Some companies have opted to design the products initially for low volume production. Once market demand justifies higher production volumes, the product is redesigned for medium production volumes. This approach incurs two design cycles and the associated costs and risks.

Other companies have opted to design the products initially for higher volume production. This approach costs more time and money up front in the design and during the low-volume phase of the product's life, thereby increasing manufacturing costs.

1.6 Migration to Higher Volume Production

If the initial design incorporated purchased embedded controllers, then migration to fully assembled custom boards is likely to be arduous. Eliminating the off-the-shelf controller from the new design will reduce production costs simply because the company is no longer paying the markup on the parts and the "value added" by the controller vendor for the purchased controller. However, if care is not taken, bugs can be introduced into the new design. Seldom do off-the-shelf controller companies provide complete IP to support this type of migration.

In many cases, the migration phase is never completed. The product is simply produced in higher quantities with the low volume design. The company loses the opportunity to capitalize on the intrinsic savings offered by higher volume production techniques.

The second approach to the migration problem has been for companies to spend the time and money up front to design the product specifically for medium production quantities. This approach often extends the initial development time—sometimes by months.

An important consideration for many projects is "time-to-market." Being able to provide a working proof of concept for investors or a "prototype" for a potential customer often will make or break a project—or company.

This approach will incur decidedly higher manufacturing costs while the product is in the low volume production phase.

The first approach puts off development issues and costs until a product reaches a sufficient level of production to warrant redesign. But design issues and reverse engineering of off-the-shelf embedded controllers can sandbag the migration phase. The second approach spends all the money and time upfront to develop a product ill suited to the manufacturing scale that it will be confined to for the first portion of its life.

Neither solution is wholly satisfactory.

Several year ago a pioneering embedded controller company, Z-World of Davis, California, took a long hard look at how to help customers develop their applications quickly and in a fashion suited to low production techniques, while affording the same customers a seamless migration path to higher quantity fully custom designs. The solution was dubbed the "Smart Core™." Thus, the core module solution was born.

A core module is a small PCB containing a CPU, memory subsystem, reset and watchdog circuits, power supervisors, and firmware. The core module provides I/O signals on headers suitable for insertion into socket strips on the customer's target system. Thus, a developer can purchase a fully functional controller and immediately plug it into the application.

When the customer decides to migrate to a medium volume production board, Z-World will license all the IP associated with the core module to the customer for inclusion in the new "single board" or integrated design. This provides a smooth low-risk transition for the designer between the low volume production and medium volume production phase.

Chapter 1

Seeing the need for improved microprocessors in the embedded market, Z-World's CEO, Norm Rogers, founded Rabbit Semiconductor to carry on innovation in the embedded controller industry at the chip level.

Today, Rabbit Semiconductor produces chips and core-module designs. Rabbit is built on Z-World's experience in the embedded systems arena. This allows Rabbit to tailor IC designs to meet the needs of the embedded system designer. Painstaking research goes in Rabbit chips to optimize them for embedded C code. The Rabbit's I/O mix is based on Z-World's decades of market experience.

Z-World continues as a force in the embedded controller industry. It offers the Rabbit core modules, but the focus is on full-featured controllers. These controllers offer the embedded system designer a smorgasbord of quarter-VGA touch-screens, high-current I/O points, protected digital inputs, analog inputs, analog outputs, industrial relays, and all manner of communication busses. These types of controllers can often be used to provide a complete control solution for low volume applications.

Embedded system designers can now purchase Rabbit microprocessors, Rabbit-based core modules or full-featured controllers from the Rabbit/Z-World fraternity. All of these products share common development tools and IP. This makes design migrations or new design variants quick, easy, low cost and low risk.

The next chapter examines how the Rabbit solution suite stacks up against the many solutions available today.

CHAPTER 2

The Basics

2.1 Evaluating Controllers

Embedded system design boils down to monitoring sensors and actuating devices. Depending on the complexity of the desired behavior, an embedded controller may not be required.

In some cases a sensor may be adequate to control the actuator. In these situations, controllers are redundant. For example, a household light switch can directly control a lamp. These types of systems are not embedded systems.

Some systems require control logic, but not necessarily a microprocessor or microcontroller. If the system's desired behavior can be implemented with simple combinatorial logic, then the system is not considered an embedded system.

If the controller requires sequential logic then the application may rightfully be called an embedded system. After all, a microprocessor is just a glorified finite state machine (FSM).

For example, a digital lock may be implemented with an FSM. For many years, a programmable logic device (PLD) was the usual tool for building an FSM. In the past, PLDs were dramatically less expensive than microcontrollers. However, microcontroller prices have dropped to the level that many engineers have switched from PLDs to small microcontrollers even for simple FSM designs. The simplicity of programming a microcontroller outweighs the ever shrinking cost advantage a PLD may have. As PLDs age, obsolescence is a concern.

Many factors bear on the selection of a controller. Performance, cost and availability are the most often bandied factors. Development tools, time-to-market and product sustainability must also be considered.

2.1.1 Performance and Cost

The overarching criterion when selecting an embedded controller is performance. If a controller cannot handle the required task, then that controller must be discarded.

A close second to performance is cost. If a controller is qualified but too expensive, then it is a poor solution. A mantra for engineers must be "If we can't afford the solution, then it's not a solution."

Comparing the performance and features of competing controllers is almost always an apples-to-oranges comparison. I/O mixes in competing models are seldom equivalent. Some controllers have internal memory, some external, and some both. Devices with internal memory may have different programming models. Comparing controllers, chip level or board level, is neither simple nor scientific. Selecting a controller is an exercise in the art of risk management.

Chapter 2

Consider the case of a simple application requiring only a few hundred assembly instructions and ten I/O pins with an anticipated production run of 1000 units. The application could be implemented in a small FPGA (Field Programmable Gate Array) as a big FSM. The application could be implemented with a small microcontroller like a PIC processor from Microchip Technologies, or with an AVR® from Atmel. The application could also be implemented with a desktop PC by using the PC's parallel printer port.

The FPGA solution is overly difficult for most engineers, and FPGAs are somewhat pricey. If the project is a "one-off," the PC-based solution is perhaps viable. Assuming that the application will require a production run of a thousand units, the PIC or AVR solution looks more attractive.

For the sake of this example, assume the AVR AT90S1200 solution is $0.75 more expensive than a PIC16C55-based solution. If each part is equally capable of performing the control task, which part should be chosen? The cheaper?

The answer is "it depends." The AVR and PIC solutions are not apples-to-apples. Table 2.1 lists some features of each part. The most glaring difference is in the programming model. The PIC solution is OTP (one time programmable) while the AVR part is Flash-based.

Table 2.1: PIC and AVR have similar features.

Feature	PIC16C55	AVR AT90S1200
Digital I/O pins	12	15
On board program space (words)	512	512
On board data space (bytes)	24	0
EEPROM (bytes)	0	64
Counter/timers	1	1
External Interrupts	0	1
Max clock speed	40 MHz	12
Machine Cycles per instruction	1 or 2	1 or 2
Clock cycles per machine cycle	4	1
In System Programmable	Yes	Yes
Programming model	OTP	Flash-based

The question that the designer must answer is: will the extra $0.75 for the AVR's Flash-based programming model buy anything useful for the project? Two facets of the product's life cycle need to be considered—product development, and production.

During product development, code must be written and loaded into a target system. If OTP parts are used, either the IC must be socketed in the development platform, or the project engineers must be experts at soldering/removing ICs in the target system. Another approach to developing for OTP devices is to use an in-circuit-emulator (ICE).

An ICE allows executable code to be loaded into it from a host platform, usually a PC. The ICE also plugs into the target system, usually in a socket intended for the actual microcontroller or microprocessor. The ICE executes the code while wiggling the signals on the target system just like the actual microcontroller would.

For simple devices like the PIC16C55, a number of excellent ICE tools are available, most under $1000. For faster processors, the authenticity of the signal wiggling can be compromised by how the ICE connects to the target's socket. For complicated processors, an ICE may not be able to emulate the target processor in real time. However, for our PIC-based example, an ICE is a reasonable solution.

If the product has multiple developers working on it, chances are that each engineer will want an ICE. This may mean spending several thousand dollars "extra" on emulators for the development team.

Even if the product only has one developer, if the ICE costs $750 then the PIC's savings over the AVR are negated. The anticipated production was only one thousand units. Of course, after the development is done, the engineering group will still have the ICE. This may or may not be useful, depending on whether a similar flavor of PIC microcontroller is used for a subsequent project.

The production phase of the product's life may benefit from the AVR's Flash-based ISP program memory if a bug is discovered. Bugs crop up in code all the time. Industry experts like to play at quantifying how many bugs can be expected in a project given the number of lines of code. Let's say that based on experience or trade publications or industry studies (or which way the wind blows in Siberia), we believe we can expect one serious bug for every three hundred lines of assembly code initially released to production.

Initially, we stated that the application would require a few hundred instructions. That implies that we believe that we can expect one serious bug to be in the code released to production. Now, that doesn't mean that there will be a bug. It just means that we think there is likely to be a bug.

If the Flash-based AVR is used in production units and a bug is found, then the existing inventory of assembled boards can be reprogrammed without having to desolder parts. Units already deployed may be field upgraded if they were designed to be. RMAs (returned materials authorized) can be reprogrammed quickly without soldering.

Reprogramming the AVR in circuit may require the addition of an extra header and a few passive components to the PCB. As of the year 2004, component insertion costs are running around $0.12. If four extra components (a header and three resistors) are required to reprogram the AVR, then the $0.75 cost increase of the AVR over the PIC is further increased by $0.12 × 4 = $0.48 plus the cost of the four extra components (assume $0.09).

For the production phase of the product to see the benefits of the AVR's ISP flash, the product cost must increase over the PIC solution by $0.75 + $0.48 + $0.09 = $1.35.

An innovative engineer might be able to eliminate the need for the extra components by designing or buying a clip-on programmer. This may add cost to the product development phase for the research to identify, purchase and test the clip-on programmer.

In this example, the engineer selecting the processor may well reason that the 75 cent or $1.35 difference in parts cost over 1,000 projected units is worth having the ability to reprogram defective controllers or upgrade existing products with new features. The engineer may also feel that the AVR's ISP flash will eliminate the need for an ICE, thus saving more money. This engineer would select the AVR.

Another engineer may decide that the debugging features of an ICE are indispensable for developers and therefore the cost of an ICE for the PIC solution would be balanced by the cost of an ICE for the AVR. This engineer may further reason that the production projection of 1000 units is likely to be wrong and that many more units will be built. Hence, reducing per-unit cost becomes more important. Furthermore, he may gamble that released code will be bug-free. This engineer would select the PIC.

In even this simplest of examples, two chains of thought lead to different conclusions. The selection of a controller is seldom a "cut and dried" process. It is an exercise in trade-offs to manage risk.

2.1.2 Performance by Benchmark

Attempts have been made to provide comparative rankings of processor performance. Benchmark algorithms exist. Most were developed to compare performance of desktop, mini or mainframe computers. The apples-to-oranges world of embedded platforms makes traditional benchmarks all but irrelevant.

Benchmarks are useful if the hardware platforms are similar. When purchasing a desktop PC it might be useful to know how fast a numeric simulation or a CAD rendering package or even a graphics intensive game will run. Embedded platforms are so dramatically different that developing a meaningful benchmark that will run "identically" on dissimilar hardware platforms is nearly impossible.

The Internet newsgroups, such as comp.arch.embedded, are riddled with tedious pedantic discussions of the pros and cons of benchmarks. Flame wars abound regarding the legitimacy of how some group or another implemented a so-called "standard" benchmark algorithm. This polemic nattering can sound all hoity-toity, but the reality is that choosing a control solution is not something that can be done with benchmarks.

Digital signal processing (DSP) is the one area where specialized and relevant benchmarks have been developed. This stems from the fact that DSPs implementing standard filter topologies are in large part computationally an apples-to-apples comparison. This is a result of the fact that the filter algorithms are straightforward. On the other hand, if the DSP is going to perform tasks other than simple filtering, then the additional features of the DSPs, such as I/O access and interrupt overhead, may come into play, and benchmarks will not be of much help.

Embedded controllers differ in feature set and implementation so dramatically that embedded benchmarking is more an art than a science. Controller selection encompasses so much more than pure computational horsepower that any benchmarks are nearly irrelevant.

About the best that can be said for benchmarks is that with careful scrutiny, in some applications, they may be of assistance in comparing computational performance.

2.1.3 Availability

Controllers, be they packaged, board-level or chip level, are as susceptible to market forces as any other commodity. Many embedded products have life cycles of a decade or even longer.

The engineer selecting a controller must consider the availability of the solution for the duration of the product's anticipated life cycle.

For example, there are numerous products on the market targeted at the cell phone industry. The low-power, small size and low cost of these parts makes them attractive to embedded designers. But before settling on a chip-de-jour, consider how fast the chip's core market is changing. New generations of cell phones are born about every six months. Does it make sense for the IC manufacturers to continue to produce an IC after it has become obsolete in the target market? No.

One of the reasons this book uses Rabbit Semiconductor's microcontrollers and modules is that Rabbit is committed to the long-term availability of their ICs. Rabbit Semiconductor grew out of Z-World, a leading manufacturer of embedded controllers. Over the years, Z-World had to manage numerous end-of-life (EOL) announcements from vendors.

A chip going EOL is always troublesome for companies using the chip. If the IC is a commodity part, like a logic chip, simply qualifying a new vendor may be fairly easy. If the IC is a single-source part, like a real-time clock or CPU, then often the best that can be done is a one-time lifetime buy of the parts. The cash outlay for this can be crippling for a small company.

Any part can be designed out of a system, but design cycles are slow and incur both financial costs and opportunity costs. These costs can be difficult to bear for any company. Nobody wants to revisit an old and successful design simply because a vendor has discontinued a part.

Another very real issue to contend with is leadtime. Some microprocessor chips have phenomenally long lead times. Certain companies have seriously bad reputations for delivering product on time. Some companies make a run of controllers only once or twice a year. If the parts have been allocated in advance to large customers, then smaller companies may be faced with six-month leadtimes.

One California-based engineer is fond of relating a story of how his firm designed an instrument around a Japanese-made microcontroller. The instrument was entering the advanced prototyping phase and the engineer was given the task of ordering parts to build a dozen new prototypes for testing. All of the parts for the instrument were readily available except the microcontroller.

The Japanese manufacturer had production runs scheduled every six months, with fixed numbers of parts scheduled to ship to the United States. US distribution channels had allocated all of the parts to large customers. The company that only needed twelve microcontrollers was quoted a 154-week leadtime (yes, three years).

To solve the immediate problem of building and testing prototypes, the engineer arranged to have some existing microcontrollers removed from older prototypes and placed in the advanced prototypes. He still was faced with the need to acquire several more microcontrollers to finish the project. Desperate as the US company was, they could not find any of the Japanese microcontrollers for sale. In the end they identified a $500 evaluation board with the $20 microcontroller on it. They purchased $3000 in evaluation boards to get the remaining $120 worth of microcontrollers they needed. An additional round of development was scheduled to replace the Japanese built microcontroller with a more available device.

Chapter 2

If an engineer settles on a packaged or board-level controller solution, then the vendor building the controller must manage the supply chain. If the controller vendor is not well capitalized, maintains poor relationships with silicon distributors, or is inexperienced in supply chain management, leadtimes on the packaged or board-level controller can vary dramatically.

In Chapter One, an example of an underwater torque tool was given. Being specialized tools, only a few of these systems were to be produced. When the design team completed the prototype and the second tool was ready to be manufactured, the torque tool company attempted to order a second PC104 motherboard. The PC104 vendor had planned poorly and was not only out of stock, but the lead-time was quoted at some ghastly number like five months. After many long and heated conversations, an engineer at the PC104 vendor found a shop-queen and shipped it to the torque tool company.

A shop-queen is a board that has spent a lot of time in "the shop." In many industries this carries the stigma of a lemon. A shop-queen in this instance was a board that was used internally for testing and experimentation in the manufacturer's facility. When the PC104 board arrived at the torque tool company, the board had bent connectors and a lot of dust. However, after a bit of cleaning, lead straightening and rigorous evaluation, it was deemed functional, placed in a production torque tool, was shipped and worked fine.

There are two morals to this tale. First, if the controller supplier is inept at managing their supply chain, the systems designer will bear the brunt. Second, even something as "standard" as a PC104 embedded PC is not always easily second-sourced.

The production departments at Z-World and Rabbit Semiconductor have nearly two decades of experience managing volatile supply chains. Customers that use Z-World or Rabbit controllers don't have to worry about leadtimes on individual parts. The folks at Z-World and Rabbit worry about that. Their commitment to product availability makes Z-World an excellent choice for packaged or board-level control solutions. The same can be said for Rabbit Semiconductor in regard to chip-level and core-module solutions. Both companies have consistently short lead times for their products, usually shipping from stock.

2.1.4 Tools and Time-to-Market

Seldom do engineers have the luxury of long development cycles. Even "big" projects pack the work into the tightest, most optimistic schedule that can be cooked up by management. There is never time budgeted for "fighting with and debugging development tools." Selecting a controller solution that has first class tools is desirable.

Tools can roughly be broken down into hardware and software tools. Hardware tools include schematic capture, PCB layout, circuit simulation, and HDL (hardware description language) compilers. Software tools include compilers that generate code for the controller's processor, assemblers, simulators, and emulators/debuggers.

The axiom "you get what you pay for," implies that good tools will be expensive. Surprisingly, this isn't always true of software development tools. The axiom, however, seems to hold fast for hardware tools.

Z-World, Rabbit Semiconductor and Softools each sell outstanding compilers for less than a thousand dollars. Technically refined yet inexpensive development tools make Rabbit-based embedded controllers a cost-effective solution for many applications.

Dynamic C™ provides a complete integrated development environment (IDE) with a C compiler, assembler and debugger. The vast libraries are provided in source form so engineers can adapt standard drivers to custom applications.

Softools offers an ANSI-compliant toolset featuring an IDE with a C compiler, assembler and debugger. Complete libraries are also provided.

Dynamic C™ and Softools both communicate with the Rabbit processors through a serial interface. Both tools offer all the debugging capability needed by developers. This simple interface means no expensive hardware programmers, emulators or debuggers are needed.

Rabbit's innovative core modules give system designers a ready-made electronic heart comprised of fine-pitch surface mount (SMT) components all on a compact board with easy to access through-hole headers. The core module designs may be freely copied to a custom PCB in production, if required.

The core modules have "development" boards available. These development boards sport a prototyping area as well as some common peripherals (LEDs, switches, power supply and the like). Rabbit core module boards differ from other manufacturers' development boards in their practicality.

Many CPU vendors offer development boards that are not easily adapted to use in control applications. Signals are only pinned out on headers, there are no prototyping areas and the boards are unreasonably expensive. Rabbit's development boards are useful tools for engineers and not simply marketing aids.

In addition to the Rabbit core modules, Z-World offers Rabbit-based board-level and packaged controllers. This one-stop-shop for inexpensive software tools and diverse hardware platforms sets Z-World and Rabbit Semiconductor apart from other embedded controller vendors.

Good hardware development tools are expensive. The more complex the circuit, the more an engineer will want to have a "good" layout tool. For example, if a circuit only has a couple of connectors on a small PCB then an inexpensive tool can produce usable PCB Gerber files and a drill file. Gerber files are CAM (computer aided manufacturing) files that describe where to put copper on a PCB. A drill file, sometimes called an NC drill file for "numerically controlled" drill file, describes where and what sized holes to place in the PCB.

If a board design includes fine pitch SMT parts and hundreds of connections, a cheap PCB layout tool will cost more in layout time, errors and debugging than initially buying a good tool. One tool that has a proven track record is PCAD™. It is perhaps the best tool available today for doing schematic capture and PCB layout of simple or complex designs for the small and medium sized company.

If complex circuit design can be avoided, then PCB design can be accomplished economically with a low-cost tool or by farming the work out to a design bureau. If a core module can be used, often the additional circuitry for an application is minimal. The application board design is often a matter of providing simple interface components and a power supply. The development boards are usually suitable for design using a low cost tool or PCB design house.

The Z-World full-grown board-level and packaged controllers often eliminate the need for any secondary PCB design. This can save thousands of dollars in tools and time. However, for medium or large production volumes, it is generally more economical to go with a Rabbit Semiconductor core module or chip-based design.

2.1.5 Libraries

When selecting a controller, an often overlooked fact is that the amount of time spent writing code for an embedded system usually dwarfs the time spent designing hardware. Selecting a control solution that minimizes the software development task is desirable.

Many companies offer controller hardware, but few offer hardware and tightly coupled software libraries. This means that software engineers may have to spend weeks writing code to talk to even the simplest devices.

Rabbit Semiconductor offers complete libraries for all of their controllers. There are drivers for simple devices like analog-to-digital converters (ADCs) as well as complex drivers for devices like ¼ VGA graphic LCDs.

Libraries go beyond hardware device drivers. Various communication protocols are provided with Rabbit-based products. TCP/IP and Modbus are two heavily used protocols. TCP/IP is usually used with Ethernet. Modbus is most often used over RS-485.

An embedded systems designer's time is best spent developing algorithms and software specific to the project, not reinventing drivers. By selecting a control solution with an abundance of tested and proven drivers, development time can be minimized.

One of the advantages of an embedded PC is the amount of software available. For instance, operating systems like Microsoft®Windows®, Linux and DOS® are readily available. Drivers for many standard devices are available. Ethernet, disk access, video and keyboards are readily available.

Care must be exercised when attempting to use desktop OSs in embedded applications. After all, none of these were designed specifically with embedded systems in mind. How do you talk to an embedded I/O device like an ADC? What do you do if you want to decode signals from a quadrature encoder? The answer in most cases is to handle the I/O directly and write your own drivers.

The experience Rabbit Semiconductor has gained over the years has allowed the company to craft feature rich drivers for embedded systems designers. For example, Rabbit drivers exist to talk to both ADCs and quadrature encoders.

The vast libraries separate Rabbit's solutions from their competition. The existence of solid drivers and a royalty-free TCP/IP stack allows the system designer to focus on the application and not on reinventing drivers or writing a protocol stack.

2.1.6 Sustainability

Products spend 90% of their life in production. The original designers often move to other companies before a product completes its lifecycle. Engineers who were not involved with the product's development will almost always be responsible for sustaining the product in production.

Modifications made to a product that has been released to production can be loosely grouped into four categories. These are bug fixes, corrections to address hardware EOL (end of life) issues, product enhancements, and improvements for manufacturability.

Bug fixes are the most urgent class of changes. Software or hardware bugs may be discovered. Selecting a controller with easy to learn software development tools will allow a future engineer to quickly come up to speed on how to modify the product's code base. If library source code is available, then even driver bugs may be repaired.

In the case of hardware bugs, selecting a microcontroller with good documentation and debugging features is prudent. A sustaining engineer will have to come up to speed quickly on the available troubleshooting tools. The desired emphasis will naturally be on understanding and fixing the product's bug, not learning complicated or poorly documented tools.

The second class of production side modifications addresses EOL issues. If the part becoming obsolete is easily second sourced then the only modification to the product will be to the product's documentation. A second source for the part will be qualified and added. However, if the part going EOL is a single source part, then hardware and software changes will likely be required.

Consider the case of a real time clock chip (RTC) going EOL. These devices are often single source. Occasionally there will be "pin compatible" devices available, but even these may not be functionally equivalent devices. So in the best case a software modification will need to be made to the RTC driver. The worst case requires both a PCB change and a new RTC driver.

As in the case of a bug fix, good software development tools, good product documentation and library source code assist in focusing on the problem not the process. This saves time and money and allows the company to address EOL issues nimbly.

The third class of post-production release modifications addresses the inevitable demands for feature enhancement. The reason for modifications to the product may be different than for the EOL modifications, but the actual work required is the same. Product enhancements will require changing software, hardware or both.

The prescient designer will initially select a controller with a few more resources than are actually needed for the product rollout. This will allow a future engineer to add product enhancements without changing the hardware.

The last class of changes relates to manufacturability. Generally these are mechanical adjustments that make assembly easier. The initial controller selection has little impact on these types of changes. One notable exception crops up when the production test department or the repair/re-work group wants to install special test or diagnostic code in the product. In this situation, an in-system programmable (ISP) controller can save the company much time and money.

Some companies opt to do functional tests on every instrument produced. If the instrument is complex or requires the interaction of many different hardware subsystems, a special functional test program (test code) may be preferred over the full application to verify that the individual hardware components are operational.

Test code seldom bears any resemblance to the full application program. A test program attempts to isolate individual hardware components and verify their operation. If sensors require calibration then the test code may also perform this task. Ideally, a test program will identify and isolate problems. The repair department will often use production's test code as a diagnostic tool.

After an instrument's hardware is qualified by the test code, the application code must be loaded. The instrument is given a cursory examination and then placed in finished goods.

If the controller is not ISP, then a socketed nonvolatile memory device will need to be designed into the system. Changing memory devices can be time consuming, and the test ROMs will be subject to physical damage from constant handling. An ISP controller allows production test code or diagnostic code to be installed and replaced easily.

2.2 Defining the Problem

Before selecting a control solution, the control problem must be well defined. This can be complicated by indecisive management or by technological hurdles that are of unknown magnitude. The former is the most frustrating for engineers. The latter can often be mitigated through research and simulation.

Management will provide a product description. Some companies have a formal process for creating a product requirement document (PRD), sometimes called a product specification, sometimes called a system requirements document (SRD). Sadly, most companies do not have a formal process. All too often, management will simply issue a short memo or a napkin sketch describing the new product's features.

In either case, the engineer must start with what is given. Obvious questions must be answered—either by managerial edict, technical prowess or educated guessing.

Next, the system architecture must be developed. Some companies have a formal process for taking an SRD and splitting out hardware and software requirements into a hardware requirements specification (HRS) and a software requirements specification (SRS). From there, hardware and software architectures can be defined.

But for most companies, at this stage the controller is simply a box on a diagram with sensors and actuators reporting to it and a blob on the side of the paper with what passes for a flowchart amongst friends.

Regardless of the formality of the process, the engineer will get an idea of the type of I/O required for the project. If ADCs are required, their resolution should be defined. If pulse width modulation is being generated then characteristics such as drive current, frequency, duty cycle range and resolution should be defined. As much as possible should be nailed down up front.

The Basics

Undoubtedly, as the project progresses, requirements will change and the initial guesses at various features will require adjustment. This is just part of the process of evolving an idea into a product. Formal processes try to avoid costly changes half way down a product development path. But front-loading a project with detailed requirements documentation is not well suited to the needs of many small companies.

Time-to-market may require playing fast and loose with the requirements phase just to get a product to market quickly. Sometimes having an imperfect plan on a placemat will be of greater use in getting a sellable product done quickly than having a perfect plan in six months.

Once a plan is in hand, technological hurdles may still complicate a project. These are problems that an engineer does not know how to solve up front. These problems are only solved through research and experimentation.

2.3 A Survey of Solutions

While this book uses Rabbit 3000-based products for all the projects detailed, there are a myriad of other controllers on the market. Many of the techniques presented are applicable to controllers other than the Rabbit 3000. So let's briefly examine some of the available microprocessors and microcontrollers and see how they compare with the Rabbit-based solutions.

2.3.1 Small Microcontrollers

There are many "small" microcontrollers on the market. Probably the most widely known is the PIC™ line of controllers from Microchip. These parts along with their hopeful brethren—AVR from Atmel, COP8™ from National Semiconductor, Nitron™ from Motorola and many others—fill a niche in the "not-too-complicated" class of embedded solutions.

These parts are available from $0.80 to about $5.00. Code is stored either in "One Time Programmable" (OTP) ROM or internal flash, making small controllers fairly self-contained.

The lower end of this style of controller finds parts like Microchip's PIC16C54 and Atmel's AT90S1200. These parts have a low pin count and subsequently low I/O capability and have been optimized for simple applications.

Table 2.2 shows how the feature sets stack up between these two parts.

Table 2.2: Low end PICs and AVRs go head-to-head.

	PIC16C54	AT90S1200	Comments
Pin count	28	20	
Maximum I/O pins	20	15	
Frequency ranges Available	DC – 40 MHz	DC – 12 MHz	
RAM	24 bytes	0	
EEPROM	ZERO	64 bytes	
General Purpose Registers	Same as RAM 24 × 8	32 × 8	PIC™ W register is accumulator AVR has no accumulator
Program Storage	512 × 12 bits (OTP)	1 KB (flash)	

Chapter 2

Table 2.2: Low end PICs and AVRs go head-to-head (continued).

	PIC16C54	AT90S1200	Comments
Code Size	512	512	PIC 12-bit instructions AVR 16-bit instructions
Counter/Timers	1	1	
External Interrupts	0	1	
Sleep mode	Y	Y	
Watchdog Timer	Y	Y	
ISP	Y	Y	
Analog Comparator	N	Y	

For systems requiring analog input, integrated analog-to-digital-converters (ADCs) are available. These are generally limited to 8-bit, 10-bit or 12-bit converters. For example, the PIC16C710 has a built-in 8-bit converter with a 4-to-1 analog multiplexer. The PIC16C710 has a similar pinout to the PIC16C54, but several pins have modified functions.

Historically, analog designers have frowned upon the performance of ADCs built on the same silicon die as the microcontroller. Noise performance on some integrated ADCs has been known to be less than stellar. This is something that must be evaluated on a part-by-part basis. Some manufacturers do an excellent job; others are not as good at keeping digital noise out of the analog portion of their designs.

Burr-Brown, now a division of Texas Instruments, is a name synonymous with excellence in analog design. The Burr-Brown/TI MSC1200 is an example of a device where a fast computational core coexists with high-resolution low-noise analog.

For around $6.50 the MSC1200 offers a lot of functionality. The following list enumerates the device's key features:

- 8051 high-speed core (4 clocks per instruction cycle)
- DC-33 MHz operation
- 4K ISP Flash
- 1K boot block
- 128 bytes of SRAM
- 24-bit delta-sigma converter
- Programmable gain amplifier (PGA) from 1 to 128
- 8 input analog MUX
- Differential or single ended inputs
- On-site burnout detection
- On-site precision reference
- On-site temperature sensor
- 8-bit current DAC
- 16 digital I/O pins
- Watchdog timer
- Two 16-bit timer/counters

- Full duplex UART
- –40C to +85C temperature range
- 2.7 to 5.25 volt allowable rails

The MSC1200 datasheet indicates that a system designer can expect to see between 5 and 22 effective bits of resolution depending on how the analog channel is configured. Many of Burr-Brown's high-resolution delta-sigma converters allow the system designer to trade-off gain and sample rate against uncertainty (noise). The MSC1200 allows the same trade-offs.

A flexible analog MUX allows the analog inputs to be either differential or single ended. The MUX offers the unusual capability to pair arbitrary inputs as a single differential channel.

The high-speed core is based on the venerable 8051 architecture. The popularity enjoyed by 8051 over decades is a testament to the original designer's insightfulness. Yet, the MSC1200 core has several improvements over the original architecture.

Notably, the MSC1200 core has only four machine cycles per instruction cycle. The traditional 8051 implementations have twelve machine cycles per instruction cycle. This, combined with a screaming 33 MHz top speed, make the MSC1200 a reasonably potent computational engine for the price.

The MSC1200 is an example of a controller that has fantastic analog noise performance married to a CPU on a single die.

Small controllers are suitable for offloading simple tasks from a larger CPU. For example, consider a controller that must implement a small servo control loop as part of a larger system. This servo loop requires sampling a position error and quickly adjusting a drive signal to a motor. If the designer moves this task from a main processor, say a Rabbit 3000, to a satellite processor like a PIC, then the software on the Rabbit becomes simpler and more of the Rabbit's resources are available for other system tasks.

Using small microcontrollers as satellite controllers is a technique that can, in some circumstances, save development time. The additional microcontroller may drive up the end product cost. This may or may not be offset by reduction in development time. If the main CPU can do the offloaded task even at the expense of more complex software, then it may be worth spending the extra time on software NRE (nonrecurring engineering) fees to reduce the per-unit cost of the hardware.

In some cases, the application of an inexpensive satellite processor can actually keep the cost of a system down. If a main processor is stretched to the limit and an additional task must be added, the incremental cost of moving from one main CPU to another may be higher than offloading the additional task to an inexpensive satellite controller.

2.3.2 Mid-sized Microcontrollers and Microprocessors

Parts costing $10 to $20 are generally considered "mid-sized" controllers. There are a large number of parts available in this class. The Rabbit 2000 and Rabbit 3000 microprocessors, the AVR ATmega128 and the Intel386™EX all fall squarely into the mid-sized class.

Different architectural philosophies drove the development of each of these processors. This is evident from each processor's feature set.

Chapter 2

For many, the name Intel conjures up images of multi-gigahertz multi-watt desktop processors. It's easy to forget that Intel has a long history of serving the embedded market. The Intel386 EX was designed specifically for the embedded market by adapting the popular desktop i386.

The Intel386 EX datasheet states, "It (the Intel386 EX) provides the performance benefits of 32-bit programming with the cost savings associated with 16-bit hardware systems." Clearly, Intel was trying to leverage the success of the x86 family while trading off performance for lower cost. By building from an existing design and maintaining binary compatibility with earlier family members, Intel ensured that a wide array of existing development tools would generate code for the Intel386 EX.

The Atmel AVR family sports a very straightforward, almost textbook, architecture. When the AVR architecture was created, all of the development tools had to be built from scratch. Atmel created a design that scaled nicely across a wide array of derivative products. The same basic architecture can be found in the least expensive AVRs (such as the AT90S1200) all the way to the mid-sized family members like the ATmega128. All of the AVRs are flash-based controllers.

Rabbit Semiconductor struck a middle road between "old" and "new." The Rabbit parts are based closely on the Zilog Z180 architecture, although they are not binary compatible with the Zilog parts. This caused Rabbit Semiconductor to have to create a suite of development tools to support the new design. At the same time, Rabbit was able to freely add instructions to, and otherwise tweak, the Z180 architecture.

So, like Intel, Rabbit built upon a widely respected and successful architecture. Unlike Intel, Rabbit did not maintain binary compatibility with the original design.

Like Atmel, Rabbit had to produce firmware development tools, and likewise, Rabbit optimized their design for C programming.

Now that we have an idea of the general philosophies behind the processor designs, let's have a look at key features of each device. We'll start with the Intel386 EX.

Intel386™EX

- Static Intel386™ core; DC – 33 MHz
- Full 32-bit internal architecture
- External 16-bit data bus allows use of less expensive memories
- 26-bit address bus
- Address pipelining allows use of slower, inexpensive memories
- MMU fully compatible with 80286 and Intel386™ DX processors
- On-chip dedicated debugging support including break point registers
- Watchdog timer
- Chip-select unit
- Timer/counter unit
- Interrupt control unit
- Parallel I/O unit (24 I/O pins)
- UARTs

The Basics

- DMA and bus arbiter unit
- Refresh control unit
- JTAG-compliant test-logic unit
- 140 mA @ 25 MHz, VCC = 3.6 volts (device held in reset with inputs held in inactive state)

By looking at the features listed above, one can tell that the CPU is a derivative of the desktop processor line. Besides the giveaway bullet of "Intel386™ Core," the processor has features such as a 32-bit architecture, a huge 26-bit external address space, DMA and a bus arbiter. All are indicative of a powerful desktop processor.

Intel was clearly targeting the embedded market with this processor. They squeezed down the external data bus to 16-bits to allow less expensive memory devices to be used in the embedded design. A counter/timer, UART and 24 I/O pins were added around the processor core as a nod to embedded interfacing. The power consumption, while low for a desktop processor, is still fairly high for low-powered embedded applications.

This reliable processor has enjoyed many years of success. The widespread availability of quality development tools is certainly a boon for the part.

Next, we consider the AVR ATmega128. The following list captures the highlights of this controller.

ATmega128

- 8-bit data bus
- Static core design; DC – 16 MHz
- 32 × 8-bit general purpose registers
- Up to 16 MIPs throughput at 16 MHz
- 128 K-Bytes of internal ISP flash
- 4 K-Bytes of internal EEPROM
- 4 K-Bytes of internal SRAM
- Up to 64 K-bytes of external memory
- Two 8-bit timer/counters
- Two 16-bit timer/counters
- Two 8-bit PWM channels
- Six PWM channels from 2 to 16 bits
- 8 channel, 10-bit ADC
- Two UARTS
- Watchdog timer
- 53 I/O pins
- 18 mA @ 16 MHz, VCC = 3.3 volts

This part is loaded with features that scream "embedded controller." The power consumption is very low. There is an eight channel 10-bit ADC on the die. The 53 pins of digital I/O is certainly targeted at the embedded market.

This controller has all of the program memory in flash and on board. A significant amount of SRAM is also included. Many small embedded applications will fit into this memory model.

Chapter 2

The 4K of onboard EEPROM can be used to store calibration constants, log a small amount of data, or even store system ID and configuration information. These are common tasks performed in embedded systems.

If a larger data memory is required, some of the 53 I/O pins can be traded off and up to 64 K-bytes of external SRAM added. For many systems, 64K of data storage is adequate. But for many applications, the ATmega128's memory model will be too limited.

Even with a RISC-like ability to execute many instructions in a single clock cycle, a 16 MHz speed is considered slow by many engineers. Depending on the application, clock speed may not be an issue.

Consider a small solar-powered controller that performs simple biophysical data acquisition. One of the major design considerations for this type of system is power consumption. There are two frontline techniques used to reduce power consumption. Run the system at a lower voltage so the V^2/R power is lowered. Run the system at a slow clock rate—power consumption is linearly proportional to clock speed.

In this type of system, a 16 MHz clock might be ten or a hundred times faster than the designer will need. In this case, the speed of the internal flash memory is not relevant when selecting a controller.

Next, we'll briefly look at the Rabbit 3000A.

Rabbit 3000A

- 8-bit data bus
- 20-bit address bus
- Static Core Design; DC – 54 MHz
- Glueless memory interface
- Clock spreader for EMI reduction
- High frequency clock and 32768 Hz clock
- Built-in clock doubler
- 4 levels of interrupt priority
- Bootable over serial port (cold boot feature)
- 56 I/O signals (shared with serial ports and other peripherals)
- Four Pulse width modulation channels
- Six UARTs
- Auxiliary I/O bus reduces loading on memory bus
- Two input-capture channels
- Two quadrature decoder channels
- Built-in watchdog timer
- Standard 10-pin programming port
- 65 mA @ 30 MHz, VCC = 3.3 volts

In some ways, the Rabbit 3000A strikes a middle ground between the Intel386 EX and the ATmega128. The 20-bit address space is much bigger than the AVR but smaller than the

26-bit space offered on the Intel part. The power consumption is higher than the ATmega128, but much lower than the Intel386.

The Rabbit offers the highest clock rate of the three devices. This does not necessarily translate into higher computational performance. Internally, the Intel core is four times as wide and the external data bus is twice as wide as the Rabbit. The Rabbit, on balance, requires more clock cycles to execute an instruction than the AVR.

One potential disadvantage of turning up the CPU clock speed is that faster external memory is required. However, as time progressed, what was once "fast" expensive memory has become "slow" cheap memory. Rabbit Semiconductor understood this when their processor was designed. This forward thinking has produced a snappy little part that today doesn't require expensive memory.

Another potential disadvantage of running a part at high-speed is the increased power consumption of the CPU and memory. As clock speed increases, so does power consumption. The Rabbit is a fully static core and can be run from DC to 54 MHz.

The Rabbit's I/O mix is a lot more complete than either the Intel or Atmel parts.

The Rabbit's feature rich I/O set is derived from Z-World's firsthand knowledge of the embedded market. Rabbit sports PWM channels, quadrature decoder channels, input capture channels and a high digital I/O count.

One of Rabbit's more useful features is the inclusion of six UARTs. At first glance, six UARTs might seem a bit extravagant, until one considers the nature of embedded communications. The advent of clocked serial peripherals means that a system designer can add many types of sophisticated I/O without loading the processor's main bus. ADCs, DACs, RTCs, EEPROMs and a host of other devices are available with clocked-serial interfaces (I^2C™, SPI™ and Microwire™). The Rabbit's UARTs make expanding the system's I/O features quick and inexpensive.

Rabbit's auxiliary I/O bus is another tool that can be used to expand the I/O feature set quickly and easily. In conventional processors, parallel I/O devices are placed on the CPU's data and address busses. This can lead to undesirable capacitive loading of the data and address busses. Rabbit's auxiliary I/O bus allows designers to place parallel I/O devices on a bus completely separate from the memory bus.

Another unique feature the Rabbit offers is a programming port that is used to program flash memory connected to the CPU's memory bus. This clever interface allows any flash memories used on the CPU bus to be treated as ISP devices. The programming port allows a PC with a serial connection and a bit of software to program flash memories through the Rabbit processor.

2.3.3 Final Thoughts

An engineer faced with selecting a microcontroller or microprocessor must be prepared to weigh a number of factors and make trade-offs. Comparing microcontrollers is an exercise in contrasting apples-to-oranges.

Finding a controller with the right mix of computational speed, integrated peripherals and development tools is an easier task than one might think. The difficulty is picking just one of the many excellent solutions available.

2.4 A Rabbit's Roots

As we saw in the last section, a company's design philosophy greatly influences the final form of its products. In the next section, we will examine the Rabbit's architecture in detail. But before we do, it will be informative to have a look at Rabbit Semiconductor's history and understand the forces that have shaped its design philosophy.

Our story starts in Silicon Valley in 1981. At the age of 40, Norm Rogers and co-worker Paul Mennen quit their jobs at GenRad and with equal investments of $34,000 formed Decmation.

Their original product was a plug-in card that allowed Digital Equipment Corporation mini-computers to run CP/M software. The idea wasn't new. There was another product already in the market that sold for around $2,500 and did about the same thing. Decmation's product sold for $700.

In the first year, the little company had a quarter-million in sales. Marketing consisted of running small advertisements in the back of electronics magazines and attending the occasional trade show. Orders were processed by telephone.

Business remained strong and in 1983, Norm mortgaged his home in San Jose, bought out half of Paul's interest in the company for $50,000 and incorporated Decmation. From 1983 on, Norm Rogers maintained the controlling interest in the company.

Norm poured time into research and development. By late 1985 he had his next big hit almost complete. This product, named Blue Lightning, did for IBM PCs what the original product did for DEC minicomputers. There were a large number of applications written for CP/M, but since IBM had thrown in with Microsoft, MS DOS PC's were in, and CP/M machines were on their way out. Blue Lightning allowed IBM users to run their legacy CP/M applications.

Again, there was already competition in the market. The competitor's well known product was called the "Baby Blue," a takeoff on the color of IBM computers. Decmation's name, Blue Lightning, was a shameless rip off.

By a stroke of luck the competitor went bankrupt the same day the Blue Lightning was introduced. The bankruptcy was caused by problems unrelated to the CP/M board business.

An improved Decmation design soon followed—Blue Thunder, which soon replaced Blue Lightning.

In an email sent to the authors of this book Norm Rogers recalls Blue Thunder:

> "My champion product was the Blue Thunder 12 MHz card. At that time no Z80 had a faster official clock speed than 6 MHz. However I discovered that NEC chips would actually run up to 14 MHz before they failed. I also discovered that one brand of Japanese dynamic RAM would run fast enough to keep up with a 12 MHz Z80. So, I put this all together, sanded off the part numbers so that no one could discover my discovery, and sold 12 MHz Blue thunders for about 3 times what the 6 MHz ones sold for. I sold a lot of them with a great profit margin." —Norm Rogers, March 2003

By 1986, the company's main business of CP/M simulators was waning. Customers were phasing out CP/M-based software applications. In search of new products, Decmation tried a number of different avenues, with varying success. The CP/M cards that plugged into PCs were redesigned with a serial port and made into "communications coprocessors" which could offload serial communications from the PC's CPU.

Another new business was development tools for Z80 based embedded systems. A great deal of very high quality CP/M based development tools were now considered obsolete and were available for purchase at low prices. Decmation licensed and rebranded these as tools for embedded systems engineers, either wrapped with a CP/M emulator or recompiled to be MS-DOS applications. This change of business direction plus threatening letters from Digital Equipment Corporation ("DEC") lawyers led to changing the name of the company from Decmation to Z-World. The "Z" stood for the Z80 microprocessor at which the development tools were aimed.

In 1987, Norm moved from San Jose to Davis, California. His young children were school age, and the public schools in Davis were better than the marginal schools in San Jose. Z-World rented a 1200 square foot office in a small business park and had four employees.

Being host to the University of California, Davis had eager young college students. A towering young computer science student who had independently written some Z80 development tools made contact with Norm. Norm recognized the energy and genius of Greg Young, and Z-World quickly hired him. Greg wrote the first integrated development environment (IDE) for Dynamic C. Greg would later go on to setup and manage the technical support group and serve as a design engineer, and today, still consults on special projects.

In 1989, Z-World brought in a professional marketing manager by the name of Carrie Maha. While she was new to the burgeoning embedded systems market, she brought an organization and polish that Z-World needed. No longer would marketing consist simply of black and white ads in the backs of magazines. Carrie built a world-class marketing and sales organization from the ground up and today serves as the Executive Vice President.

The first Z-World embedded platform, dubbed the Little Giant, combined a Z180 microprocessor and practical I/O designed to control and supervise real-world embedded systems. Z-World's Dynamic C development system, written by Norm and Greg, made it easy to program the new board. Bundled with Dynamic C, this product was an instant hit when it was introduced in 1990. The Little Giant still sells well today as a legacy product to the original OEMs. The vertical integration of controller hardware and development tools would propel Z-World through the decade and into the next century.

By 1997, the company had moved twice to accommodate its rapid growth. Revenue was up, profit was up and the stage was set for another technological leap. Norm Rogers had spent the better part of two decades working around the shortcomings of the Z80 and Z180. Ten years of tweaking compilers and building hardware had given Norm a well-rounded and intimate view of the relationship between hardware, the firmware that runs on it and the software tools used to create the firmware. Norm was ready to build a new processor.

In 1997, Norm was introduced to Pedram Abolghasem, an ambitious electrical engineering student at Sacramento State University. Upon graduating, Pedram joined the company to spearhead the R & D efforts behind the first microprocessor that would address the many weaknesses of the Z180 design.

Chapter 2

Embarking on a methodical search for a partner who had already ridden the ASIC pony, Pedram located a one man consulting company, Systemyde International, owned by former Zilog designer Monte Dalrymple. With 15 patents credited to his name and experience as Principal Architect and Designer on Zilog products such as the Z16C30, Z16C32 and Z16C35, Monte had the experience needed to bring Norm's ideas to fruition.

In September 1997, the development of the Rabbit 2000 began in earnest. Norm, Pedram and Monte worked tirelessly to move the project forward. Norm supplied the architecture. Monte, as a consultant, wrote the Verilog code and provided mature advice to Norm and Pedram.

Initially, Pedram was the only onsite engineer assigned to the project. This meant he was the catch-all for work. Pedram designed printed circuit boards, using the largest and most state of the art Altera's FPGAs, with which to test Monte's Verilog code. Pedram found numerous bugs. Monte fixed them. In the end, Pedram and Monte turned Norm's ideas into silicon.

The entire development costs of the Rabbit 2000 amounted to less than $500,000. This was only $1/40^{th}$ the cost that some other companies were spending at the time in their microprocessor development efforts—many of which failed.

In late 1999, Rabbit Semiconductor was launched to continue and support the microprocessor development. In November 1999, the Rabbit 2000 processor hit the streets. Its success was followed by the Rabbit 3000 in 2001, and the Rabbit 3000A in 2002. The Rabbit 4000 is in the works.

Norm Rogers is still majority shareholder and company president of both Z-World and Rabbit Semiconductor. Pedram is the Director of Engineering. Monte continues to work closely with Norm and Pedram on Rabbit's new designs.

With the advent of the Rabbit line of processors, Z-World and Rabbit Semiconductor have attained a degree of vertical integration not seen in the embedded market to date. Customers can choose from a plethora of off the shelf solutions from processors to fully packaged controllers. The Dynamic C platform, now a 32-bit Microsoft Windows-based application, supports C and assembly language development as well as debugging on all the target hardware. The Rabbit processors have enjoyed wide industry acceptance and are supported by third party tools.

2.5 Rabbit in Detail

As the rest of the book uses the Rabbit 3000 microprocessor for all projects, an examination of the chip's architecture is in order. Figure 2.1 shows a top-level block diagram of Rabbit 3000.

2.5.1 The CPU

At the heart of the Rabbit 3000 is the CPU block. As seen in Figure 2.1, all external data is transferred to and from the external buses to the CPU. The CPU registers act as a source or destination for all external bus transfers.

The CPU communicates with the internal peripherals using an internal bus. This bus consists of an 8-bit internal data bus, 8-bit internal address bus and sundry internal arbitration signals. The 8-bit internal address bus means the Rabbit 3000 is limited to 256 internal peripheral I/O registers. It also means access to the registers doesn't require a two byte operand fetch to

The Basics

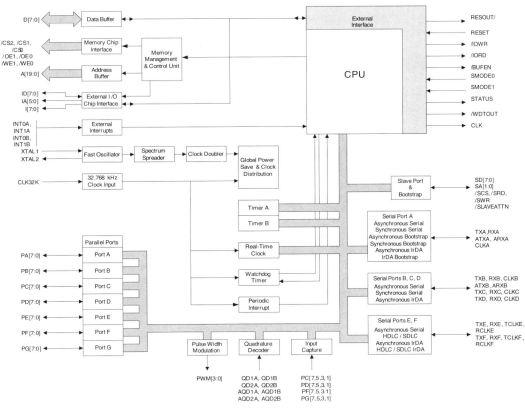

Figure 2.1: A bird's-eye view of the Rabbit 3000 shows a feature rich architecture.

acquire a complete 16-bit address. This increases the Rabbit's efficiency.

The CPU block, while similar to the Z180 CPU, has some important differences. A diagram of the registers available in the CPU block is shown in Figure 2.2.

Figure 2.2: The Rabbit CPU has an abundance of registers.

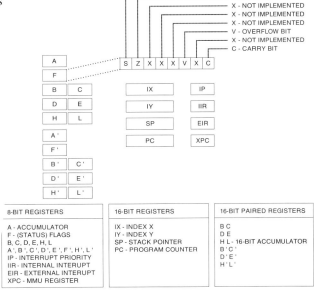

Chapter 2

Those familiar with the Z80 and Z180 will immediately recognize the familiar general purpose registers (A, F, B, C, D, E, H, L) and the alternate register set (A', F', B', C', D', E', H', L'). Rabbit Semiconductor has added some new registers and extended the functionality of the traditional complement.

The alternate register set was originally used to store the contents of the regular register set when a special function was called—often an interrupt service routine (ISR).

The EXX instruction exchanges the contents of HL, BC, DE with HL', BC', DE'. The instruction EX AF, AF' treats the resisters A and F as a register pair and exchanges the contents with AF'. In the Z80/Z180 the exchange instructions were the only way to access the contents of the alternate registers. This made using the alternate registers for data manipulation cumbersome at best. Rabbit has added a couple of instructions that make using the alternate registers as general purpose registers easier. Rabbit states that this "effectively doubles the number of registers that are easily available."

The alternate registers are still a bit cumbersome to use for computation or data manipulation as compared to their regular register counterparts. They are by no means "easy" in the sense that one might assume that the regular registers and alternate registers are orthogonal from an opcode point of view. However, the Rabbit instructions have breathed new life into an awkward legacy feature.

The Rabbit CPU, like its predecessors, is still an accumulator-based architecture. The 8-bit A register is the accumulator for 8-bit operations. For 16-bit operations, the HL register pair is the accumulator. In some circumstances, the IX and IY registers can be used as a 16-bit accumulator.

The IIR and EIR registers point to interrupt tables for internal and external interrupt generators. The Z80 R register is logically analogous to the Rabbit's IIR register.

The IP register is new. The 8-bit register is divided into four 2-bit fields. The fields are used to hold a history of the Interrupt priority. This is an important tool for managing the 4 levels of interrupt priority the Rabbit sports.

The XPC register is also new, and there is some debate as to whether this register should be treated as part of the CPU core architecture or as part of the Memory Management Unit (MMU). The register establishes where the XPC Segment maps in physical memory. However, instructions operate on XPC like a CPU register rather than an I/O mapped peripheral register. We, like Rabbit, will treat XPC as a CPU register.

Rabbit's designers explain that the XPC was created as a CPU register so that XPC can be accessed quickly. When we talk about the MMU, we'll see that the XPC register is used to locate the Extended Memory Segment. Since this is the primary avenue for accessing the bulk of the Rabbit's external 20-bit address space, the Extended Memory Segment is remapped frequently. Saving a few clock cycles on each access to the XPC register will favorably impact code execution on Rabbit-based systems using Dynamic C.

The Basics

Other MMU registers are implemented as I/O mapped peripheral registers and do not show up in the CPU. The only MMU related register afforded the privilege of being a proper CPU register is the XPC register.

The IX and IY registers are index registers. They are commonly used as pointers to memory locations. The HL register pair can also be used as a pointer into memory.

2.5.2 Parallel I/O

The Rabbit 3000 has seven 8-bit digital I/O ports. The 56 I/O pins share functions with other on-chip peripherals. For example, if a system design requires a serial port then some number of digital I/O pins will be traded-off. This is common in microcontrollers.

The Rabbit 3000 I/O ports have several desirable features. One of the biggest practical concerns is the ability to interface to 5-volt logic. The Rabbit 3000 is designed to operate on a 1.8 – 3.6 volt supply. However, all of the Rabbit's inputs (with the exception of the oscillator and power pins) are fully 5-volt tolerant.

Rabbit I/O pins are powered from a separate set of power supply pins than the CPU core. The VDDIO pins supply current to the I/O pins. VDDCORE pins supply current to the CPU core. A savvy designer can reduce overall EMI by confining the CPU's noisy harmonic-rich power supply to a very limited physical area on the PCB. The I/O pins can be powered from a separate (filtered) supply rail. The EMI generated as part of the fast oscillator and CPU switching will not be passed to the I/O pins and therefore not be conducted on the I/O lines. Compliance with FCC and CE standards will be much easier to achieve with this arrangement.

The Rabbit I/O pins on ports D, E, F and G have the option to synchronize their output updates with a timer. Figure 2.3 shows a conceptual diagram of an output with the timer synchronization option.

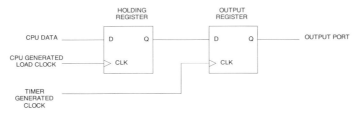

Figure 2.3: Timer synchronized outputs allow for precise timing of output pulses.

For systems that need to generate precisely timed pulses, timer synchronization is an excellent feature. The timer that updates the port can also be used to generate an interrupt. The ISR can load the next scheduled value into the first flip-flop holding register in the output chain. The next time the timer trips, the output will be updated and the ISR called again. This cycle can be repeated indefinitely.

Timers A1, B1 and B2 can be used to provide the synchronized update clock.

Another clock source available to update the output registers synchronously is the peripheral clock (PCLK). The "global power save and clock distribution", as seen in Figure 2.1, generates PCLK from either the fast clock crystal (divided by 2, 4, 6 or 8) or the 32.768 KHz clock.

Applications that can take advantage of this feature include Pulse code modulation (PCM) audio playback. Pulse width modulated (PWM) DACs can also benefit from synchronized bit transitions.

2.5.3 Rabbit Serial Ports

Rabbit has six serial ports. At first glance, this may seem like an over abundance. However, this is one of the features that can trace roots back to Z-World. Over the years, one of the most often requested features for Z-World's embedded controllers was "more serial ports." The Rabbit 3000 delivered.

All six serial ports may be used as simple three wire asynchronous serial ports with a maximum bit rate of the system clock divided by eight. This mode is suitable for driving RS-232 transceivers if no hardware handshaking is required.

In asynchronous mode, a 7-bit or 8-bit data length can be selected, with optional parity. A special 9-bit protocol is also supported in which a 9th bit is set or cleared to mark the first byte of a block transfer.

Embedded systems commonly employ protocols other than RS-232. One of the most common is the three wire RS-485 interface.

RS-485 has an A and B line in addition to the ground-reference. This half-duplex protocol is sensitive to the difference in voltage between A and B. If the voltage A-B is positive and greater than 200 mV then the data is a one. If the voltage A-B is negative and greater than 200 mV then the data is a ZERO.

RS-485 is designed so that many nodes can share the same A and B signals. Only one node is allowed to transmit at a time. This feature is why RS-485 is referred to a "multi-drop protocol."

One issue that Z-World discovered early on with the Z180 UART is that detecting when the last bit was finally shifted out of the output register and onto an RS-485 bus was very difficult. This made turning around the RS-485 bus a bit tricky. If the RS-485 transmitter were disabled too quickly, data would be lost. If too much time elapsed before the bus was released, then the RS-485 transmitter might cause bus contention with another node.

Building on Z-World's experience, Rabbit Semiconductor enabled the Rabbit 3000 UARTs to generate an interrupt when the last bit is finally shifted out on to the bus. This is how the Rabbit can quickly and safely turn off the RS-485 transmitter, freeing the bus for traffic originating from another node.

Ports A, B, C and D can also be operated in synchronous mode. This is sometimes called "clocked serial" mode.

When operating in synchronous mode, either the Rabbit 3000 or an external device can generate the clock. If the Rabbit 3000 generates the clock, the maximum clock rate is equal to the system clock divided by two. If an external device generates the clock, then the maximum clock rate is limited to the Rabbit's system clock divided by six. In this mode, SPI devices may be attached to the UARTs.

Ports E and F support HDLC/SDLC communication. This protocol supports frame checking, clock recovery and error checking. HDLC has found use in high-end router back-channel communication interfaces, and internal mainframe computer busses. HDLC is not commonly supported by microprocessors in Rabbit's price range. If HDLC is required for an application, a designer often has to purchase special USARTs to accommodate the requirement.

The Rabbit 3000 supports NRZ, NRZI, Biphase-level (Manchester), Biphase-space (FM0) and Biphase-mark (FM1) encoding. A digital phase-locked loop (DPLL) is used to recover the data clock from the received signal.

A standard CRC-CCITT polynomial ($x^{16} + x^{12} + x^5 + 1$) is used to generate the cyclic redundancy check (CRC) code. Frame flags are generated automatically. Framing errors are also detected automatically.

The HDLC protocol is used for high-speed communications using infrared IrDA transceivers in FIR mode.

Port A can be used to cold-boot (also called bootstrap) the Rabbit processor. Normally, upon reset, the processor begins executing code from the memory located at physical address 0x00000. As a general rule, Rabbit-based systems will have a ROM or flash memory located at 0x00000. In the case that the flash memory is not programmed, or in systems that operate purely from RAM, the processor can be forced to acquire data from a bootstrap port.

SMODE1 and SMODE0 pins control the bootstrap operation. When they are both zero, the processor boots normally. When SMODE1 and SMODE0 are {1,0}, the Rabbit will boot from serial port A in a synchronous mode. The Rabbit requires an external device to provide the clock for the synchronous boot. This precludes the use of an EEPROM as a boot device.

When SMODE1 and SMODE0 are {1,1}, the Rabbit will boot from serial port A in asynchronous mode. The communication rate is 2400 baud and is derived from the 32.768 kHz clock.

When the Rabbit is bootstrapping, it expects to see three-byte messages, called triplets. Each triplet is composed of a 2-byte address and one byte of data. Each triplet causes the Rabbit to store the data byte at the 16-bit address with the exception that if the most significant bit (MSB) of the address is a 1, then the data will be written into the addressed onboard peripheral register.

This allows a bootloader program to be loaded into SRAM and the MMU registers to be configured so that the SRAM appears at physical address 0x0000.

The sequence 0x80, 0x24, 0x80 will cause 0x80 to be written to the internal peripheral register 0x24. This will terminate bootstrap operation; it allows the newly loaded code to begin executing at address 0x00000.

2.5.4 Slave Port

In the past, processor-to-processor communications was somewhat difficult. To overcome this problem, dual-ported RAM has often been used to provide a shared space accessible to two separate processors. Even with silicon prices on an exponentially decreasing cost curve, dual-porting is expensive. Lead times for such specialty items can be long.

Chapter 2

Dedicated serial channels are another scheme often employed to communicate between processors. This approach requires few wires and UARTs are fairly inexpensive. For many applications, a dedicated serial port is a reasonable solution to processor-to-processor communication. The Rabbit, as we have just seen, sports six serial ports.

Sometimes, a designer may not have an extra serial port. Or perhaps the processors are running on slow-speed clocks to conserve power and the available serial port bandwidth is overly limited. In such cases, the Rabbit processor offers a processor-to-processor parallel communication channel called a *slave port*.

The chosen moniker is a bit unfortunate. One often thinks of "slave" devices being dumb. The Rabbit "slave port" is best thought of as a specialized communications port for high-efficiency processor-to-processor communications.

The slave port allows a Rabbit processor to be placed on a second Rabbit processor's I/O bus and appear as a peripheral. The slave port implements six registers. Three of these can be written from the external bus and read by the "slave" processor's internal CPU. The remaining three registers can be written by the slave processor's CPU and read from the external bus.

These six slave port registers don't act exactly as a dual port RAM, but they do provide a means of high-speed parallel communication between two Rabbits. Although the following examples show Rabbit-to-Rabbit connections, there is nothing that prohibits the Rabbit slave port from connecting to another processor or even a CPLD or FPGA.

Figure 2.4 shows a diagram of the slave port signals. The data path is 8-bits wide. The two address signals act as register selects. The /SWR and /SRD determine which way data will be flowing on the bus. The /SCS signal enables the port. /SLAVEATTN can be used by the Rabbit to signal the "master" that the "slave" requests a transaction.

The slave port is versatile; the CPUs may be "masters" or "slaves" or even both. Figure 2.5 illustrates how such creativity might manifest in a design.

The slave port opens up the possibility of building very smart peripherals yet having them communicate with other system processors over the I/O bus. For example, consider a Rabbit "slave" processor that acts as a communication subsystem. The slave might have serial streams from six separate I/O devices converge on the six serial ports. The slave could not only handle the bit level serial transactions, but implement higher level protocols, data compression/decompression, error detection/corrections and parsing. The system "master" processor may be attending to local I/O tasks, control-loops and MMI responsibilities. Offloading the bulk of the communication burden to the slave processor allows the master to only have to deal with the highest level details of the messaging.

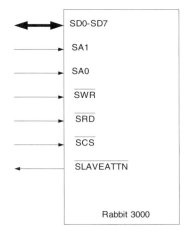

Figure 2.4: The Slave port presents a seamless interface to many CPU I/O busses.

The Basics

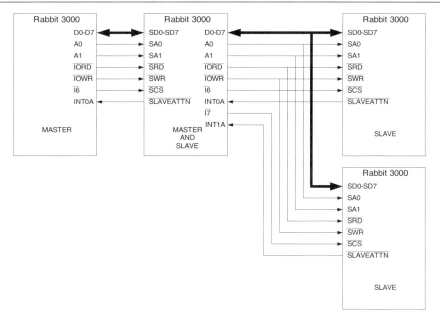

Figure 2.5: Slaves can have slaves—the possibilities are endless.

In addition to fostering processor-to-processor communications, the Rabbit's slave port may be used to cold-boot the processor. Upon reset, if the SMODE1 and SMODE0 pins are held in a {0,1} configuration, the Rabbit will accept boot code over the slave port in much the same way that boot data can be provided over serial channel A.

2.5.5 Input Capture

One of Rabbit's more unique features is the dual channel Input Capture module. This clever bit of engineering allows the Rabbit 3000 to precisely measure the time between external digital events. This can be used for measuring pulse width and frequency. The module's ability to generate interrupts allows the design engineer to add up to four external interrupts to the Rabbit-based system even if the module's event timing capabilities go unused.

Figure 2.6 shows a conceptual model of a single input capture channel. The Source Register (ICSxR) determines which of the bits from Ports C, D, F or G will be the source of the start and stop conditions.

The Trigger Register (ICTxR) determines whether the start and stop conditions will be rising-edge or falling-edge triggered or triggered on any edge. This register also determines whether a "start" or "stop" event will latch the value of the 16-bit counter into the Capture Register. ICTxR controls options for generating an interrupt and starting and stopping the counter.

The Input Capture Control Register (ICCR) specifies what interrupt priority is to be used by the Input Capture module. Both Input Capture channels share this register.

A program may reset the counter through the Control and Status Register. This register also contains a "disable/enable interrupts" bit for the Input Capture module. By reading the ICCSR, a program is able to determine if a start or stop condition has occurred or if the 16-bit Input Capture Counter has rolled over. The ICCSR is shared between both Input Capture channels.

The only registers shared between the two channels are the ICCSR and the ICCR registers. Each Input Capture Channel has a 16-bit counter as well as support registers.

In Figure 2.6, we have used an 'x' to denote a position in a register name where a '1' or a '2' may be placed to identify a register with a specific Input Capture channel. For example, the ICT1R is the Input Capture Trigger Register for channel 1, while ICT2R is the Trigger register for channel 2.

The system peripheral clock (PCLK or PCLK/2) feeds timer A8. The A8 timer can divide its source by n + 1, where n is 0 to 255. This means the highest time resolution of the Input Capture is, best case, one-half period of PCLK/2.

Rabbit Semiconductor suggests three operating modes for the Input Capture module:

1. The counter is enabled by the start condition and halted by the stop condition. The stop condition also latches the counter value into the Input Capture Registers.
2. The counter runs continuously and the start and stop conditions latch the counter value into the Capture Registers.
3. The counter is started by software instructions and halted by the stop event.

Mode one is great for measuring pulse width or for measuring the time between external events. For example, a positive going pulse can be measured by configuring the Input Capture module to accept a rising edge as a start event and the falling edge as the stop event. The Input Capture Counter can be cleared and the Input Capture module configured to start the counter when the start event (rising edge of our pulse) occurs. There is one caveat: the system will stop counting even if a stop signal occurs before a start signal, so you do need to synchronize properly.

Once a stop event happens, the counter is halted, and the contents are latched into the Input Capture Register. The application program need only look at the Input Capture Register to see how long the pulse was.

Provisions are included to contend with counter rollover and interrupt generation, to alert the application program that the stop event has occurred.

Mode two can be used to timestamp events. Each start event and each stop event latch the contents of the counter into the Input Capture Registers. In most cases, the Input Capture module will also be programmed to generate an interrupt to inform the application program that an event has occurred and the data in the Input Capture Registers needs to be processed before a future event occurs and overwrites the timestamp in the Input Capture Registers.

Since start and stop events do not have to occur on the same external pin, this mode can be used simply to generate interrupts. This gives a system designer the ability to have up to four more external interrupt signals (a start and stop event on two channels). This is a simple way of expanding the available external interrupt pins.

The Basics

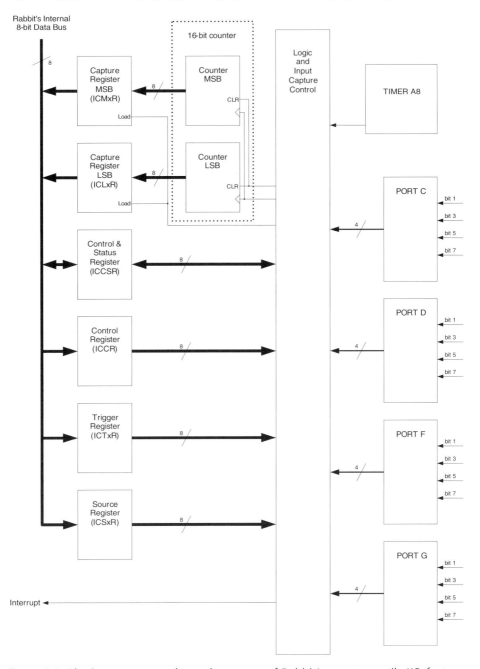

Figure 2.6: The input capture channels are one of Rabbit's most versatile I/O features.

Mode three allows the timer to free-run, and only the stop event will halt the timer. This is useful for measuring the time between the application program clearing the Input Capture timer and an external event occurring.

Chapter 2

The input pins for the start and stop events need not be the same pin. They needn't even be on the same port. The two Input Capture channels can share an external pin as an event generator. The Rabbit Input Capture configuration registers allow versatile and creative configuration of the Input Capture module. For bit level details on each register, refer to the Rabbit 3000 Microprocessor User's Manual included on the CD packaged with this book. (The latest copy is available on the Rabbit web site.)

2.5.6 Quadrature Decoder

The quadrature decoder module is another Rabbit Semiconductor innovation that fell out of Z-World's long experience in the embedded market. Z-World has known for years that optical quadrature encoders are commonly used in rotating machinery as feedback for shaft position. In keeping with Rabbit's philosophy of "let the silicon do the work," the Rabbit 3000 sports an 8-bit quadrature decoder module.

Before looking at the details of the Rabbit's quadrature decoder, let's quickly review what quadrature encoding is and where it is used.

Quadrature encoding uses a two-bit Gray code to indicate the direction of travel of a mechanical device. Mechanical and optical encoders are available that implement quadrature encoding on rotary shafts. These range in price based on the angular resolution of the device as well as the quality of the mechanical construction.

All quadrature encoders contain some sort of encoder wheel, such as the one shown in Figure 2.7. The encoder wheel is read optically or mechanically and an electrical signal is produced on two wires. The two electrical signals are designated as the In-phase (I) signal and Quadrature (Q) signal. In Figure 2.7, the I-signal corresponds to the outer ring and the Q-signal to the inner ring.

The idea behind a quadrature encoder is that the bit sequence produced will present unique transitions depending on the direction of rotation. Unit distance binary codes, such as the two-bit reflective Gray Code shown in Figure 2.7, are ideally suited for this purpose. The code sequence shown in the table at the bottom of Figure 2.7 illustrates how the In-phase and Quadrature signals vary as a function of angular displacement.

The code sequence in the table progresses from the left to the right and this corresponds to a clockwise traversal of the wheel. Careful inspection of the code sequence will reveal that no transition that occurs on a clockwise traversal will appear in a counter-clockwise traversal. For example, the transition of IQ from {1,1} to {0,1} will only appear on a clockwise traversal. Similarly, the sequence {1,1} to {1,0} will only appear on a counter-clockwise traversal.

Figure 2.7 picked an arbitrary rotation of the ring to correspond to 0° (IQ={1,0}). The encoder wheel was structured such that a clockwise traversal (counter clockwise rotation) will create a code sequence that will cause the Rabbit's quadrature decode module's 8-bit counter to increment. Likewise, counter-clockwise traversal will cause Rabbit's counter to decrement.

Rabbit's quadrature decoder is easy to use. There are two channels. They share control and status registers, but each channel has a unique 8-bit counter to keep track of the encoder's position.

The Basics

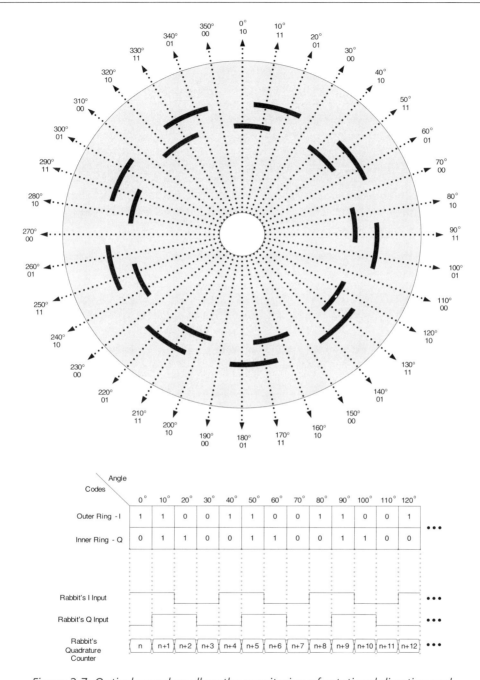

Figure 2.7: Optical encoders allow the monitoring of rotational direction and with some additional processing, speed and position can also be obtained.

Chapter 2

Figure 2.8 shows a block diagram of the quadrature decoder channels. The I and Q inputs can be selected from Port F. Some care must be taken when hooking up a quadrature encoder to port F, since only predefined pin pairs can be used. Port F bits "0 and 1" or bits "4 and 5" may be used to connect to channel 1. Port F bits "2 and 3" or bits "6 and 7" may be used to connect to channel 2. The odd number bits always correspond to the I-signal.

The control registers allow application code to determine if a counter has rolled over. The Quadrature Decoder may also be configured to generate an interrupt when a counter makes a transition between the 0xFF and 0x00 states. The interrupt, if enabled, will occur regardless of the transition direction between the two states (0xFF $\Rightarrow$ 0x00 or 0x00 $\Rightarrow$ 0xFF).

The QDCSR (Quadrature Decode Ctrl/Status Register) allows an application program to reset either channel's counter to 0x00. This allows an application program to "home" the quadrature encoder wheel to correspond to a "count" of 0x00.

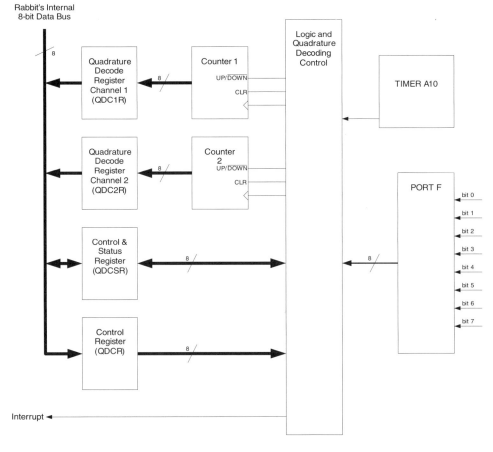

Figure 2.8: Two independent channels of quadrature decoding give the Rabbit great flexibility in controlling rotary systems.

The Basics

It is no accident that the Input Capture Module and the Quadrature Decoder module both allow inputs from Port F. The pulse width of either the In-phase or Quadrature signal will be proportional to the velocity of the encoder wheel. The Input Capture Module, shown in Figure 2.6, only allows measurements from the odd numbered bits of Port F (or C, D or G). This will correspond to the I-signal if a quadrature encoder is connected.

A clever engineer will combine the rotational displacement information from the Quadrature Decoder Module with velocity data gathered through the Input Capture Module to get a complete picture of the behavior of any rotating machinery the Rabbit is monitoring or controlling.

2.5.7 Pulse Width Modulation (PWM) Channels

Pulse Width Modulation is one of the most versatile tools available to the embedded system designer. Everything from audio generation to motor control can be implemented with PWM. Recognizing this, Rabbit Semiconductor developed a PWM module that provides four independent channels of PWM output. Rabbit went one step further than many companies—the Rabbit's PWM module allows the system designer optionally to spread the "on-state" power over the output waveform's period. In some applications, such as audio output or a DAC implemented with a low pass filter, the ripple in the final waveform is reduced.

Let's first look at what PWM is and then see how Rabbit's PWM module operates.

Figure 2.9 shows a pulse and the definitions for some common terms. Assuming that energy in a pulse is contained completely in the "on time" portion of the waveform, the duty cycle is directly proportional to the energy. 100% duty cycle implies a steady state high voltage. 0% duty cycle implies a "pulse" that is off all the time. By varying the duty cycle of the pulse, we can cause the energy in the waveform to go from 0 to 100%.

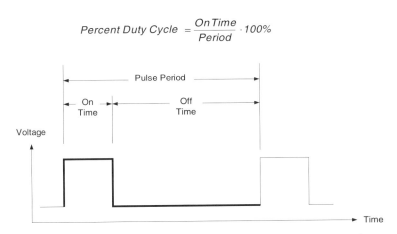

Figure 2.9: Pulse Width Modulation is nothing more than varying duty cycle.

The term "pulse width modulation" derives from an assumption that the pulse period is fixed. In this case, the "pulse" is thought of the high part of the waveform in Figure 2.9. "On time" is equivalent to pulse-width when we are talking about PWM.

Chapter 2

Some newcomers confuse "pulse width" with "pulse period." Careful attention to nomenclature and definitions is important.

The Rabbit PWM module is straightforward—a single 10-bit free-running counter counts from 0x000 to 0x3FF (0 to 1023) and rolls over. The application program loads a PWM "width" register with a value "n." Out of 1024 counts, the PWM's output pulse is high for n+1 counts. This gives the application the ability to vary the duty cycle from about 0.1% to 100%. 1024 ticks from timer A9 define the overall period.

Figure 2.10 shows a block diagram of the PWM module. In the register names PWMxR and PWLxR we are using an "x" to indicate that a number between 0 and 3 can be substituted to control PWM channel 0 to 3.

To connect the PWM module to a pin on Port F, the Port F function register (PFFR is part of Port F and is not explicitly shown in Figure 2.10) must be programmed. If an application needs to run a PWM output all the way down to 0% duty-cycle, PFFR can be programmed to disconnect the PWM module from the Port F pin and the corresponding Port F pin can subsequently programmed to be off.

Figure 2.10 shows that only three bits of the PWM LSB register (PWLxR) actually go to the PWM state machines. Two of these bits are concatenated with the 8 bits from the PWMxR register to form the 10-bit value of *n*. The third bit in PWLxR determines the Rabbit's PWM energy spreader is enabled.

The PWM frequency can be calculated from the formula

$$F = F_{cpu} / 2 / 1024 / (TAT9R + 1)$$

and it is the same for all four PWM signals.

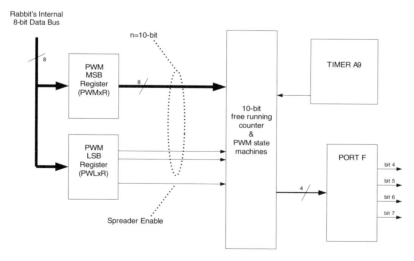

Figure 2.10: Two registers are all that are needed to control each PWM channel.

The Basics

There are a couple of ways to think about the PWM spreader. The easiest, but not completely accurate, way is to simply think of the spreader as a 4× frequency multiplier. The reason this is not accurate is that the actual timer ticks, and therefore resolution, do not change based on the spreader being enabled or not.

Figure 2.11, borrowed from Rabbit's User's Manual, shows how the PWM spreader affects a series of sample configurations. The interesting bit is that the Rabbit's spreader automatically adjusts the pulse widths to be different lengths totaling up to "$n + 1$" complete ticks of A9 so the total energy in a given 1024 tick period is the same regardless of whether the spreader is enabled or not. Notice in Figure 2.11 when $n = 257$, the spreader automatically alternates the short (64 count) and long (65 count) pulses.

The most accurate way to describe the PWM spreader is as a device that breaks up a 1024 count period into four quadrants and distributes $n + 1$ total counts of "on time" across the four quadrants as evenly as possible.

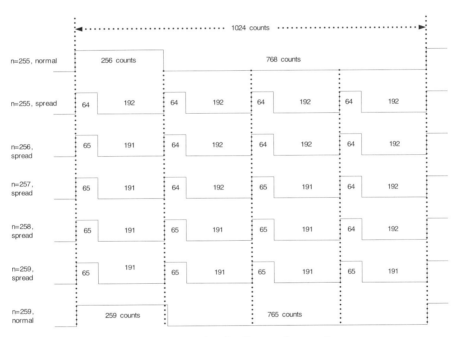

Figure 2.11: The spreader distributes the on-time energy across four quadrants in the 1024 count period.

2.5.8 Auxiliary I/O Bus

The Auxiliary I/O Bus is stroke of genius! As Z-World learned over the years, interfacing a bunch of I/O devices to a memory bus, especially a high-speed memory bus, causes all manner of problems associated with the capacitive loading associated with the I/O devices. The Rabbit 3000 gives system designers the option to keep the CPU's memory bus separate from the I/O bus. This occurs at the sacrifice of 14 I/O pins from Ports A and B. Furthermore, the use of the Slave Port and the Auxiliary I/O bus is mutually exclusive.

Chapter 2

The Auxiliary I/O bus uses the 8 bits of I/O Port A as a data bus. 6 bits from Port B provide the least significant 6 bits of the 16-bit I/O address. 6 bits is only 64 unique addresses, but for many systems this is more than adequate.

When the Auxiliary I/O bus is used, I/O transactions occur simultaneously on the main system bus. If additional address lines are needed, they can be pulled from the main bus and buffered. If only one or two signals are needed, a designer might consider using the UHS family Tiny Logic™ buffers from Fairchild Semiconductor. These offer a small SOT-23 footprint and only cost pennies. Texas Instruments has a similar line of products.

When using a Rabbit Core module in a design, the Auxiliary I/O Bus is often the only I/O bus available. This keeps the trace lengths of the high-speed memory bus short and unwanted EMI minimized. The Auxiliary I/O bus is one of Rabbit's strongest features for real world embedded designs.

2.5.9 Timer A

Rabbit has two timer modules—Timer A and Timer B. Of all or Rabbit's modules, the timer models catch the most flak from users. Unfortunately, the transistor fairy has come and gone. We are stuck with the timers as they sit in the Rabbit 3000. So let's see how they operate and how Rabbit intended each to be used. From there, the system designer can make the most of what's available.

The Timer A module is actually fairly straightforward. Figure 2.12 shows a block diagram of the Timer A module. Each timer, A1-A10, is an 8-bit down-counter with an accompanying reload register. This allows each counter to divide the incoming clock by 1 to 256.

A quick glance at Figure 2.12 shows that the output of every timer has a preferred function. In the cases of A8, A9 and A10 the timers cannot generate an interrupt and are only useful for their assigned task.

The A8, A9 and A10 timers may a first glance look as generic as A2–A7, but alas they are not. These three resources serve only their I/O modules and cannot generate interrupts. They cannot

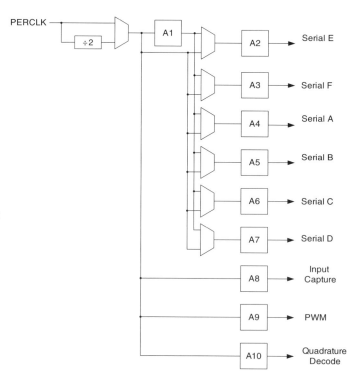

Figure 2.12: Timer A consists of ten 8-bit timers.

be used as prescalers or be cascaded with A1. They are intended to produce relatively fast clocks for their respective I/O modules. They fulfill their purpose well. Some engineers find it less painful to think of these timers as part of their assigned modules.

A1 can be used as a prescaler for A2–A7. However, unless ultraslow bit rates are used, A1 isn't usually needed as a prescaler for baud rate generation. A1 can generate an interrupt when it counts down to 0x00. Unless A1 is being used as a prescaler, it is available for general-purpose interrupt generation.

A2–A7 can generate an interrupt when their counts reach 0x00. If these are not being used to generate baud rates, they can be used as general-purpose time bases. Figure 2.12 shows that these timers may not be cascaded to produce low frequency interrupts. Furthermore, the Timer A module shares a single interrupt among all the timers with interrupts enabled. The Interrupt Service Routine (ISR) is responsible for determining which timer(s) tripped the interrupt.

Rabbit Semiconductor has been very careful in their design regarding the bits that indicate if a particular timer (A1–A7) has rolled over. Under no circumstances will a bit be dropped if multiple timers simultaneously rollover. When an ISR reads the register (TACSR—Timer A Control and Status Register) containing the rollover bits, the ISR must be exhaustive in the inspection of these bits as they are cleared in TACSR upon read. If multiple timers have expired, then the ISR must execute the code required to service each timer that expired.

If additional timers expire while the ISR is running, a new interrupt will be generated once the ISR has completed and the interrupt priority is lowered. Under no circumstances will the Rabbit Timer A Module lose rollover information.

The timers A1–A10 may not be read directly. They may not be used to timestamp events. However, the Input Capture Module, driven by A8, can be used as a timestamp. Timer B may also be read directly by the CPU.

2.5.10 Timer B

Figure 2.13 shows how the two "timers" B1 and B2 are arranged. Not shown in Figure 2.13 are two control registers that determine from which source the 10-bit counter is clocked and whether a "match" will generate an interrupt or not.

There is no method to reset the 10-bit counter. This means that to use B1 or B2 to generate an interrupt on a schedule other than 1/1024 of the selected clock source, the ISR must compute the next "match" value required to get the next IRQ at the desired time. While unorthodox (to be kind), this arrangement is functional.

To read the value of the 10-bit counter, a special procedure, described in the Rabbit Microprocessor User's Manual, must be followed. This is required because the 10-bit counter can't be represented in a single 8-bit byte. Since two bytes have to be read to acquire the counter's state, the possibility exists that in the time between the two reads, a counter rollover may have occurred thus invalidating the first byte read.

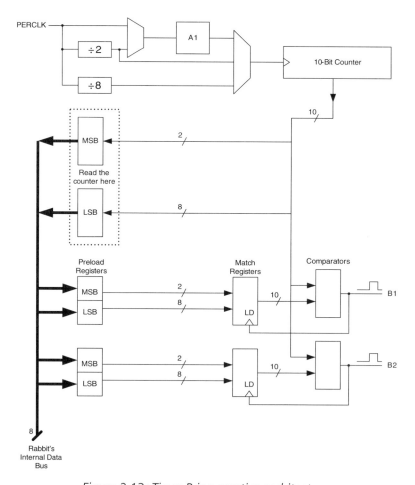

Figure 2.13: Timer B is a creative architecture.

Besides being useful for generating periodic interrupts, B1 and B2 can both be used to synchronously update parallel I/O.

By most standards, Timer B is a bit of a pain to use. The repeated computation of reload values is annoying. The fact that B1 and B2 share a common counter and time base is a little less than optimal. However, what the hardware lacks, the firmware can make up. The trade-off is simply CPU cycles.

2.5.11 Clock Spreader

Electromagnetic regulatory compliance is a very real issue for companies. Testing is expensive. Redesigning circuits in a product that is already complete, just to pass government EMI tests, is expensive and time consuming.

Rabbit Semiconductor looked at Z-World's experience with EMI and created a Clock Spreader to reduce EMI at the source. Most EMI issues occur at harmonics of the system crystals or CPU bus periods.

The Clock Spreader works by introducing a seemingly random modulation on the crystal oscillator, thus spreading out the energy associated with the oscillator's fundamental and harmonic frequencies over a greater bandwidth. This reduces peak emissions and makes passing government EMI tests much easier.

The Clock Spreader has three settings: OFF, NORMAL and STRONG.

The settings operate as one would intuitively expect. OFF allows the system crystal to ring at the most pure frequency possible. This has the benefit of ensuring the most predictable system wide timing. Baud rates derived from the crystal will be as accurate as possible. The disadvantage is that in these days of ever-increasing EMI regulations, the system will also have the highest emissions.

The NORMAL and STRONG modes introduce a small or modest amount of modulation (spreading). The more modulation, the more the undesired harmonic energy is spread out making EMC compliance easier.

In some extremely rare situations if the Clock Spreader is set to STRONG and the highest baud rates are being used with asynchronous serial channels, some bit errors may occur. Only very rarely will a designer run into a baud rate issue. If one does crop up, the Clock Spreader can be dialed back from STRONG to NORMAL or even to OFF.

2.5.12 Clock Doubler

The Rabbit 3000 sports a Clock Doubler. Slow speed external crystals can be used and doubled by the Rabbit 3000. One application for the Clock Doubler is to allow the CPU the option of trading power consumption against computational horsepower.

A system's power consumption is roughly linearly proportional to the system clock rate. For example, turning on the Clock Doubler, the Rabbit 3000 can approximately halve the time required to compute arithmetic operations. However, when the application isn't "computing," the Clock Doubler can be disabled, allowing the power consumed by the CPU to be reduced by 50%.

Chapter 2

2.5.13 MMU and MIU

The memory management unit (MMU) and memory interface unit (MIU) combine to give Rabbit 3000 the ability to directly address six megabytes of memory. That's a huge memory space for a little 8-bit processor.

Currently available compilers only handle up to a 1-megabyte address space. This is in keeping with the Rabbit's 20-bit physical address bus. The overhead of swapping memory pages through the Chip Select, Output Enable and Write Enable or selective inversion of A18 and A19 lines is not incurred. Nonetheless, clever engineers could use the Rabbit to address a full six megabytes if they are willing to do the bookkeeping manually.

Being an 8-bit processor, Rabbit's external and internal data busses are 8 bits wide. This does not theoretically constrain the size of the external address bus. Supporting a wide address bus requires the CPU to do multiple fetches across the 8-bit wide data bus to acquire instruction address operands.

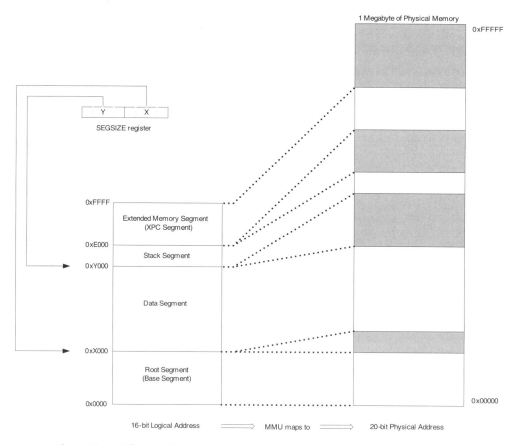

Figure 2.14: The MMU creates segments in the 16-bit logical address space that allow access to windows in the 20-bit physical address space.

For example, a 16-bit wide address bus requires the CPU to perform three read cycles over the 8-bit data bus to fetch a complete instruction that loads a register from a memory location. The first read acquires the "load" instruction. The two following read cycles acquire the address.

A 16-bit (64 KB) address space is perfectly good for many applications. The Tandy TRS-80™ Model III business computer ran with only 48 KB of stock memory. The 1980s are well behind us and modern systems are expected to support larger memory models.

Rabbit Semiconductor was unwilling to trade the overhead needed to fetch a third byte for each operand containing an address. To eliminate this inefficiency, Rabbit adopted an MMU. This module maps a 16-bit logical address into a 20-bit physical address. The CPU instructions generally operate on 16-bit addresses and therefore only two read cycles are required to fetch addresses. Rabbit has also included some special instructions that operate directly on 20-bit addresses. These instructions, while useful, are slower than instructions that operate on 16-bit addresses.

The MMU maps the small 16-bit logical address space into the larger 20-bit physical space by creating windows, called "segments." Figure 2.14 illustrates the idea of mapping with segments.

The segments are named more or less for how Dynamic C uses them. For this discussion, we are not concerned with how Dynamic C manipulates the MMU (this is covered in Chapter 3). Here we are only concerned with how the Rabbit's hardware treats the address space.

The first segment, called the Root Segment or Base Segment, is forced to start at logical address 0x0000 and physical address 0x00000.

The size of the second and third segments is determined by the contents of the SEGSIZE register. The four least significant bits of SEGSIZE specify the logical address where the second segment, called the Data Segment, starts. The four most significant bits from SEGSIZE specify the logical address where the third segment, called the Stack Segment, starts. The fourth segment, called the XPC Segment or Extended Memory Segment always starts at logical address 0xE000.

Upon receipt of a 16-bit logical address, the MMU decides to which segment the address belongs, based on SEGSIZE. Then the MMU computes a 20-bit physical address by adding the 16-bit logical address to a segment mapping register unique to each segment.

Figure 2.15 shows the three mapping registers accessible by the user and how they are added to the 16-bit logical address to form a 20-bit physical address. The XPC, STACKSEG and DATASEG registers are 8 bits. The XPC is actually a proper CPU register so that it can be changed quickly by an application. STACKSEG, DATASEG and SEGSIZE are internal peripheral I/O registers.

Figure 2.15 shows that a change of the least significant bit in XPC will move the XPC segment in physical memory by 4K. The reasoning behind the fixed 8K size of the XPC is now explainable.

The XPC segment is mapped into physical memory with a granularity that is one-half of its size. This allows the XPC segment to be slid over large slices of code or large data structures in memory without completely losing sight of the image just viewed.

Chapter 2

For example, if a 12K space in physical memory contains code, initially the XPC segment could be mapped over the first 8K. As the code execution progresses to the 6K mark, the XPC segment could be adjusted to view the 4K–12K portion. This allows the currently executing code (near 6K) to be visible in both mappings.

As it turns out, this ability to slide a segment gradually across physical memory makes life easier for compiler designers. The Z180 MMU did not have this feature—Rabbit Semiconductor added it to their MMU.

Rabbit's memory management is split between two modules. As we have seen, the MMU is responsible for mapping a 16-bit logical address into a 20-bit physical address. The MIU contains potent mojo to implement separate instruction and data spaces, generate external chip selects, output enables and write enables. The MIU also is where the Rabbit obtains the capability to access a full six megabytes of memory.

Figure 2.15 shows that the MIU has the ability to selectively invert A16 or A19 depending on whether an instruction or data fetch is occurring. The MMIDR (MMU Instruction/Data Register) controls this behavior. This is how the Rabbit implements separate instruction and data space (I & D space).

Separate I & D space only applies to the first two segments. The Stack Segment and XPC segment are unaffected by the selective inversion of A16 and A19

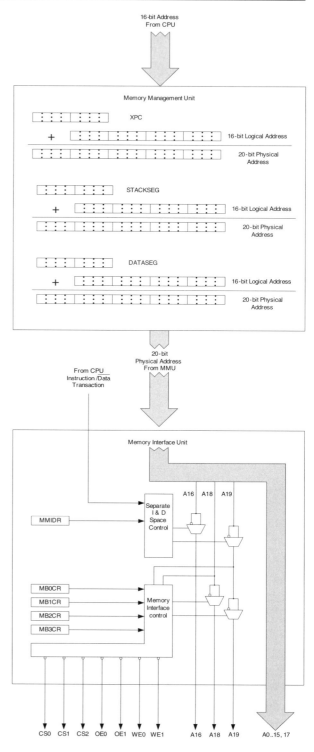

Figure 2.15: The MMU and MIU form a complete memory management structure.

The Basics

dictated by the MMIDR. This arrangement allows the XPC Segment (and Stack Segment) to map into the physical memory without complications from the MIU.

The MIU is also responsible for generating the memory interface signals (/CS0, /CS1, /CS2, /OE0, /OE1, /WE0, /WE1). This is done through the four Memory Bank Control Registers (MBxCR).

The one megabyte memory space is divided into four 256K quadrants. Each quadrant has a corresponding MBxCR. These registers allow the application program to specify the assertion of the memory interface signals as a function of the physical address being accessed. This allows the Rabbit to accommodate a variety of memory devices.

For example, consider a system with a 512K SRAM chip and a 512K flash chip. A logical approach to memory interface would be to configure MB0CR and MB1CR to assert the /CS0. Connect /CS0 to the flash chip. Configure MB2CR and MB3CR to assert /CS1 for accesses to top 512K of the memory space. Then connect /CS1 to enable the SRAM. Both the flash and SRAM could share the Write Enables and Output Enables.

Next, consider a system with the same 512K SRAM chip, but only a 256K flash chip. In this system, we probably still want to place the flash chip down at 0x0000 (as that's where the Rabbit will want to boot from). We may still want to place the SRAM at the top of the 20-bit address space. We configure MB0CR to assert the /CS0, then connect /CS0 to the flash chip. Next, configure MB2CR and MB3CR to assert /CS1 for accesses to the top 512K of the memory space and connect /CS1 to the SRAM. Now we must decide what to do if the application attempts to access the second quadrant (256K–512K). We might map that quadrant back onto the flash memory, or we might map it to the SRAM. Another approach would be to configure MB1CR to assert /CS2, then use /CS2 to generate an external interrupt. The ISR could diagnose the attempt to access a nonexistent memory device and set the system up for a graceful recovery.

Now that we understand how an application program can modify the MBxCR registers to assign a /CSx to specific address range, let's look at how we can use this to connect six 1 MB memory devices to the Rabbit. Figure 2.16 shows such an arrangement.

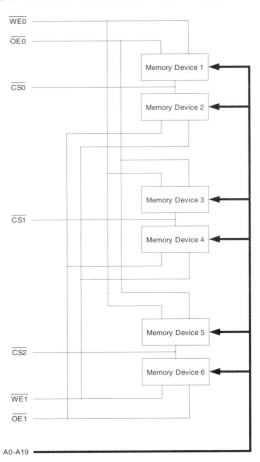

Figure 2.16: The Rabbit 3000 can directly connect to six memory devices, each of which can be up to one megabyte.

61

Chapter 2

The astute engineer will, at this point, be squinting and wondering how the four Memory Bank Control Registers (MBxCRs) will generate three chip-selects across six megabytes of address space. The answer is, "not automagically." The application program must do the bookkeeping necessary to keep track of which chip selects are set up at any given time in which MBxCR. The application program must also keep adjusting the MBxCR configuration to swap in the desired memory device at the desired time. Keep in mind that this is not normally required, since products manufactured by Rabbit Semiconductor have at most one megabyte of memory.

The Rabbit MIU offers an additional feature that is useful for managing memory sizes greater than 1 MB. This is the selective inversion of A18 and A19 depending on the MBxCRs.

Figure 2.17 shows how 4 MB of memory can be connected to the Rabbit and, through the use of selective inversion, 256 KB quadrants in each memory device may be paged into the

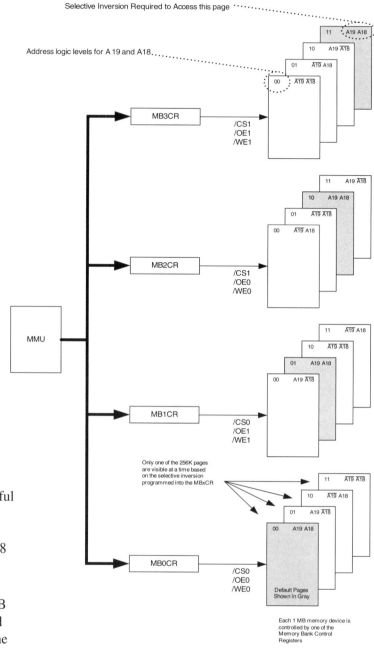

Figure 2.17: Selective inversion of A18 and A19 allow the Rabbit to page 256 KB quadrants in 1 MB external memory devices manually.

visible space. Again, the application program is responsible for the paging, and careful housekeeping will be required.

In Figure 2.17, it is important to understand that the selective inversion of A18 and A19 occur after the MBxCR register has been chosen. In Figure 2.15, as well as in Figure 2.17, the address lines A18 and A19 are brought from the MMU into the Memory Interface Control block. The Memory Interface Control block uses A18 and A19 to determine which MBxCR is applicable to the current address. If selective inversion is required, then the inversion is accomplished on the A18 and A19 line inside of the Memory Interface Control block shown in Figure 2.15.

Figure 2.15 shows that A19 may be inverted as part of the separate I & D functionality before being delivered to the Memory Interface Control Block. For the sake of understanding the Rabbit's ability to index large memory blocks by paging, through the use of selective inversion of A18 and A19, the separate I & D space functionality should be ignored initially.

As a practical matter, Rabbit Semiconductor and Z-World do not offer any commercial products with more than 1 MB of memory on the main processor bus. Dynamic C does not support the bookkeeping to track the manual paging of external memory devices configured as shown in Figures 2.16 and 2.17.

Rabbit Semiconductor has built engineering prototypes sporting large memory footprints such as those shown in Figures 2.16 and 2.17. The technology exists for large memory support. It only awaits an application.

Figure 2.15 omits several MIU related registers. The Memory Timing Control Register (MTCR) tells the MIU how to align the memory interface signals against the system clock, thus allowing the Rabbit maximum versatility in interfacing to a large array of available memories.

The Breakpoint/Debug Control Register (BDCR) allows the MUI to control how RST 28 instructions are handled. The RST 28 instruction can act as a software interrupt (commonly called a trap) for debugging. In final production code, the RST 28 instructions can be left in the application code to ensure overall system timing is the same between "debugging" and "runtime" execution. The BDCR can command RST 28 to behave as a NOP. Dynamic C uses this feature.

2.5.14 Glueless Memory Bus

Our discussion of the Rabbit architecture started with the CPU and has progressed outward through the array of onboard peripherals and MMU. At last we come to the microprocessor's largest boundary—the memory bus.

Once again, Rabbit's designers pulled out all the stops to create an easy-to-use interface. The system designer can plunk down just about any old 8-bit memory device and the Rabbit 3000 will talk to it. The Rabbit generates all the control signals necessary to talk to memory devices—Output Enables, Chip Selects, Write Enables and of course the Address and Data buses are all provided. Memory interfacing is glueless.

Figure 2.16 shows just how easy it is to interface memory to the Rabbit 3000.

The Rabbit Semiconductor web site maintains a Technical Note (TN226) with a list of viable flash devices: http://www.rabbitsemiconductor.com/support/techNotes_whitePapers.shtml

2.6 In Summary

When picking a processor for a project, many issues need to be considered. Some are technical. Some are economic. Some are related to risk management. In all cases, evaluating processors or controllers will involve apples-to-oranges comparisons.

Rabbit Semiconductor grew out of Z-World's vast experience building embedded controllers for a wide array of applications. Rabbit's designers created a processor capable of simplifying many of the tasks commonly found in embedded system design. Rabbit and Z-World, through the development of core modules and packaged controllers, have hardware solutions from the chip level to packaged controllers.

Z-World supports software development on Rabbit-based systems with highly integrated tools dubbed Dynamic C™. Third party tools such as those offered by Softools are also available.

Next, we will look at how to get started with a Rabbit-based core module using Dynamic C as the development environment.

CHAPTER 3

Starting Out

In Chapter 1, we covered the concept of using microprocessors versus cores. We will start this chapter with a brief description of the RCM3200 Rabbit core and then get into the Rabbit development environment. We will cover development and debugging aspects of Dynamic C and will jump into our first Rabbit program. We will highlight some of the differences between Dynamic C and ANSI C. Knowing these differences will make our life easier as we learn to program with Dynamic C.

3.1 Introduction to the RCM3200 Rabbit Core

A processor does not mean a lot by itself. The designer has to select the right support components, such as memory, external peripherals, interface components, etc. The designer has to interface these components to the CPU, and design the timing and the glue logic to make them all work together. There are design risks involved in undertaking such a task, not to mention the time in designing, prototyping, and testing such a system.

Using a core module solves most of these issues. Buying a low-cost module that integrates all these peripherals means someone has already taken the design through the prototyping, debugging, and assembly phases. In addition, core manufacturers generally take EMI issues into account. This allows the embedded system builder to focus on interface issues and application code.

As discussed in Chapter 1, there are several advantages to using cores. The greatest advantage is reduced time-to-market. Instead of putting together the fundamental building blocks such as CPU, RAM and ROM, the designer can quickly start coding and focus instead on the application they are trying to develop.

To illustrate how to use a core module, we will setup an RCM3200 core module and step through the code development process.

The RCM3200 core offers the following features:

- The Rabbit 3000 CPU running at 44.2 MHz
- 512K of Flash memory for code
- 512K of fast SRAM for program execution
- 256K of battery backed SRAM for data storage
- Built in real-time clock

Chapter 3

- 10/100Base-T Ethernet
- Six serial ports
- 52 bits of digital I/O
- Operation from 3.15 V to 3.45 V

During development, cores mount on prototyping boards supplied by Rabbit Semiconductor. An RCM3200 prototyping board contains connectors for power and I/O, level shifters for serial I/O, a reset switch, and a pair of switches and LEDs connected to I/O pins. A useful feature of the prototyping board is the prototyping area that has both thru-holes and SMT pads. This is where designers can populate their own devices and interface them with the core.

The Rabbit Semiconductor prototyping boards are designed to allow a system developer to build preliminary designs and write code on the prototyping board. This allows initial system development to occur even if the application's target hardware is not available.

Once final hardware is complete, the core module can be moved from the prototyping board to the target hardware and the system software can then be finalized and tested.

3.2 Introduction to the Dynamic C Development Environment

The Dynamic C development system includes an editor, compiler, downloader, and in-circuit debugger. The development tools allow users to write and compile their code on a Windows platform, and download the executable code to the core. Dynamic C is a powerful platform for development and debugging:

Development:

- Dynamic C includes an integrated development environment (IDE). Users do not need to buy or use separate editors, compilers, assemblers or linkers.
- Dynamic C has an extensive library of drivers. For most applications, designers do not need to write low-level peripheral interface code. They simply need to make the right API calls. Designers can focus on developing the higher-level application rather than spend their time writing low-level drivers.
- Dynamic C uses a serial port to download code into the target core. There is no need to use an expensive CPU or ROM emulator. Users of most cores load and run code from flash.
- Dynamic C is not ANSI C. We will highlight some of the differences as we move along.

Debugging:

Dynamic C has a host of debugging features. In a traditional development environment a CPU emulator performs these functions. However, Dynamic C performs these functions, saving the developer hundreds or thousands of dollars in emulator costs. Dynamic C's debugging features include:

- Breakpoints—Set breakpoints that can stop program flow where required, so that the programmer can examine and change the state of variables and registers or figure out how the program got to a certain part of the code

- Single stepping—Step into or over functions at a source or machine code level. Single stepping will let the programmer examine program flow, or values of CPU registers, program variables, or memory locations.
- Code disassembly—The disassembly window displays addresses, opcodes, mnemonics, and machine cycle times. This can help the programmer examine how C code got converted into assembly language, as well as calculate how many machine cycles it may take to execute a section of code.
- Switch between debugging at machine code level and source code level by simply opening or closing the disassembly window.
- Watch expressions—This window displays values of selected variables or even complex expressions, including function calls. The programmer can therefore examine or evaluate values of selected variables during program execution. Watch expressions can be updated with or without stopping program execution and can be used to trigger the operation of hardware devices in the target. Use the mouse to "hover over" a variable name to examine its value.
- Register window—All processor registers and flags are displayed. The contents of registers may be modified as needed.
- Stack window—Shows the contents of the top of the stack.
- Hex memory dump—Displays the contents of memory at any address.
- STDIO window—**printf** outputs to this window, and keyboard input on the host PC can be detected for debugging purposes.

3.3 Brief Introduction to Dynamic C Libraries

Dynamic C provides extensive libraries of drivers. Low-level drivers have already been written and provided for common devices. For instance, Dynamic C drivers for I²C, SPI, various LCD displays, keypads, file systems on flash memory devices, and even GPS interfaces are already provided. A complete TCP stack is also included for cores that support networking.

There are some differences between Dynamic C and ANSI C. This will be especially important to programmers porting code to a Rabbit environment. As we cover various aspects of code development, we will highlight differences between Dynamic C and ANSI C.

Source code for Dynamic C libraries is supplied with the Dynamic C distribution. Although the Dynamic C library files end with a ".LIB" extension, these are actually source files that can be opened with a text editor.

For example, let us examine the LCD library. If Dynamic C is installed into its default directories, we find an LCD library file at DynamicC\Lib\Displays\LCD122KEY7.LIB:

The library file defines various variables and functions. Because it is an LCD library, we find functions that initialize a display and allow the programmer to write to an LCD.

Looking at the function descriptions, the programmer can quickly understand how Rabbit's engineers implemented each function. The embedded systems designer can tailor the library functions to suit particular applications and save them in separate libraries.

Chapter 3

> **Quick Summary:**
> - Dynamic C is not ANSI C
> - Dynamic C library files end with a ".LIB" extension, and are source files that can be opened with a text editor

3.4 Memory Spaces in Dynamic C

In Chapter 2, we looked at various memory spaces used by the Rabbit. Here we see how Dynamic C manipulates the MMU to provide an optimal memory usage for the application.

The Rabbit has an external 8-bit data bus. This allows the processor to interface to inexpensive 8-bit memory devices. The trade-off with a small data bus is the multiple bus accesses required to read large amounts of data. To minimize the time required to fetch operands containing addresses while still providing a useful amount of address space, the Rabbit uses a 16-bit address for all instruction operands.

A 16-bit address requires two read cycles over the data bus to acquire an address as an operand. This implies an address space limited to 2^{16} (65536) bytes. A 16-bit address space, while usable, is somewhat limiting.

To achieve a usable memory space larger than 2^{16} bytes the Rabbit's designers gave the microprocessor a memory management unit (MMU). This device maps a 16-bit logical address to a 20-bit physical address.

The Rabbit designers could have simply made the Rabbit's instructions accept 20-bit address operands. This would require 3 bytes to contain the operands and would therefore require three fetches over the 8-bit data bus to pull in the complete 20-bit address. This is a 50% penalty over the 2 fetches required to gather a 16-bit address.

Many programs fit quite handily in a 16-bit address space. The performance penalty incurred by making all the instructions operate on a 20-bit address is not desirable. The MMU offers a compromise between a large address space and efficient bus utilization. Good speed and code density are achieved by minimizing the instruction length. The MMU makes available a large address space to applications requiring more than a 16-bit address space.

The Rabbit 3000™ Designer's Handbook covers the MMU in exacting detail. However, most engineers using the Rabbit only need understand the rudimentary details of how Dynamic C uses the feature-rich Rabbit MMU.

3.4.1 Rabbit's Memory Segments

The Rabbit 3000's MMU maps four segments from the 16-bit logical address space into the 20-bit physical address space accessible on the chip's address pins. These segments are shown in Figure 3.1.

In Chapter 2, we examined the MMU registers in some detail. Further information is also available in the Rabbit 3000 documentation available on the CD included with this book. We will

Starting Out

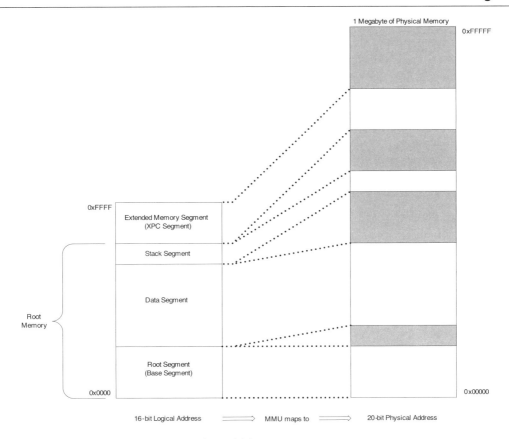

Figure 3.1: The Rabbit 3000 MMU segments.

not rehash the details of how the MMU performs the mapping of a 16-bit logical address to a 20-bit physical address. Our objective here is to understand how Dynamic C uses the Rabbit's MMU to create a framework for embedded applications written in C and assembly language.

Dynamic C uses the available segments differently depending on whether separate instruction and data space is enabled. First we will consider the case without separate I & D space enabled.

3.4.2 Dynamic C's Memory Usage Without Separate I & D Space

Dynamic C's support of separate I & D space allows much better memory utilization than the older model without separate I & D space. This section is included for the benefit of engineers who may have to maintain code written for the older memory model. New applications should be developed using separate I & D space. The newer memory model almost doubles the amount of root memory available to an application.

Dynamic C uses each of the four segments for specific purposes. The Root Segment and Data Segment hold the most frequently accessed program code and data. The Stack Segment is where the system stack resides. The Extended Memory Segment is used to access code or data that is placed outside of the memory mapped into the lower three segments.

Chapter 3

A bit of Rabbit terminology worth remembering is the term "root memory." Root memory contains the memory pointed to by the Root segment, the Data Segment and the Stack Segment (per Rabbit 3000 Microprocessor Designer's Handbook). This can be seen in Figure 3.1.

Another bit of nomenclature to keep in mind is the word "segment." When we use the word "segment" we are referring to the logical address space that the MMU maps to physical memory. This is a function of the Rabbit 3000 chip. Of course, Dynamic C sets up the MMU registers, but a "segment" is a slice of logical address space and correspondingly a reference to the physical memory mapped.

Segments can be remapped during runtime. The XPC segment gets remapped frequently to access extended memory, but most applications do not remap the other segments while running.

The semantics may seem a little picky, but this attention to detail will help to enforce the logical abstractions between Dynamic C's usage of the Rabbit's hardware resources and the resources themselves.

An example is the phrase "Stack Segment" and the word "stack." The "Stack Segment" is just a mapping of a slice of physical memory into logical address space. There is no intrinsic hardware requirement that the system "stack" be located in this segment. The "Stack Segment" was so named because Dynamic C happens to use this third MMU segment to hold the system stack. The "Stack Segment" is a piece of memory mapped by the MMU's third segment. The "stack" is a data structure that could be placed in any segment.

The Root Segment is sometimes referred to as the Base Segment. The Root Segment maps to BIOS code, application code and Dynamic C constants. In most designs the Root Segment is mapped to flash memory. The BIOS is placed at address 0x00000 and grows upward. The application code is placed above the BIOS and grows to the top of the segment. Constants are intermixed with the application code.

Dynamic C refers to executable code placed in the Root Segment as "Root Code." The Dynamic C constants are called "Root Constants" and are also stored in the Root Segment.

The Data Segment is used by Dynamic C primarily to hold C variables. The Rabbit 3000 microprocessor can actually execute code from any segment; however, Dynamic C uses the Data Segment primarily for data. Application data placed in the Data Segment is called "Root Data."

Some versions of Dynamic C do squeeze a few extra goodies into the Data Segment that one might not normally associate with being program data. These items are nonetheless critically important to the proper functioning of an embedded system. A quick glance at Figure 3.2 will reveal that at the top 1024 bytes of the data segment are allocated to hold watch-code for debugging and interrupt vectors. Future versions of Dynamic C may use more or less space and may place different items in this space.

Dynamic C begins placing C variables (Root Data) just below the watch-code and grows them downward toward the Root Segment. All static variables, even those local to functions placed in the extended memory, are located in Data Segment. This is important to keep in mind as the Data Segment can fill up quickly.

Starting Out

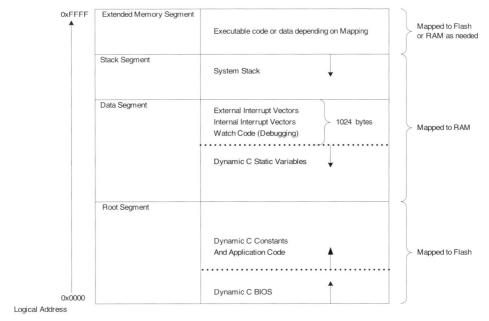

Figure 3.2: Dynamic C's usage of the Rabbit 3000 segments.

Dynamic C's default settings allocate approximately 28K bytes for the Data Segment and 24K bytes for the Root Segment spaces. The macro DATAORG, found in `RabbitBios.c`, can be modified, in steps of 0x1000, to change the boundary between these two spaces. Each increase of 0x1000 will gain 0x1000 bytes for code with an attendant loss of 0x1000 for data. Each incremental decrease of 0x1000 will have the opposite effect.

The Stack Segment, as the name implies, holds the system stack. The stack is used by Dynamic C to keep track of return addresses as well as to pass some variables between functions. Variables of type "auto" also reside on the stack. The system stack starts at the top of the stack segment and grows downward.

The XPC Segment, sometimes called the Extended Memory Segment, allows access to code and data that is stored in the physical memory devices outside of the areas pointed to by the three segments in "Root Memory." Root Memory is comprised of the Root Segment, the Data Segment and the Stack Segment.

The system's "extended memory" is all of the memory not mapped into the Root Memory as shown in Figure 3.1. Extended Memory includes not only the physical memory mapped into the XPC segment, but all the other physical memory shown in Figure 3.1 in gray.

When we refer to extended memory, we are not referring just to memory mapped into the XPC Segment. The XPC segment is the tool (MMU segment) that Dynamic C uses to access all of the system's extended memory. We will use XMEM interchangeably with "extended memory" to mean all physical memory not mapped into Root Memory.

Chapter 3

Generally, functions can be placed in XMEM or in root code space interchangeably. The only reason a function must be placed in root memory is if the function is an interrupt service routine (ISR) or if the function modifies the MMU mapping of the XPC register.

If an application grows large, moving functions to XMEM is a good choice for increasing the available root code space. Rabbit Semiconductor has an excellent technical note TN219, "Root Memory Usage Reduction Tips." For engineers with large applications, this technical note is a must read.

An easy method to gain more space for Root Code is simply to enable separate I & D space, but for when that is not an option, moving function code to XMEM is the best alternative.

3.4.3 Placing Functions in XMEM

Assembly or C functions may be placed in root memory or extended memory. Access to variables in C statements is not affected by the placement of the function, since all variables are in the Data Segment of root memory. Dynamic C will automatically place C functions in extended memory as root memory fills.

Functions placed in extended memory will incur a slight 12 machine cycle execution penalty on call and return. This is because the assembly instructions LCALL and LRET take longer to execute than the assembly instructions CALL and RET. If execution speed is important, consider leaving frequently called functions in the root segment.

Short, frequently used functions may be declared with the root keyword to force Dynamic C to load them in Root Memory. Functions that have embedded assembly that modifies the MMU's special function register called XPC must also be located in Root Memory. It is always a good idea to use the root keyword to explicitly tell Dynamic C to locate functions in root memory if the functions must be placed in root memory.

Interrupt service routines (ISRs) must always be located in root memory. ISRs will be discussed in detail in Chapter 7.

Dynamic C provides the keyword xmem to force a function into extended memory. If the application program is structured such that it really matters where functions are located, the keywords root and xmem should be used to tell the compiler explicitly where to locate the functions. If Dynamic C is left to its own devices, there is no guarantee that different versions of Dynamic C will locate functions in the same memory segments. This can sometimes be an issue for code maintenance.

For example, say an application is released with one version of Dynamic C, and a year later the application must be modified. If the xmem and root keywords are contained in the application code, it doesn't matter what version of Dynamic C the second engineer uses to modify the application. The compiler will place the functions is the intended memory—XMEM or Root Memory.

3.4.4 Separate Instruction and Data Memory

The Rabbit 3000 microprocessor supports a separate memory space for instructions and data. By enabling separate I & D spaces, Dynamic C is essentially given double the amount of root memory for both code and data. This is a powerful feature, and one that separates the Rabbit 3000 processors and Dynamic C from many other processor/tool combinations on the market.

The application developer has control over whether Dynamic C uses separate instruction and data space (I & D space). In the Dynamic C integrated development environment (IDE) the engineer need only navigate the OPTIONS $\Rightarrow$ PROJECT OPTIONS $\Rightarrow$ COMPILER menu and use the check box labeled "enable separate instruction and data spaces".

When Separate I & D space is enabled, some of the terms Z-World uses to describe MMU segments and their contents are slightly altered from the older memory model without separate I & D spaces. Likewise, some of the macro definitions in RabbitBios.c have altered meanings.

For example, the DATAORG macro in the older memory model tells the compiler how much memory to allocate to the Data Segment (used for Root Data) and the Root Segment (used for Root Code) and Root Constants. In a separate I & D space model, the DATAORG macro has no effect on the amount of memory allocated to code (instructions), but instead, tells the compiler how to split the data space between Root Data and Root Constants. With separate I & D space enabled, each increase of 0x1000 will decrease Root Data and increase Root Constant spaces by 0x1000 each.

The reason for the difference in function is an artifact of how Dynamic C uses the segments and how the MMU maps memory when separate I & D space is enabled. For most software engineers, it is enough to know that enabling separate I & D space will usually map 44K of SRAM and flash for use as Root Data and Root Constants and 52K of flash for use as Root Code.

The more inquisitive developer may wish to delve deeper into the memory mapping scheme. To accommodate this, we will briefly cover how separate I & D space works, but the nitty-gritty details are to be found in Chapter 2 and the accompanying Rabbit Semiconductor CD.

When separate I & D space is enabled, the lower two MMU segments are mapped to different address spaces in the physical memory depending on whether the fetch is for an instruction or data. Dynamic C treats the lower MMU two segments (the Root Segment and the Data Segment) as one combined larger segment for Root Code during instruction fetches. During data fetches, Dynamic C uses the lowest MMU segment (the Root Segment) to access Root Constants. During data fetches the second MMU segment (the Data Segment) is used to access Root Data.

When separate I & D space is enabled, the lower two MMU segments are both mapped to flash for instruction fetches, while for data fetches the lower MMU segment is mapped to flash (to store Root Constants) and the second MMU segment is mapped to SRAM (to store Root Data).

This is an area where it is easy to become lost or misled by nomenclature. When separate I & D space is enabled, the terms Root Code and Root Data mean more or less the same thing to the compiler in that "code" and "data" are being manipulated. But the underlying segment mapping is very different than when separate I & D space is not enabled.

Chapter 3

When separate I & D space is not enabled, the Root Code is only to be found in the physical memory mapped into the lowest MMU segment (the Root Segment).

When separate I & D space is enabled, the Root Code is found in both the lower MMU segments (named "Root Segment" and "Data Segment"). Dynamic C knows that the separate I & D feature on the Rabbit 3000 allows both of the lower MMU segments to map to alternate places in physical memory depending on the type of CPU fetch. Dynamic C sets up the lower MMU segments so that they BOTH map to flash when an instruction is being fetched. Therefore Root Code can be stored in physical memory such that Dynamic C can use the two lower MMU segments to access Root Code.

This may seem contrary to the segment name of the second MMU segment, the Data Segment. The reader must bear in mind that the MMU segments were named based on the older memory model without separate I & D space. In that model, the CPU segment names were descriptive of how Dynamic C used the MMU segments. When the Rabbit 3000 came out and included the option for separate I & D space, the MMU segments were still given their legacy names. When separate I & D space was enabled, Dynamic C used the MMU segments differently, but the segment names on the microprocessor remained the same.

This brings us to how Dynamic C uses the lower two MMU segments when separate I & D space is enabled and a data fetch (or write) occurs. We are already familiar with the idea of "Root Data," and this is mapped into physical memory (SRAM) through the second MMU segment—the Data Segment.

Constants are another type of data with which Dynamic C must contend. In the older memory model without separate I & D space enabled, constants (Root Constants) were intermixed with the code and accessed by Dynamic C through the lowest MMU segment (the Root Segment). In the new memory model with separate I & D space enabled, Dynamic C still uses the lower MMU segment (the root segment) to access Root Constants. But with separate I & D space enabled, when data accesses occur, the lowest MMU segment (root segment) is mapped to a space where code is not stored. This means there is more space to store Root Constants as they are not sharing memory with Root Code.

Root Constants must be stored in flash. This implies that the lowest MMU segment is mapped into physical flash memory for both instruction and data accesses. Root Code resides in flash, as do Root Constants.

Given this overview, we can consider the effect of DATAORG again. DATAORG is used to specify the size of the first two MMU segments. Since Dynamic C maps the first two MMU segments to Root Code for instruction accesses, and treats the first two MMU segments as one big logical address space for Root Code, changing DATAORG has no effect on the space available for Root Code.

Now consider the case when separate I & D space is enabled and data is being accessed. The lowest MMU segment (the Root Segment) is mapped into flash and is used to access Root Constants. The second MMU segment (the Data Segment) is mapped into SRAM and is used to access Root Data.

Starting Out

Changing DATAORG can increase or decrease the size of the first two segments. For data accesses, this means the size of flash mapped to the MMU's first segment is either made larger or smaller while the second segment is oppositely affected. This means there will be more or less flash memory mapped (through the first MMU segment) for Dynamic C to use for Root Constants with a corresponding decrease or increase in SRAM mapped (through the second MMU segment) for Dynamic C to use as Root Data.

When separate I & D spaces are enabled, the stack segment and extended memory segment are unaffected. This means that the same system stack is mapped regardless of whether instructions or data are being fetched. Likewise, extended memory can still be mapped anywhere in physical memory to accommodate storing/retrieving either executable code or application data.

For most engineers it is enough just to know that using separate I & D space gives the developer the most Root memory for the application. In the rare circumstance in which the memory model needs to be tweaked, the DATAORG macro is easily used to adjust the ratio of Root Data to Root Constant space available. For the truly hardcore, the Rabbit documentation has all the details.

3.4.5 Putting It All Together

We have spent a considerable amount of time going over segments. In Section 3.7.3, we will look at code that will reinforce these concepts.

> **Quick Summary:**
> - Logical addresses are 16-bits
> - Physical addresses exist outside the CPU in a 20-bit space
> - The MMU maps logical addresses to physical addresses through segments
> - Depending on application requirements such as speed and space, it may be important to control where code and data are placed. Dynamic C's defaults can be overridden, allowing the programmer to decide where to place these code elements in memory

Chapter 3

3.5 How Code is Compiled and Run

Let's look at the traditional build process and contrast it with how Dynamic C builds code:

3.5.1 How Code is Built in Traditional Development Environments

- The programmer edits the code in an editor, often part of the IDE; the editor saves the source file in a text format.
- The programmer compiles the code, from within the IDE, from command line parameters, or by using a make utility. The programmer can either do a "Compile All," which will compile all modules; or the make utility or IDE can only compile the modules that were changed since the last time the code was built. The compiler generates object code and a list file that shows how each line of C code got compiled into one or more lines of assembly code. Unless specified, each object module has relative memory references and is relocatable within the memory space, meaning it can reside anywhere in memory. Similarly, each assembly module gets assembled and generates its own relocatable object module and list file.
- If there are no compilation or assembly errors, the linker executes next, putting the various object modules together into a single binary file. The linker converts relative addresses into absolute addresses, and creates a single binary file of the entire program. Almost all linkers nowadays also have a built-in locator that locates code into specific memory locations. The linker generates a map file that shows a number of useful things, including where each object module resides in memory, how much space does the whole program take, etc. If library modules are utilized, the linker simply links in pre-compiled object code from the libraries.
- The programmer can download the binary file into the target system using a monitor utility, a bootstrap loader, using an EPROM emulator, or by simply burning the image into an EPROM and plugging in the device into the prototyping board. If a CPU emulator is being used, the programmer can simply download the code into the emulator.

Starting Out

Figure 3.3 illustrates how code is built on most development environments:

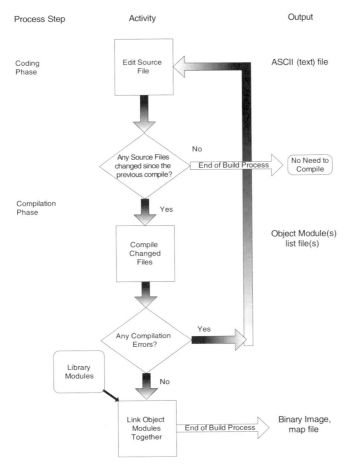

Figure 3.3: The traditional build process.

3.5.2 How Code is Built with Dynamic C

- The programmer edits the code in the Dynamic C IDE, and saves the source file in a text format.
- The Dynamic C IDE compiles the code. If needed, the programmer can compile from command line parameters. Unlike most other development environments, Dynamic C prefers to compile every source file and every library file for each build. There is an option that allows the user to define precompiled functions.
- There is no separate linker. Each build results in a single binary file (with the ".BIN" extension) and a map file (with the ".MAP" extension).
- The Dynamic C IDE downloads the executable binary file into the target system using the programming cable.

Chapter 3

Figure 3.4 illustrates how code is built and run with Dynamic C:

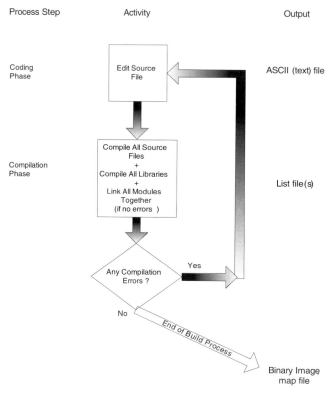

Figure 3.4: How Dynamic C builds code.

Quick Summary:

- Dynamic C builds code differently from the traditional edit/compile/link/download cycle
- Each time code is built, Dynamic C always compiles each library file and each source file
- Each time code is run, Dynamic C does a complete build.
- Within the Dynamic C IDE, executable images can be downloaded to a target system through a simple programming cable

Starting Out

3.6 Setting Up a PC as an RCM3200 Development System

Before we start using Dynamic C to write code, we need to set up an RCM3200 core module and prototyping board. This simple process only takes a few minutes.

Setting up an RCM3200 development system requires fulfilling the following steps:

1. Using the CD-ROM found in the development kit, install Dynamic C on your system.
2. Choose a COM (serial) port on your PC to connect to the RCM3200.
3. Attach the RCM3200 to the prototyping board.
4. Connect the serial programming cable between the PC and the core module.
5. Provide power to the prototyping board.

Now that the hardware is setup, we need to configure Dynamic C. Some Rabbit core modules are able to run code from fast SRAM instead of flash. This feature can be enabled from the Dynamic C "Options ⇒ Project Options ⇒ Compiler" menu. The RCM 3200 will run programs from fast SRAM instead of flash.

For our simple examples, it really doesn't matter whether we configure Dynamic C to generate code that will run from fast SRAM or from flash. However, for the sake of consistency, we always configure Dynamic C to enable code to be run from fast SRAM for the examples in this text that use the RCM3200.

3.7 Time to Start Writing Code!

Now that the RCM3200 system is ready for software development, it is time to roll up the sleeves and start writing code. The first program is very simple. The intent of this exercise is to make sure the computer (the host PC) is able to talk to the RCM3200. Once we are able to successfully compile and run a program, we will explore some of Dynamic C's debugging features, as well as some differences between Dynamic C and ANSI C.

3.7.1 Project: Everyone's First Rabbit Program

It has been customary for computer programmers to start familiarizing themselves with a new language or a development environment by writing a program that simply prints a string ("Hello World") on the screen. We do just that—here's the program listing:

Program 3.1: helloWorld.c

```
main()
{
    printf ("Hello World");         // output a string
}
```

Chapter 3

Here's how to compile and run the Rabbit program:

1. Launch Dynamic C through the Windows Start Menu—or the Dynamic C Desktop Icon.
2. Click "File" and "Open" to load the source file "HELLOWORLD.C." This program is found on the CD-ROM accompanying this book.
3. Press the "**F9**" function key to run the code.

After compiling the code, the IDE loads it into the Rabbit Core, opens a serial window on the screen, titled the "STDIO window," and runs the program. The text "Hello World" appears in the STDIO window. When the program terminates, the IDE shows a dialog box that reads "Program Terminated. Exit Code 0."

If this doesn't work, the following troubleshooting tips maybe helpful:

- The target should be ready, indicated by the message "BIOS successfully compiled..." If this message did not appear or a communication error occurred, recompile the BIOS by typing **<Ctrl+Y>** or select **Reset Target/Compile BIOS** from the **Compile** menu.
- If the message "No Rabbit Processor Detected" appears, verify the target system has power and the programming cable is connected between the PC and the target.
- The programming cable must be connected to the controller. The colored wire on the programming cable is closest to pin 1 on the programming header on the controller. Make sure you use the connector labeled as PROG and not the connector labeled DIAG. The other end of the programming cable must be connected to the PC serial port. The COM port specified in the Dynamic C **Options** menu must be the same as the one to which the programming cable is connected.
- To verify the correct serial port is connected to the target, select **Compile**, then **Compile BIOS**, or press **<Ctrl+Y>**. If the "BIOS successfully compiled ..." message does not display, try a different serial port using the Dynamic C **Options** menu. Don't change anything in this menu except the COM number. The baud rate should be 115,200 bps and the stop bits should be 1.

A Useful Dynamic C Shortcut

"**F9**" causes Dynamic C to do the following:

- Compiles the project source code
- Assuming that there were no compilation errors,
    - Loads the code into flash on the target board
    - Begins execution of the application code on the target system

Starting Out

Although the program terminates, the IDE is still controlling the target. In this mode, called debug or run mode, the IDE will not let the programmer edit the code. For the IDE to release the target and allow editing, we need to close the debug session by clicking on "Edit" and "Edit Mode." Alternatively, pressing "**F4**" will enter Edit Mode.

> **Auxiliary Files Created by Dynamic C During Compilation**
>
> **helloWorld.BDL** is the binary download image of the program
>
> **helloWorld.BRK** stores breakpoint information. It can be opened with a text editor to see the number of breakpoints and the position of each breakpoint in the source file
>
> **helloWorld.HDL** is a simple Intel format Hex download file of the program image
>
> **helloWorld.MAP** shows what the labels (variables and code references) resolve to. In addition, it shows the length and origin of each module and which memory space (Root Code, Root Data, or XMEM Code) in which the module resides
>
> **helloWorld.ROM** is a program image in a proprietary format

3.7.2 Dynamic C's Debugging Features:

Dynamic C offers powerful debugging features. This innovation eliminates the need for an expensive hardware emulator. This section covers the basics of using Dynamic C's debugging features.

Program 3.2 (`watchDemo.c` on the enclosed CD-ROM) is the simple program that will be used to illustrate Dynamic C's debugging features.

Program 3.2: watchDemo.c

```
void delay ()
{
   int j;
   for (j=0; j<20000; j++);        // create a delay
}

main() {
   int count;                       // define variable
   count = 0;                       // initialize counter
   while (1)                        // start an endless loop
   {
       count++;                     // increment counter
       delay();                     // wait a bit
       printf("count = %d\n", count);// print something useful
   } //end of endless loop
} // end of program
```

The code will print increasing values for count in the STDIO window.

3.7.3 Dynamic C Help

Dynamic C makes obtaining help for its library functions simple. Placing the cursor over a function name in the source file and pressing **<Ctrl+H>** will bring up a documentation box for that function.

3.7.4 Single Stepping

To step through a program, we need to compile and download the program to the RCM3200 without running the program. This can be accomplished by any of the following methods:

- pressing <F5>
- selecting Compile from the menu bar
- selecting the lightning bolt icon on the Tool Bar

The IDE highlights (in green) the first character of the program's first executable statement. This green highlighting is always used to indicate the current execution point.

The **F7** and **F8** keys will single step the statements—each time one of these keys is pressed, one program statement will be executed. The difference is that if the statement is a call to another function, pressing **F8** will completely execute the called function, while pressing **F7** will execute the called function one statement at a time.

To summarize, **F8** allows the user to "step over" functions, while **F7** allows the user to "step into" functions.

Pressing **F7** to single step watchDemo.c will execute the following sequence of statements,

`count = 0;`	The first statement of the program
`while`	The "`while`" statement
`(1)`	The statement that must be evaluated for the "`while`" branch
`count++;`	The first statement in the "`while`" body
`for (j=0; j<20000; j++);`	The first statement in `delay()`;

The "`for`" loop in `delay()` has a conditional statement (j<20000), the loop control variable adjustment statement (j++) and a null statement (;) in the loop body. The programmer would have to press F7 another 60,000 times to complete all of the statements in the `delay()` function.

Using step-over, **F8**, Dynamic C would execute the following sequence of statements:

`count = 0;`	The first statement of the program
`while`	The "`while`" statement
`(1)`	The statement that must be evaluated for the "`while`" branch
`count++;`	The first statement in the "`while`" body
`delay();`	The `delay()` function (all 20,000 loops)
`printf("count = %d\n", count);`	The function that prints to STDOUT
`(1)`	The statement that must be evaluated for the "`while`" branch
`count++;`	The first statement in the "`while`" body

Starting Out

```
delay();           The delay() function (all 20,000 loops)
printf("count = %d\n", count);   The function that prints to STDOUT
```

and so on.

F8 is useful for stepping over functions that are believed to be good, while allowing the programmer to carefully examine the execution path of suspect code.

3.7.5 Adding a Breakpoint

Breakpoints are useful for controlling program execution during debugging. Placing the cursor over the statement where the breakpoint is desired and pressing **F2** will assign a breakpoint to the statement. **F2** can be used to remove existing breakpoints. Placing the cursor over an existing breakpoint and pressing **F2** will remove the breakpoint. Dynamic C indicates that a breakpoint exists by changing the background color of the first character in the statement to red.

Alternatively, breakpoints may be placed or removed using the **Toggle Breakpoint** option from the **Run** menu.

Breakpoints are used to halt program execution so the programmer can inspect the state of the target system's hardware, CPU's registers, and program variables. When the program halts due to a breakpoint, step-into and step-over can be used to observe the execution path of suspect code.

Once breakpoints are set, the program is run (using **F9**). The program will advance until it hits a breakpoint. While the program is paused, the IDE allows breakpoints to be added or removed from the target.

Breakpoints may also be set in a program running at full speed. This will cause the program to break if the execution thread hits a breakpoint.

Another technique for using breakpoints will allow software developers to determine if a particular segment of code is being executed. A breakpoint can be placed in the segment of interest and program execution started. If the program never halts, then the segment of interest was not executed. This is a useful technique for determining if code branches occur as expected.

"Normal" breakpoints allow interrupts to continue being serviced even when the breakpoint has been reached. "Hard" breakpoints can be used to disable all execution while the breakpoint is being serviced. This type can be especially useful when debugging ISRs.

3.7.6 Watch Expressions

Once a program is halted, examining the contents of variables is simple. The easiest way to examine a variable is to hover the mouse over the variable of interest. Dynamic C will pop up a small box showing the value. An expression may be evaluated in a similar manner by highlighting the expression and then hovering over it.

Dynamic C also provides a tool called a watch window. The programmer can view expressions added to the watch window.

Chapter 3

For example, adding the integer "count" from watchDemo.c to a watch window allows the developer to observe how "count" changes as the code is single stepped.

Expressions in watch-windows are updated whenever code execution is halted. Breakpoints halt code execution, as do the step-into and step-over tools.

Expressions can be updated while code is running at full speed by pressing **<Ctrl+U>**. The Dynamic C help menu for "Watch Windows" explains that runWatch() should be periodically executed in the program to support **<Ctrl+U>** requests for updating the IDE Watch Window.

Expressions can be added to a watch-window by selecting **Add/Del Watch Expression** from the **Inspect** menu, or by using the **<Ctrl+W>** shortcut.

You can use the Watch Window to change the values of variables as well as execute functions.

Summary of Some Dynamic C Shortcut Keys

Edit, Compile and Run
F4: Enter Edit Mode
F5: Compile code and load executable on target, but don't begin execution
F9: Compile and load executable on target if not already done and then begin execution

Debugging Functions
Ctrl+W: Add or Delete a Watch Expression
Ctrl+U: Update Watch Expression Window
F2: Set or Remove a Breakpoint
Ctrl+A: Clear All Breakpoints
F7: Single Step walking into functions (step-into)
F8: Single Step walking over functions (step-over)

Fast Help
Ctrl+H: Provide help on a function
Alt+F1: Instruction Set Reference

3.7.7 Dynamic C is Not ANSI C

The American National Standards Institute (ANSI) maintains a standard for the C programming language. Code that conforms to the ANSI standard is theoretically more easily ported between operating systems.

Embedded systems have unique performance demands. Different companies have adopted different approaches to meeting the unique challenges presented by embedded systems.

Z-World has enjoyed nearly two decades of unrivaled success with the approach of carefully adapting the C programming language to suit the demands of embedded controllers. Language extensions and creative interpretation of standard C syntax make Dynamic C a desirable environment for embedded systems development.

Starting Out

As universities teach ANSI C with a bent toward desktop application development, it is worth pointing out some of the differences between Dynamic C and ANSI C. This will help new comers to Dynamic C avoid common pitfalls while being able to take full advantage of the enhancements Dynamic C offers.

Chapter 10 will compare and contrast ANSI C and Dynamic C in detail. For now, let's examine a pitfall that new comers often fall into—declarations with initializations.

Most programmers will look at the code presented in Program 3.3A and expect "count" to be an integer that is initialized to be 0. Dynamic C takes a slightly different view of the code.

Program 3.3A: oops.c

```
void main( void ) {
     int count=0;
     printf("count = %d\n", count);
}
```

Dynamic C assumes a variable that is initialized when declared is a constant. This is a common enough situation that Dynamic C will generate a warning when the above code is compiled. The warning states, "line 2 : WARNING oops.c : Initialized variables are placed in flash as constants. Use keyword 'const' to remove this warning."

Changing the declaration to,

```
const int count=0;
```

causes the compiler to generate the same executable code, but without the warning.

If the desired result is to declare an integer named count and initialize it to zero then the code shown in Program 3.3B should be used.

Program 3.3B: better.c

```
void main( void ) {
     int count;
     count=0;
     printf("count = %d\n", count);
}
```

Chapter 3

Initializing global variables may at first glance appear impossible. Dynamic C offers a ready solution with the compiler directive #GLOBAL_INIT.

The Dynamic C documentation explains, "`#GLOBAL_INIT` sections are blocks of code that are run once before `main()` is called."

The code shown in Program 3.3C shows how a global variable can be initialized.

Program 3.3C: GlobalVarInit.c

```
int GlobalVarCount;
void main( void ) {
#GLOBAL_INIT
{
 GlobalVarCount = 1;
}
    printf("GlobalVarCount = %d\n", GlobalVarCount);
}
```

The #GLOBAL_INIT directive can be used to initialize any static variables. If a function declares a static variable and needs that variable initialized only once, then #GLOBAL_INIT can be used to accomplish this. Program 3.3D shows how to do it. The output generated is,

    LocalStaticVar = 1
    LocalStaticVar = 2
    LocalStaticVar = 3

The static integer LocalStaticVar is initialized to 1 before `main()` is executed.

Program 3.3D: LocalVarInitializedOnce.c

```
void xyzzy (void) {
static int LocalStaticVar;
#GLOBAL_INIT
{
 LocalStaticVar = 1;
}
printf("LocalStaticVar = %d\n", LocalStaticVar++);
}
void main( void ) {
xyzzy();
xyzzy();
xyzzy();
}
```

Program 3.3E shows how "not to" initialize a static variable. The static integer LocalStaticVar is assigned the value of 1 every time xyzzy() is called. This is generally not a desired behavior for static variables, which are intended to retain their value between function calls. The output generated is,

    LocalStaticVar = 1
    LocalStaticVar = 1
    LocalStaticVar = 1

Program 3.3E: LocalVarAlwaysInitialized.c

```
void xyzzy (void)
{
static int LocalStaticVar;
LocalStaticVar = 1;
printf("LocalStaticVar = %d\n", LocalStaticVar++);
}
void main( void )
{
xyzzy();
xyzzy();
xyzzy();
}
```

As useful as the compiler directive #GLOBAL_INIT is, it can become a source of confusion.

The key point to remember when using #GLOBAL_INIT is that the order of execution of #GLOBAL_INIT sections is not guaranteed!

All #GLOBAL_INIT code sections are chained together and executed before main() is executed.

Global variables can be modified in multiple #GLOBAL_INIT code segments. If this is done, the compiler will not generate any warnings or errors. If the coder is careless, a global initialization may be overwritten by a subsequent #GLOBAL_INIT code segment. Since the order of execution of #GLOBAL_INIT sections is not guaranteed, the global variable is not guaranteed to have been initialized by the intended #GLOBAL_INIT code segment. To further complicate matters, a source file may have the order of execution of #GLOBAL_INIT sections altered by different versions of the compiler.

#GLOBAL_INIT is a useful and reliable compiler directive. Like any tool, the powerful #GLOBAL_INIT directive must be used within the compiler's constraints. Do not initialize global variables in multiple #GLOBAL_INIT sections. Realize that the order of execution of #GLOBAL_INIT sections is not guaranteed. Know that different versions of Dynamic C are free to re-order the execution of #GLOBAL_INIT sections. Code accordingly.

Chapter 3

> **Summary: Other Differences Between Dynamic C and ANSI C**
>
> Variables initialized upon declaration are constants to Dynamic C and placed in flash memory.
>
> GLOBAL_INIT is a convenient way to initialize variables in functions.

3.7.8 Dynamic C Memory Spaces

Section 3.4 discussed where Dynamic C places variables. We will now re-examine the placement of code and data and how we can force Dynamic C to put code and data where we want.

We compiled Program 3.4 (memory1.c) using Dynamic C version 7.33, and then examined the associated map file (memory1.map). Here are the pertinent excerpts from the source code and the map file:

Program 3.4: memory1.c

```
int my_function (int data)
{

    static int var_func_static;
    int var_func;

    var_func_static = 3;
    var_func = var_func_static*data;

    printf ("%d multiplied by %d is %d\n",data,var_func_static,var_func);

    return var_func;
}


void main()
{

    static int var_static1;
    int var_not_static1;
    static const char my_string[]="I like what I have seen so far!\n";

    var_static1 = 0xA;
    var_not_static1 = 0x5;

    var_not_static1 = my_function (var_static1);

    printf ("%s",my_string);
}
```

Starting Out

The top section of the map file shows origin and sizes of various segments:

```
//Segment      Origin          Size
Root Code      00:0000         0055d5
Root Data      00:bfff         000899
Xmem Code      ff:e200         001716
```

Excerpts of the map file show us where in memory we will find `my_string`, `my_function()`, and `main()`:

```
// Global/static data symbol mapping and source reference.
//   Addr      Size  Symbol
    b857       2     my_function:var_func_static
    b855       2     main:var_static1
 10:022c      33     main:my_string
// Parameter and local auto symbol mapping and source reference.
//Offset Rel. to       Size    Symbol
    4         SP        2      my_function:data
    0         SP        2      my_function:var_func_not_static
    0         SP        2      main:var_not_static1
// Function mapping and source reference.
//   Addr     Size           Function
    1c26      63             my_function
    1c65      58             main
```

Looking at the addresses above and comparing them to the global static data symbol addresses, we can see that the static variables got placed in the Root Code space, while the string got placed in XMEM.

We can see that Dynamic C lumped together the static variables from `main()` and `my_function` with the string constant, and kept the nonstatic variables in the stack. Notice that the stack has reserved two bytes for my_function:data; this is how the lone integer parameter gets passed from `main()` to `my_function()`.

Also notice that the function and main got placed in Root Code.

Chapter 3

Now that we are beginning to get comfortable with where Dynamic C places code and data by default, let's play with it a little—let's try to save root space and move as much as we can to XMEM. We think the program may take just a little longer to execute since Dynamic C and the MMU will have to convert all physical memory accesses to the internal logical representation, but we will save on the precious root space. Changing the above program to work differently, we get memory2.c in Program 3.5:

Program 3.5: memory2.c

```
xmem int my_function(int data)
{
    static int var_func_static;
    int var_func_not_static;

    var_func_static = 3;
    var_func_not_static = var_func_static*data;

    printf ("%d multiplied by %d is %d\n",data,var_func_static,var_func_not_static);

    return var_func_not_static;
}


xmem void main()
{

    static int var_static1;
    int var_not_static1;
    static const char my_string[]="I like what I have seen so far!\n";

    var_static1 = 0xA;
    var_not_static1 = 0x5;

    var_not_static1 = my_function (var_static1);

    printf ("%s",my_string);
}
```

Starting Out

We can expect to see some differences in the map file; we should find the code for main() and my_function() in xmem space. Let's look at the map file and find out if that is the case:

```
//Segment    Origin      Size
Root Code    00:0000     005561
Root Data    00:bfff     00089f
Xmem         ff:e200     001792


// Function mapping and source reference.
//   Addr      Size  Function
     e420      64        my_function
     e460      60        main

// Global/static data symbol mapping and source reference.
//   Addr      Size  Symbol
     b857      2         my_function:var_func_static
     b855      2         my_function:var_func_not_static
     b853      2         main:var_static1
     b851      2         main:var_not_static1
  10:022c     33     main:my_string

// Parameter and local auto symbol mapping and source reference.
//Offset Rel. to    Size    Symbol
     3       SP      2       my_function:data
```

This time we can see that the function and main got placed in XMEM space. The variables, except for the one used for parameter passing between main and the function, got placed in Root Code.

The keywords `xmem` and `root` allow the engineer to force the compiler to locate functions in specific areas of memory. The map file can be used to verify that Dynamic C did what the engineer intended.

Dynamic C versions 8.01 and higher are quite smart about how they locate functions. Most software engineers need not worry about manually locating functions. This is especially true when separate I & D space is enabled, as that gives plenty of root space for both code and data for most applications. However, in the cases when engineers want to tweak the compiler's choices, `xmem` and `root` give the engineer full control.

3.8 What's Next?

In this chapter, we have explored Dynamic C's development environment and memory usage. We have written several small programs that run on the RCM3200 core module. We have examined some of the powerful debugging features offered in the IDE.

In the next chapter, we will continue our exploration of debugging techniques.

CHAPTER 4

Debugging

When developing a software application, bugs will invariably be introduced in the code. Dynamic C has several resources that can assist a developer in locating, isolating and eliminating bugs.

In the last chapter we saw how an engineer can force Dynamic C code to single step through code and set break-points. In this chapter, we will explore Dynamic C's debugging features in more detail as well as some generic debugging techniques.

4.1 The Zen of Embedded Systems Development and Troubleshooting

Troubleshooting is puzzle solving. When it comes to troubleshooting, an engineer's mind-set is eighty percent of the game. The techniques and tools in the toolbox are the remaining twenty percent.

Before discussing the nitty-gritty technical details of debugging in Dynamic C, we will touch on some concepts to help the reader develop a state of mind conducive to embedded systems development and trouble shooting.

While developing an embedded system, an engineer wears three distinct hats. They are inscribed,

- DEVELOPER
- FINDER
- FIXER

Wearing the DEVELOPER hat implies a responsibility to find or create a cost effective solution to the control problem at hand.

An engineer dons the FINDER hat when a malfunction or bug is observed. The responsibility of the FINDER is to delve deeply into the bug and determine the root cause of the malfunction.

Once a bug is identified, the engineer slips on the FIXER hat. The FIXER, much like the DEVELOPER, must find a cost effective solution to a problem. However, unlike a DEVELOPER, a FIXER is usually constrained to an existing design and seldom has the leeway a DEVELOPER does.

4.1.1 The DEVELOPER Hat

Embedded systems developers have many details with which to contend. An enormously helpful philosophy is that of "baby steps for the engineer." This philosophy derives from the fact that diagnosing runtime errors in embedded systems is often significantly more difficult than diagnosing problems in more controlled environments.

Software engineers writing middleware code for SAP or C++ or Perl for Web servers will often knock out a few hundred lines of code in a module before testing it. In embedded systems, a good rule of thumb is to try to test the code every 20 to 50 lines. Start simple. Progress in small steps. Be sure that each function, each module and each library operates correctly both with "normal data" as well as with data outside the boundary conditions.

4.1.2 Regression Testing—Test Early and Test Often

Another bit of philosophy admonishes, "Test early and test often."

When testing embedded systems, it is useful to develop a suite of tests that can be run repeatedly throughout the code development. These tests might be as simple as a sequence of button presses on the MMI, or as complex as replacing hardware drivers with functions (often by substituting a library file) that, instead of acquiring live sensor data, report simulated sensor data.

If the test suite runs successfully on a previous code build, but not on the current code build, and the programmer has only made incremental changes between the two code builds, then we have a good idea where to start looking for defective code.

This test suite is also useful when upgrading a compiler. If the code compiles and successfully completes the tests with a previous version of the compiler, but not with the current version, then we have both an indication that something is wrong, and that changing compiler versions caused the bug.

This technique of running the same old tests on new pieces of code is called regression testing.

4.1.3 Case Study—Big Bang Integration and No Regression Test Suite

Not every system will have difficult to find bugs, but when one does crop up, the time it takes to reproduce can be staggering. Here's one company's experience with an obscure bug.

The company made equipment that printed numbers in magnetic ink on the bottom of checks. The engineering group was long overdue releasing the new high-speed document printer. Marketing had already announced the new product and sales of the older products had subsequently tanked. Customers were holding out for the announced, but unreleased, high capacity, high-speed printer.

The new printer had a hopper that could hold 5,000 documents and could process the documents as fast as the fastest human operator could enter the information to be printed.

The document handler generally operated as expected. With the exception that every 16,000 to 47,000 documents, the code would "lock-up." After a few seconds, the onboard watchdog would reset the system. The log files stored in RAM would invariably be corrupted.

Chapter 4

The software team had opted for a big-bang integration of software modules written by five talented engineers for the system's microprocessors. Each engineer had his or her own style and method of "testing" their code. No collaborative regression tests existed.

The hardware design team was confident in the design. Simple bits of code were used to verify that the motors and solenoids were properly under software control. Digital storage oscilloscopes were used to verify noise levels and transients were well within design tolerances.

The software engineers were sure it was an issue associated with the big bang integration. The software team burned many a gallon of midnight oil looking for solutions and spent months developing simulators and regression tests to try and find the bug.

Through robust testing, they found and fixed many bugs that would have eventually caused customers grief. However, through software testing alone, they were unable to reproduce or identify the cause of the "it just crashes every now and again" bug.

After several months, several hundred thousand dollars, and a 30% company wide reduction in force due to lack of sales of existing product, the problem turned out to be ESD. As the checks were pulled along by little rubber rollers, ESD built up on the documents and was accumulated on a plastic photo-sensor. If conditions were dry enough, eventually the charge would discharge into a nearby PCB trace and would addle the CPU.

The solution was to add a couple of grounded little tinsel brushes to wipe off the static build up from the documents.

This simple little problem almost killed the company. It did cost forty people their jobs. What could have been done?

The hardware group had proceeded with their design in incremental steps. Each piece of hardware was tested. The integrated units were tested. The careful testing and development gave the hardware group, and the engineering management, confidence in the hardware design.

The software group had been much more cavalier. Bright people had coded whole modules quickly—sometimes overnight. The rapid development, coupled with the lack of formal regression testing or integration testing, inspired a lack of confidence in the final code base.

Not only did management feel the issue was a software problem, so did the software engineers. They just *knew* that some buffer was overflowing or some pointer was errant.

The fact that as they developed tests for each module, and for the integration of modules they found bugs, further enforced the belief that the code base was unstable.

It wasn't until months had passed and the software group had found and fixed many bugs in the code base that the company developed enough confidence in the code base to begin to seriously look at potential hardware issues again. Which is where the "show stopper" bug was found.

The lesson to be learned here is that not all problems that appear to be software actually are. Additionally, if one is not confident with the code base, a lot of time and money may be spent looking in the wrong place.

4.1.4 The FINDER Hat

Troubleshooting requires a very methodical mindset. Never assume anything. Never assume that the tools are working flawlessly. Never assume that the hardware is good. Never assume that the code is without bugs.

When troubleshooting, don't look at anything on the bench as being black or white. Think in terms of gray. Each piece of code or hardware or test gear should be assigned a degree of confidence.

For example, consider a digital multimeter (DMM) that is telling us that we have a low supply voltage. Don't assume that the DMM is correct. Crosscheck it.

We could measure a fresh 9-volt battery. If we have confidence in the freshness of the 9-volt battery, then our confidence has increased in the DMM's accuracy (assuming the DMM displayed about 9 volts when connected to the battery).

We could cross check the DMM's measurement of the "low" supply voltage with another DMM or better yet, an analog meter. If we get the same measurement from both instruments, the confidence level improves in both the test instruments (reading similarly) and that the measured supply rail might be low.

But we're not done yet. We can hook up an oscilloscope to the supply rail in question. Even a crummy old slow scope will do in most situations. What we want to eliminate is the possibility that the DMM is giving us a false measurement due to AC noise on the supply rail—which would still be a problem, but a different one than a "low voltage" rail.

Each of the above steps gives us a clue. Each step helps us build confidence in our understanding of the situation. Never assume—always double check.

When debugging a system, there are two distinct hats that must be worn. The first hat is inscribed FINDER, the second FIXER. We should always be cognizant of which hat we are wearing.

The FINDER hat is worn during the diagnostic phase. Before we can properly "fix" a problem, we must understand it.

Some engineers sometimes approach a malfunctioning system with shotgun solutions like these:

- Just make the stack bigger
- Just add an extra time delay here or there
- Just put a few capacitors on the power supply or in the feedback loop
- Just ground the cable shields
- Just disable interrupts
- Just disable the watchdog
- Just make everything a global variable

People that don't necessarily understand the distinction between FINDER and FIXER take this sort of shotgun approach.

Chapter 4

Shotgun solutions don't always address the root cause of the problem. They may mask a problem in a particular unit or prototype, or under a given set of conditions, but the problem may re-assert itself later in the field.

For example, an engineer may see a problem in a function's behavior, and determine that "all this passing parameters by address is overly difficult—I'll just make this function's variables global." This engineer may have fixed this problem for the function in question, but nothing was done to address the root cause of the problem. The engineer that wrote the code may not have understood how to pass parameters into and out of functions. Quite likely another place in the code has similar problems. These problems may just not be asserting themselves at the present time.

A much better solution would have been for the engineer to recognize the deficiency, correct the function that was exhibiting the odd behavior and then carefully comb through the code base looking for similar problems.

The FINDER must carefully amass clues, generate a hypothesis for the problem and then build confidence in the hypothesis through experimentation.

Part of the process of amassing clues involves NEVER CHANGING MORE THAN ONE THING AT A TIME. If two or three tweaks are made to a system as part of the exploration of a problem, there is no way to tell which change affected the behavior of the problem.

If we make a change to a system and the change doesn't seem to change the system's behavior, in most circumstances, the next move should be restore the system to the original configuration (undo the change) and verify the problem still exists.

In many situations, we may make a change that we feel should be incorporated into the final project, even if it didn't affect the problem at hand. If that occurs, be disciplined, undo the change and proceed with the FINDER's duty. Once the root cause of the problem is found, we can always come back (wearing the DEVELOPER's hat) and make additional changes that we might consider good engineering practice.

For example, consider a target system that is exhibiting difficulty communicating with another system. We notice that the communications cable's shield is ungrounded. We ground the shield, but the communications problem still exists. Even though we might consider it a good engineering practice to ground the shield, the system should still be placed back into the original state until we get to the bottom of the communications problem. Always make just one change at a time. Always go back to the initial configuration that caused the problem.

After making a change to a system and observing the behavior, a useful practice is to write a few notes about what change was made and what behavior was observed. The reasons for this are simple. Humans forget. Humans get confused.

An experiment run a couple of minutes ago might be clear in one's mind, but after another four hours of tracking down a difficult bug, the experiment and results will be difficult to recall with confidence. Time is the enemy of recollection. Take notes. All the best detectives have a little notepad.

Bob Pease is an internationally revered engineer and author. In his book "Troubleshooting Analog Circuits," Pease introduces Milligan's Law—"When you are taking data, if you see something funny, Record Amount of Funny." This is as important in software troubleshooting or system level troubleshooting as it is in analog troubleshooting. Take copious and clear notes.

4.1.5 The FIXER Hat

Once the root cause of a problem is understood, the engineer wearing the FIXER hat can devise a solution. The duty of the FIXER is the same as that of any design engineer—balance the cost and timeliness of a solution with the effectiveness of the solution.

Beyond repairing the bug, the FIXER has an institutional responsibility to provide feedback to the engineer or engineering group that introduced the bug. Without this feedback, the same bug may be introduced into future products.

Depending on the production status of the defective product, the FIXER may have the additional burden of devising material dispositions for existing stock, work in progress, and systems deployed at customer sites. In some situations, the FIXER may be called upon to do as much diplomacy as engineering.

4.2 Avoid Debugging Altogether—Code Smart

These are some guidelines that are useful to both the DEVELOPER and the FIXER. As with any guideline, following these to the extreme is probably not going to be either possible or desirable for embedded systems development. An engineer should keep the spirit of these guidelines in mind.

For example, a guideline might say a NULL pointer check before every pointer access is a good idea. On a PC application this might be acceptable, but a NULL pointer check and special handling of NULL pointer cases can bloat and slow down code too much for an 8-bit system. The programmer must take extra care to make sure the situation doesn't happen in the first place. The spirit of the guideline is clearly "be careful that pointers are initialized correctly."

4.2.1 Guideline #1: Use Small Functions

Keep functions to a page or less of code when possible. Minimize their side effects; for example, don't modify global variables whenever possible. Test functions by themselves to make sure they work as expected for various input values, especially boundary conditions. Use and check return values where invalid input is a possibility. Remember:

- Baby Steps
- Test early. Test often.

4.2.2 Guideline #2: Use Pointers with Great Care

An uninitialized or badly initialized pointer can point to anywhere in memory and therefore corrupt any location in RAM during a write. Be careful that pointers are initialized correctly.

4.2.3 Guideline #3: Comment Code Well

Write good descriptions for functions that include what inputs are used for and what output is to be expected. Comment code lines where the purpose of the code isn't obvious. The code that the programmer writes today may not be so familiar in six months when a bug needs to be fixed.

4.2.4 Guideline #4: Avoid "Magic Numbers"

Use a single macro to define a constant value that is or may be reused so that we only have to change it in one place. For example, the following code uses a macro to define BIGARRAY-SIZE, which is then used in more than one place:

```
#define BIGARRAYSIZE 500

char bigarray[BIGARRAYSIZE];

...

memset (bigarray, 0, BIGARRAYSIZE);
```

The following code segment is an example of how NOT to define arrays. Use of magic numbers (like 500) often leads to confusion—especially during future code maintenance or debugging.

```
char bigarray[500];

...

memset (bigarray, 0, 500);
```

4.3 Common Problems

There are common problems that technical support staff at Z-World see time and time again. Here are some common mistakes that programmers can avoid when programming in Dynamic C:

4.3.1 Mistake #1: Omission of ioe, ioi

Omitting the prefix when intending an internal or external I/O read or write in assembly code is a common mistake people make. This will cause an unintended memory read or write.

4.3.2 Mistake #2: Off by One Tests

When writing loops, be sure the terminus condition is exactly what is intended. Remember, an N sized array is indexed from 0 to N-1:

In the following example, we intend to declare a character array with 128 elements. The array will be indexed 0 to 127. We want to initialize the array with zeros.

```
char str[128], i;
// initialize str[n] to n
for(i=0; i < 128; i++)  { str[i] = 0;}    // correct
for(i=0; i < 129; i++)  { str[i] = 0;}    // incorrect
for(i=1; i < 128; i++)  { str[i] = 0;}    // incorrect
for(i=0; i <= 128; i++) { str[i] = 0;}    // incorrect
```

4.3.3 Mistake #3: Failure to Check Return Values

Many functions have return values that specify success or failure of the operation. It usually matters that a function failed to do what was expected, so check the return value and handle the case of failure!

4.3.4 Mistake #4: Unterminated Strings

The standard library functions `strcat()`, `strcpy()`, `strlen()`, and `strcmp()` all assume that the input string arguments are NULL (zero byte) terminated. A common error is to construct a string and forget to null terminate it, then use it as an argument to one of these functions. The results can be minor, sporadic errors or no error, or catastrophic to program execution, depending on which function is used and what happens to be in the memory adjacent to the string at the time.

This type of error sometimes causes a working program to fail with a different version of the compiler. It can also happen when a change in another part of the program changes what happens to be in adjacent memory.

This type of error can be especially dangerous because it may produce no apparent failure in test conditions, then fail in the field because of changed content of adjacent memory. The functions `strncat()`, `strncpy()`, and `strncmp()` take an extra argument of maximum length to aid in coding defensively against this kind of error.

Using strncpy() and strncat() to prevent buffer overruns is not a panacea. In particular, the programmer must take into account what happens when the string is too large for the buffer. In this case, the destination buffer (and hence the resulting string) will *not* be NULL-terminated. For example:

```
#define BUFFER_SIZE 10
char buffer[BUFFER_SIZE];
strncpy(buffer, "This string is too long", BUFFER_SIZE);
```

Note that the source string in the strncpy() call is longer than 10 bytes. After the strncpy(), buffer will contain "This strin" (the first 10 chars of the long string), but it will not be

Chapter 4

NULL-terminated. Therefore, if we later attempt to print the string, it will output the string followed by garbage characters. To prevent this, we can always explicitly NULL-terminate the string after the strncpy() like this:

```
buffer[BUFFER_SIZE - 1] = '\0';
```

The above will work even when the string being copied is shorter than BUFFER_SIZE—it will just add an extra NULL character at the end of the buffer.

This sort of "defensive programming" technique adds little overhead to the code's size or execution, but can prevent hard to diagnose and hard to isolate bugs. The worst that will happen if a string too long for the buffer is copied with strncpy(), followed by an explicit NULL termination, is that the data in the buffer will be truncated. If this is noticed, meaning it is a bug, then at least the bug will be traceable to buffer manipulation. Without the explicit termination, all manners of odd behavior can crop up, depending on the program's structure and memory contents. Some of these "odd" behaviors, such as "rebooting", can be very hard to trace back to poor string handling. Defensive programming is smart programming.

4.3.5 Mistake #5: Misuse of strlen() and sizeof()

The function call sizeof(N) evaluates to a constant that is the size of the argument, N. If N is an array, then sizeof(N) evaluates to the number of bytes allocated to the array. In particular, this is true for character arrays, which are often used to store strings. However, if N is a pointer, then it will always evaluate to 2. Thus, depending on how a string is passed to sizeof(), the value returned could vary. Hence, we should use strlen() or strnlen() instead of sizeof() to measure the length of NULL-terminated strings. For example:

```
const char str1[] = "01234";
char str2[6];
char *str3;
str3 = str1;

sizeof(str1);       // this equals 6
sizeof(str2);       // this equals 6
sizeof(str3);       // this equals 2
strlen(str3);       // this equals 6
```

4.3.6 Mistake #6: Comparison of Floats for Equality

Never use floats in comparison tests for equality! For example:

```
float x, y;
x = 33.0 ; y = 11;
if (x/y == 3.0){ ...
```

Debugging

Floats are not guaranteed to be accurate to the last represented decimal place. The value of x/y is represented in hex as 0x403FFFF, whereas 3.0 gets loaded as 0x40400000, so the comparison fails to work as expected.

4.3.7 Mistake #7: Portability and Undefined Behavior

Dynamic C differs from standard C in its implementation of modules and has several non-ANSI extensions to assist in real-time programming, but it has relatively few *syntactic* differences. However, there are many situations and constructs in C code—too many to list here—that are unspecified, undefined or left to the implementation of the compiler by the standard. Here is a small sample of unspecified behavior from Annex J of the 1999 Standard, which deals with these portability issues:

- The order in which sub-expressions are evaluated and the order in which side effects take place, except as specified for the function-call (), &&, ||, ?:, and comma operators (6.5).
- The order in which the function designator, arguments, and sub-expressions within the arguments are evaluated in a function call (6.5.2.2).
- The order in which the operands of an assignment operator are evaluated (6.5.16).

Unless C code is written carefully to avoid dependencies and these types of constructs, the code is not as portable as most people might think. We should not assume that code ported from another compiler to Dynamic C will work correctly. Once it is compiled successfully, test it! Test early. Test often.

As much as mindset and good coding techniques are important, good debugging tools and techniques are time savers. Dynamic C provides an array of ready to use debugging tools.

4.4 Dynamic C Debugging Tools

Dynamic C has several standard debugger features to assist in analyzing program execution.

- Watch expression—evaluate variables and expressions.
- Breakpoints—cause program execution to halt when it reaches them.
- Assembly Window—shows machine code and other information about code, allows debugging at the machine code level.
- Stack Window—shows location and contents of the top of the stack.
- Register Window—shows contents of processor registers and allows the contents of some to be edited.
- Dump Window—allows the user to display the contents of arbitrary memory locations to the screen or to files.

With the exception of watch expressions and printf(), these features are useful only when single stepping or when program execution is stopped at a breakpoint, because the information they display changes too quickly to watch or analyze otherwise. Except for breakpoints, which are accessed from the Run menu, or using shortcut keys, these features are accessed from the Inspect menu or shortcut keys. The windows that show their output can be opened and closed from the Windows menu. Many debugging windows have right mouse button functionality for configuration.

4.4.1 Watch Expressions

Watch expression allow the user to view the value of C variables, complex C expressions that can include function calls, or constants. They are very useful for analyzing the flow of an algorithm while single stepping through it, but can also be updated while the program is running by hitting ctrl+U.

Watch expressions are compiled to a reserved area in memory when they are entered into the watch list. Watch expressions will be executed on each step when single stepping, or at the first RST 28h executed when ctrl+U is hit while running.

4.4.2 Breakpoints

Breakpoints can be set in C code unless it is declared "nodebug" and they can be set in assembly code sections that begin with "#asm debug." Breakpoints can be "soft," that is interrupt priority is left unchanged when stopped at the breakpoint, or "hard," where interrupts are turned off at breakpoints. A common use for breakpoints is to set one at the point in code where we want to start analyzing the behavior

4.4.3 Breakpoint Internals

RST 28h instructions are inserted between C statements in Dynamic C. These software traps allow the debugger to single step through the code. If "nodebug" is placed before a function declaration then Dynamic C omits the RST 28h instructions. The code in a "nodebug" segment cannot be stepped into or paused. The code will, however, run slightly faster as the debugger no longer burdens each statement in the code.

To accommodate flash chips with very large sectors, setting a breakpoint works by clearing a bit in an RST 28h instruction so that it becomes an RST 20h. This is because it is easy to clear a bit in byte writable flash without erasing the sector. Logic on the target and host PC keeps track of whether the breakpoint has been cleared, since the bit cannot be reset without erasing the sector. The additional logic means that breakpoints will continue to use additional resources even after they have been cleared. The Clear All Breakpoints commands will make the debug kernel treat the RST 20Hs like RST 28Hs and eliminate the extra overhead, but once another breakpoint is set, it and all previously set breakpoints revert to the extra overhead state until such time as the program is recompiled and downloaded.

4.4.4 Breakpoints in Assembly Code

Blocks of assembly code are marked by the compiler directive #asm, and RST 28HS are not normally inserted between assembly code instructions. Single stepping will still work in blocks of assembly, but we cannot set breakpoints, a block of assembly code unless the beginning delimiter is modified like so

```
#asm debug
```

This causes the assembler to insert RST 28HS in between assembly instructions. This is necessary *only* if we wish to use breakpoints in the assembly code block. Adding the debug keyword to an assembly block can have the effect of making relative jumps—which are

Debugging

limited to 128 bytes—out of range. If we add the debug keyword and see the assembler error message "relative jump out of range" during compilation, it means that we should change the JR instruction to a JP instruction.

If we want to put a breakpoint in assembly code without adding RSTs between all instructions, we can embed a null C statement (a semicolon) in assembly code. C statements may be placed in assembly code blocks by placing a "c" character in column 1. Here is a simple example,

```
#asm
// some assembly instructions
c ;     // set a breakpoint on the semicolon (a Null C statement)
// some more assembly instructions
#endasm
```

Prior to Dynamic C 7.10, breakpoints could be hard coded by putting RST 20Hs or RST 18Hs in code. When a hard coded breakpoint was hit, program execution could not continue. This no longer works in later versions; these instructions will just be treated as breakpoints that have been toggled off. Use the embedded C null statement method above.

4.4.5 Assembly Window

The assembly window shows the current machine code and mnemonics being executed when single stepping. C source code lines can be seen in the window in versions 8.01 and later of Dynamic C. The F8 and F7 keys are used to single step through the disassembly window. We can switch back and forth between stepping at source level and assembly level simply by opening and closing the assembly window. With later versions of Dynamic C that show C code in the disassembly window, alt-F7 and alt-F8 can be used to step at the C source level while the assembly window is open.

The assembly window shows machine cycle times (assuming 0 wait states) for each line of instruction, and cycle times can be totaled by selecting a block of code with the left mouse button. Later versions will show the cycle sum of selected assembly blocks in the bottom status bar of the assembly window. The total in the far right column does not take into account looping or branching, it only gives the raw, total of the cycle times for the selected instructions.

4.4.6 Stack Window

At the time of this writing, Dynamic C has no symbolic stack tracing; only raw hex data appears in the stack window, not function or parameter names. The window is still useful for seeing what parameter values are loading before a function call, for seeing how deep into the stack current execution is, and for detecting stack imbalance.

The stack generally starts just below E000h (µC/OS-II tasks use separate stacks which may be mapped lower). The address on the left of the stack window is the current value in the stack pointer register (SP). The value decreases when something is pushed on the stack, and increases when something is popped. A function call will always push the address of the next statement onto the stack; a function return will pop the address off. On entering a function, both the value and the address showing at the top of the stack should be the same as just before returning from the function. Stack imbalance should not happen in C code, but can be caused by extra or omitted pushes or pops in assembly code. When single stepping over C statements with F8 at source code level, the SP value shouldn't change *in between* C statements. The stack is used for temporary storage while evaluating complex expressions, so the SP value may change *within* the execution of C statements. However, compiler errors are not unheard of, so we shouldn't rule them out.

4.4.7 Register Window

The register window allows the programmer to view the contents of the processor registers. Its contents only display accurate information when a program is stopped at a breakpoint or for single stepping. The register window is especially useful for debugging assembly code. The values contained in registers can be changed using the right mouse button (expect for SP and PC).

4.4.8 Memory Dump

The memory dump feature allows the programmer to view the contents of any memory location in the target. You can view the information on the PC screen or even write it to a file. Printable ASCII characters in memory show up as ASCII characters on the right of the raw hex data, making it especially useful for debugging algorithms that construct strings. We can also send memory dumps to files.

4.4.9 printf()

Printf outputs data to the Dynamic C `stdio` window by default. Output can be sent to a file simultaneously (we can set this up in the Options ⇒ Environment Options ⇒ Debug Windows.) Starting with Dynamic C version 7.25, `printf()` output can also be redirected to a serial port[1]. Printf is not normally used in a finished application.

Printf should not be overused. The code generated by the printf function is big. The CPU overhead of transferring data to the Dynamic C `stdio` window or redirecting it to a serial port is nontrivial. Printf usage can mask or introduce timing relating errors. Never use `printf()` in an ISR!

String literals used by printf statements use root code space. Some users pepper their code with printf() calls to do tracing. One customer sent his program to Z-World Tech Support, desperately seeking help in finding more root space. Commenting out several hundred debug `printf()` calls quickly freed up more than enough root space.

On solution to this problem is to use a small and fast assembly language macro that squirts data out a serial port for diagnostics. Z-World has provided a set of compact and fast assembly macros for outputting simple data to a serial port. These macros can be found on

---

[1] See the sample program \SAMPLES\SERIAL_STDIO.C for an example of how to do this.

the accompanying CD in FASTERMACS.LIB. There is also a short C file TESTSERMACS.C that illustrates how the macros are used.

4.4.10 How the Debugger Adds Overhead

Debugging in Dynamic C source code works largely by inserting RST 28h instructions in between C statements. A RST 28h works like a software interrupt; when it is executed, program execution is transferred to the vector table. When in debug mode (running with programming cable attached), the vector table transfers program execution into the debug kernel. The debugger kernel does a lot of processing to handle the target-end logic of debugging, and turns off interrupts at some points. RST 28s are inserted into functions compiled as debug only. Functions may declared debug or nodebug; the default being debug.

To declare a function as debug or nodebug, simply insert the compiler directive before the function declaration. Here are two examples.

```
nodebug int foo(){ /* some code... */ }
debug int foo2(){ /* some code... */ }
```

When the program starts up and no programming cable is attached, the vector table entry in RAM for RST 28h is changed to a ret instruction for Rabbit 2000 programs. For Rabbit 3000 and later processors, an 80h is written to the Breakpoint Control Debug Register (BDCR: 0x001C) that makes the RST 28h function as a NOP.

4.5 Isolating the Problem

This, of course, is most or all of the battle. Before an engineer can modify code to eliminate a bug, the bug must be clearly identified and understood. Ninety percent of the debugging effort will be spent identifying the mechanism in the code that produces the problem.

Some bugs are easy to identify. For example, an equation with a typo will yield erroneous results. This sort of bug is fairly easy to isolate. One must only work backwards from results to source data looking for where the weirdness happens. The printf() statement is a classic tool for inserting debug messages into the code.

Using a watch window to examine the variable that ultimately holds the erroneous result as well as watching the variables with the source data while single stepping through the algorithm will also quickly identify the problem.

Once a problem is identified, a solution can be formed and tested.

Unfortunately not all bugs are easy. The renowned Bob Pease once offered the following blessing to one of the authors, "May all your troubles be middle-sized, so you can find them." We offer the same good wish to the reader. But for the times that the bugs are subtler, we offer the following sections.

Chapter 4

4.5.1 Reproducibility

If the problem is easily repeatable, the programmer is lucky and may get to the bottom of it quickly. Generally, the more sporadic and infrequent the problem is, the harder it is to find and fix. Perhaps the most common cause of inconsistently reproducible problems is memory corruption. Memory corruption can result from any of the following causes:

- Uninitialized or incorrectly initialized pointers or variables
- Buffer (array bounds) overflow
- Stack overflow/underflow

We should always run with run-time pointer checking and array bounds checking enabled when debugging. Runtime error checking features only work in C code. Read the *Unterminated strings* and *Off by one errors* sections for more information on array bounds errors.

Inspect and test assembly routines carefully for array bounds overflows and pointer errors. Runtime errors in assembly routines are difficult to diagnose once integrated into larger frameworks. When initially developing an assembly module, test early and test often.

One type of run time error that can occur is a stack overflow. Runaway recursion can cause this, but other more subtle factors are often at fault.

When using µC/OS-II, one source of stack overflow could be that the stack is too small for one or more µC/OS-II tasks. If the program uses µC/OS-II, try increasing the task sizes of the tasks. The default storage class of Dynamic C is "auto" starting with Dynamic C 8.00, and "static" prior to that. The storage class can be changed with `#class auto/static`. When using `#class auto`, try using `#class static` to reduce stack usage, but keep in mind that this change can make reentrant functions nonreentrant. Switching the other way will break any functions that depend on local variables retaining their previous value on reentry into the function. Dynamic C library functions use auto and static specifiers explicitly on local variable declaration and so are not affected by a change to the default storage class.

Other reasons that a problem may not be consistently reproducible include:

- An ISR modifies but fails to save a register that is only infrequently used by the rest of the program, so it rarely causes a problem.
- Interrupt latency that only occasionally interferes with other parts of the program. Dynamic C library ISRs are designed not to keep interrupts turned off for too long. When writing ISRs, write them in assembly, and make them efficient.
- Borderline timings issues can happen. Serial ports may run at a high bit rate that handles the throughput *most* of the time, but occasionally not, etc.
- Electromagnetic interference (EMI) may cause unexpected trouble.

4.5.2 Increasing Reproducibility

Unlike standard C, static variables are not initialized to 0 by default on program start-up if no explicit initialization is performed. Starting with Dynamic C version 7.30, a BIOS macro, `ZERO_OUT_STATIC_DATA`, is provided to force the static data area to be zeroed out on start-up.

Debugging

If the bug is sporadic, try running the program with and without this option enabled to see if it makes a difference for reproducibility. A bug that may cause odd behavior without this option may produce a run-time pointer error with it, which isolates the problem to a line of C code when debugging. The initial zeroing of static data out only occurs when the BIOS starts, so will not normally happen in a debug session unless the BIOS is recompiled and restarted. Compiling and running a program for the first time, hitting Ctrl-y, editing the BIOS file, or editing a library used by the BIOS all force this to happen.

If the problem is dropped data with serial communications, try slowing the bit rate down to see if the problem goes away and speeding it up to see if happens more consistently. If clock interrupts are being used, try turning them off to see if data loss still occurs.

4.5.3 Minimize the Failure Scenario

Once we have achieved reproducibility, the next step is to minimize the failure scenario. A minimal scenario means the smallest piece of code that exhibits the problem in the shortest amount of time. A minimal scenario is useful for several reasons. Most obviously, reducing the error to a minimal scenario will frequently isolate and identify the problem precisely. If not, it may at least give a smaller piece of code to compile, debug, or have another pair of eyes look at.

4.5.4 Single Stepping

Single stepping works by hitting the F7 or F8 key while a program is stopped in debug mode. F8 steps *over* function calls or assembly language calls. F7 steps *into* the calls. An attempt to step into a "`nodebug`" function will simply step over the function call. If the disassembly window is open, stepping occurs machine instruction by machine instruction. If it is not, then stepping occurs C statement by C statement. Functions must be "`debug`" for stepping to work in them at the source level. Later versions of Dynamic C allow single stepping by C statement with the assembly open by using Alt+F7 and Alt+F8. Using F7 and F8 when stopped at a µC/OS-II task statement that causes the task to suspend works slightly differently—F7 will cause the program to stop at the next statement in the next active task (assuming it is a debug function), while F8 will cause the program to stop at the next statement in the current task after program execution gets back there.

Single stepping is useful in a variety of ways, including:

- One of the common symptoms for bugs in a Dynamic C program is the "`Target communication error`" message box. More often than not, this not really a target communication problem so much as a program-crashing-and-taking-the-debug-kernel-with-it due to a programming error problem. If other programs such as sample programs don't have this problem on the target board and PC, then it most likely the latter. Single stepping can be a good way to see how far the program can execute before a crash.
- If the programmer uses other methods to determine that a program works up to a certain point, then it may be useful to set a breakpoint at that point, and start single stepping from there.

Chapter 4

- If a program consistently fails quickly after starting, single stepping from the beginning may identify the failure point quickly.
- Using watch expressions and/or the register and/or dump windows in combination with single stepping assists in careful analysis of algorithms when creating or debugging them. Watch expressions are updated with each step when single stepping.

4.5.5 Binary Search

Another technique is what we can loosely term a "binary search." This method of reducing the failure scenario is to comment out half the code and see if the other half fails. Repeat this process of splitting the problem in "halves" until the problem is isolated. Of course, a program cannot usually be neatly split in two at any point, so this method requires care.

4.5.6 Analyze the Behavior under Various Conditions for Clues

When debugging, clues can be gleaned from observing a buggy system under a variety of conditions. There are physical conditions like a low voltage rail that may be causing odd behavior. There may be software-controlled conditions like masking interrupts.

When possible, test the system under a variety of conditions. Determine if a particular set of set of conditions is required for the bug to assert itself. Clues are what are required to determine what is wrong in a system and how to fix it.

4.5.7 Eliminate Debugger/Target Communication Overhead

When debugging, the debug kernel and target communication use a fair amount of CPU cycles to handle debugging logic and interaction between target and host PC. A CPU intensive program that works flawlessly in stand-alone run mode may have problems such as dropped characters in serial transfers when run in debug mode. Conversely, a program may work better in debug mode because the extra overhead masks a timing problem. It is a good idea to see if a program failure happens in both stand-alone run mode and debugging run mode.

The quickest way to *nearly* eliminate debugger overhead is to disconnect the programming cable and cycle power to run stand-alone. When a board powers up without the programming cable attached, the vector table entry for RST 28h is replaced with a RET instruction, so the overhead for an RST 28h becomes 16 clock cycles. In the Rabbit 3000, the overhead of an RST 28h is reduced to just 2 clock cycles because the Rabbit 3000 has a feature to make the RST 28 instruction function as NOPs. Code is made "nodebug," by using the compiler directive `#nodebug` at the top of the main source file. C Functions can be made debug or nodebug on a function-by-function basis by using the `debug/nodebug` keywords. The default is `debug`. Once functions are debugged it is suggested that they be made `nodebug`. Library files released with Dynamic C contain only `nodebug` functions.

4.5.8 Change Timing

When a bug occurs, sometimes the cause is related to how many CPU cycles the microprocessor has to devote to a particular task. Sometimes changing the speed of the processor can affect the repeatability of a bug.

For example, if a program is losing characters during serial communications, try lowering the bit rate.

If it seems that a bug might be related to the processor speed, try changing the bug's behavior by changing clock speeds. The rabbit offers several software configurable options for generating the CPU's clock. The CPU maybe clocked at the high-speed crystal's fundamental frequency, or the twice that frequency, depending on the clock doubler. To slow things down, the CPU can be clocked from the 32,768 Hz crystal.

One issue to consider when making adjustments to the speed at which the CPU runs, is that when the CPU clock is derived from the 32,768 Hz crystal, the Dynamic C debugger is not available. Behavioral changes will have to be observed with tools other than Dynamic C talking over a programming cable. External LEDs might be an option worth considering.

If changing clock speeds changes the behavior or frequency of occurrence, then we have another clue.

4.6 Run-Time Errors

Dynamic C for Rabbit has built-in array bounds and pointer checking, but no stack checking at the time of this writing. These features are enabled in the compiler options dialog box. It is recommended that these options be enabled during program development. They will make the program slightly larger and slower. Once a program is tested and deemed ready for deployment, the bounds and pointer checking can be turned off, the faster code regression tested and then deployed.

The default behavior of run-time errors when debugging is for a dialog box to pop-up specifying the file and line number where the error is detected followed by a program reset. The default behavior in stand-alone-run-mode is for a tight loop to be entered that causes a watch-dog time-out and board reset.

Other types of runtime errors include math range and domain errors. The Dynamic C User's Manual has a list of defined run-time error codes.

4.6.1 Array Bounds Check

Array bounds check provides strict boundary checking, so that any array bounds overflow will generate run-time error messages that identify the offending line and file. This will, however, only work on direct array assignments.

Program 4.1 shows some simple examples of when and why runtime errors will occur.

Chapter 4

Program 4.1: Examples of Runtime Errors.

```
// given the following declarations and assignment, the subsequent
// C statements will generate errors.

int x[20];   // Reserve space for twenty two-byte integers (40 bytes)
int i;
i = 20;

x[20] = 1;   // This will generate a compiler error even if array bounds
             // checking is turned off because the array is indexed from
             // 0 to 19, and the compiler can detect this because a
             // constant is used for the array index.

x[i] = 1;    // This causes an array bounds error at run-time,
             // because the compiler generates code to check the
             // assignment (if that option is turned on.)

memset(x,0,41);   // This call tells the memset function to load 41
             // bytes of zeros starting at the beginning of the
             // array x.  This will not generate a run-time error
             // because the function does not know that x is an
             // array and not just a pointer.  Therefore, it will
             // write past the end of the array, which is only 40
             // bytes long.  Depending on what is stored there,
             // this could cause problems in the program's
             // behavior.
```

4.6.2 Pointer Checking

Pointer checking will only generate run-time errors if an attempt is made to write to a memory location via a dereferenced pointer that points to a location not in the stack segment or the data area of the data segment.

Pointer checking will be more effective if ZERO_OUT_STATIC_DATA is enabled. Upon power up, uninitialized pointers may happen to contain valid (but logically wrong) values if they are not zeroed out. Consider the following code segment:

```
int * xptr;
    *xptr = 0;
```

The code will definitely cause a detectable run-time error if the `ZERO_OUT_STATIC_DATA` macro is enabled. This happens because the pointer is initialized with the address 0x0000. The address 0x0000 is in the root code segment.

If, however, the `ZERO_OUT_STATIC_DATA` is not enabled, `xptr` would not be initialized to 0x0000 and `xptr` could possibly contain an address that would not cause a run-time error.

4.6.3 Defining Your Own Run-Time Error

Defining one's own run-time error is as simple as making a call to `exit(N)` where N is a one-byte integer. The exit call will put up a dialog box that specifies the exit code in debug mode. This can be used in the same way a `printf()` is used to bring attention to an exceptional condition. The `exit()` function has a thousand times less overhead than `printf()`.

In stand-alone run mode, the `exit()` call will function the same as all intrinsically defined run-time errors.

Programmers can define custom runtime error handlers. This approach will allow the programmer to customize the system's response to run-time errors that occur. Dynamic C ships with a sample program called `Define_error_handler.c` that shows how to do this.

Dynamic C also provides the ability to log run-time errors in a RAM buffer. The log only works with battery backed RAM. Dynamic C ships with a sample program called `DISPLAY_ERRORLOG.C` that shows how to set up the buffer and recover the log.

4.7 Miscellaneous Advanced Techniques

Given below are some additional techniques that programmers will find useful.

4.7.1 RAM Trace Log

Sometimes the regular Dynamic C debug tools are either too slow or too intrusive for an application. However, an engineer still might need a tool to trace execution of the program. This method can be useful for tracing program execution when regular Dynamic C debug tools are unsuitable, such as tracing execution in an ISR.

The sample program `TRACEBUF.C` shows a way of setting up a program trace in a 256-byte buffer.

As written, the TRACE macro destroys the contents of the alternate DE register, DE', so `TRACE()` should be inserted where DE' doesn't need to be preserved.

An output function is provided, but the buffer could be viewed in the Dynamic C Dump Window. Simply use the Dump Window to examine the addresses following the initialization address of `traceBufPtr`.

Chapter 4

Program 4.2: Setting up a program trace with TRACEBUF.C.

```c
#define TRACEBUFSIZE 0x513
char traceBuff[TRACEBUFSIZE];
char * traceBufPtr;
// the following macro should be on one line - do not split this
// macro across multiple lines - regardless of how this shows
// up on "the printed page" in this book.
#define TRACE(x) asm ex de',hl$ ld hl,(traceBufPtr) $ ld (hl),x $ inc \
 L $ ld (traceBufPtr),hl $ex de',hl

int INITTRACE(){
     // zero out buffer
   memset(traceBuff,0,TRACEBUFSIZE);
   // point to 256 byte boundary
   traceBufPtr =  (void*)((((unsigned)traceBuff+256) & 0xff00));
}

// Print out n buffer entries
PrintTrace(int n){
   int i;
   char * bptr;

   bptr = (void*)((((unsigned)traceBuff+256) & 0xff00));
   for(i=0;i<n;i++) { printf("%d ",(int)*(bptr++));   }
}
main(){
  int i;
  INITTRACE();
  TRACE(1);
  TRACE(2);
  for(i=0;i<4;i++)   {
      TRACE(3);
  }
  TRACE(4);
  TRACE(5);
  PrintTrace(8);
}
```

4.7.2 Assert Macro

As of Dynamic C 8.50, Z-World introduced a standard assert() macro. Asserts are useful to check for conditions that *must* be true. For example, if a function takes a pointer argument, and that pointer must not be NULL, then the programmer can add an assert like this:

```
assert(ptr != NULL);
```

If `ptr` is `NULL`, then a message will be printed to the stdio window indicating that an assertion failed, along with the filename and line number of the failure. If asserts are used liberally enough, they can detect errors and faulty assumptions before we even suspect that there is a problem.

The developer might not want asserts to be present in production code. In this case, defining the macro `NDEBUG` will disable all asserts.

For readers that may be developing with Dynamic C below version 8.50, the following sample program ASSERT.C shows an implementation of the standard `assert(x)` macro. If x is false, print x, file and line where the assert macro is, then exit with exit code 1.

Program 4.3: Implementing the assert(x) macro.

```
#define DEBUG   // comment out to ignore assert
#ifdef DEBUG
#define assert(x) if(!(x)) \
 printf("assert " #x " in file %s at line %d\n",__FILE__,__LINE__); \
 exit(1);
#else
#define assert(x)
#endif
main() { assert(1==2); }
```

4.7.3 Evaluate Expression and the Function Call Trick

Dynamic C has a debugging tool that can be found under the INSPECT menu, called "Evaluate Expression." This tool operates much like a watch expression, except that it only occurs once—that is, it does not get evaluated at every break point and when single-stepping. The expression that is evaluated may be either a mathematical expression or a function call. One particularly interesting technique that uses "Evaluate Expression…" is executing a debugging function from the "Evaluate Expression" window. The development engineer can command the function to be executed at any time. Consider the code in Program 4.4:

Chapter 4

Program 4.4: Evaluating functions.

```
int var1;
long var2;

void printinfo(void) {
    printf("***************\n");
    printf("var1: %d\n", var1);
    printf("var2: %ld\n", var2);
}


void main(void) {
    long count;
    var1 = 0; var2 = 0;
    printinfo;          // Simply referencing the function name
    for (count = 0; ; count++) {
            var1++;
            if (count % 3) var2++;
    }
}
```

The `printinfo()` function is simply a debugging function that outputs the values of a couple variables. The `main()` routine continuously increments these variables. We can run the program and then select the "Inspect/Evaluate Expression..." option. Type in "`printinfo()`" into the dialog, and select "Evaluate". If we look in the stdio window, we will see the output from the `printinfo()` function. Note that the `main()` routine contains a simple reference (not a call!) to the `printinfo()` function. This is not strictly necessary for this sample; however, if `printinfo()` were contained in a separate library, then this reference would be necessary to ensure that it is linked into the program.

This technique has been used extensively by Z-World to develop Ethernet drivers. A function called `prt_nicreg()` queries the Ethernet chip for the values of its internal registers and then displays them. These values are not accessible via simple C variables, so being able to call the `prt_nicreg()` function to access them greatly eases development.

4.8 Final Thoughts

Debugging, like taxes, is a fact of life. Engineers are called upon to debug systems. The better one gets at debugging, the less time one spends doing it.

Eighty percent of debugging is the proper mindset. After all, debugging is puzzle-solving. Finding clues, piecing them together and creating a solution is the order of business.

Good tools and techniques for applying them comprise the remaining twenty percent of the task. Dynamic C has many resources available for debugging Rabbit-based systems. We have outlined a collection of techniques for applying these tools.

Over the last two chapters, we have covered how to create programs in Dynamic C and how to debug systems using Dynamic C's features. In the next chapter, we will examine hardware issues that often arise in embedded system design.

CHAPTER 5

Interfacing to the External World

5.1 Introduction

This chapter is concerned with the practicalities of attaching sensors and actuators to digital controllers. While the RCM3400 prototyping board will be used for all the examples in this chapter, the concepts covered are applicable to most embedded systems.

5.2 Digital Interfacing

There are many books devoted to digital design. Most are concerned with formal methods for logic reduction or techniques used to implement sequential logic. Even with all the available material, device manufacturers are compelled to publish application notes and white papers describing the practical application of their devices.

The working engineer will seldom refer to textbooks discussing canonical equations and logic reduction by Karnaugh map[1]. Engineers are often too busy trying to figure out how to prevent their circuits from being damaged by ESD, or being overheated from driving too much current.

We will cover these issues here. We begin with a look at how to bridge the gap between 3.3-volt systems (such as most Rabbit-based designs) and 5-volt systems.

5.2.1 Mixing 3.3 and 5 Volt Devices

Not so long ago, TTL-based digital systems were designed to operate on 5-volt rails. As new CMOS logic technologies have become mature and robust, there has been a natural migration to lower voltage systems.

Power consumption is proportional to the SQUARE of the voltage. In simple DC circuits, we know that,

$$Power = \frac{voltage^2}{resistance}$$

---

[1] If you happen to be in need of a truly excellent textbook on combinatorial logic, sequential state machine and asynchronous state machine design, Richard Tinder's book *Digital Engineering Design: a Modern Approach* (Prentice Hall) will be an excellent addition to your library.

Interfacing to the External World

In AC systems, effects of capacitance and operating frequency also enter into the equation. CMOS devices have very small quiescent currents (very high resistance), but the energy stored in their internal parasitic capacitors is governed by,

$$Energy_{CAP} = \frac{1}{2} \cdot C \cdot V^2$$

As digital states change, these parasitic capacitors must be charged and discharged. The resistive paths through which this charge is moved dissipate power. The more capacitive nodes involved in a system level state change mean more energy that must be moved and power dissipated. The faster the state changes occur means more power is dissipated over a given time interval.

This brings us to the equation,

$$Power_{CONSUMED} \propto k \cdot \frac{1}{R} \cdot C \cdot F \cdot V^2$$

where:
    k is a catchall constant
    F is the system's frequency of operation
    R is derived from quiescent currents
    C is determined from dynamic currents
    V is the switching voltage

The important bit is that if we drop the voltage by half, we decrease our power about four times.

Energy is related to power by,

$$Energy = \int_{time} Power \cdot dt$$

For the simple case of a static system, we can simply multiply watt (W = J/s) by time to get energy.

So by reducing the rail voltage of a system, the power consumed is reduced by an inverse square and so, therefore, is the energy required to operate the system.

In this age of laptop computers, PDAs and cell-phones, energy storage directly translates to weight (and volume). Ultimately, the push for smaller, lighter, portable, energy efficient devices has pushed the digital world to lower supply rails.

Older 5-volt systems are still ubiquitous. Design engineers are often faced with the challenge of interfacing newer 3.3-volt technology to legacy 5-volt systems.

Chapter 5

There are two main issues to consider. When driving 3.3-volt inputs with 5-volt outputs, the CMOS inputs will be driven above their 3.3-volt supply rail. If the 3.3-volt device has a high-side ESD diode this will lead to smoke—see Figure 5.5b. This will be discussed further in the next section when we discuss input protection diodes.

The Rabbit 3000 has 5-volt tolerant inputs. A Rabbit powered from 3.3-volt rails may have its inputs driven with either 3.3 or 5 volts. No damage will occur.

The second issue to consider is how the 3.3-volt outputs will drive the inputs of a 5-volt device. With a large array of logic families from which to choose the 5-volt CMOS device, the issue of noise margin is easily solved.

Since CMOS devices drive their outputs very close to ground, the $V_{IL(MAX)}$ characteristic of the 5-volt powered device is seldom a concern. However $V_{IH(MIN)}$ is often a concern. The following table shows a comparison of $V_{IH(MIN)}$ for some common logic families.

	Logic Families						TinyLogic™ single gate devices		
	HC	HCT	VHC	VHCT	LVT	LVX	HS	HST	UHS
$V_{IH(MIN)}$	0.7·VCC	2.0 V	0.7·VCC	2.0 V	2.0 V	2.0 V	0.7·VCC	2.0 V	0.7·VCC

The "T" families, such as HCT, VHCT and HST have input stages optimized for interfacing to older TTL devices. This is perfect for interfacing to 3.3 volt CMOS systems.

Other families exist with input thresholds fixed at 2.0 volts regardless of supply voltage. Fairchild's LVT and LVX families fall into this category.

The potential problem with devices that have a minimum High Input Voltage of 0.7*VCC is that with a supply rail of 5 volts, the input threshold is only guaranteed valid if it exceeds 3.5 volts. This says if a 3.3-volt device is driving a 5-volt device with a 0.7*VCC input threshold the configuration will not work.

For most CMOS families, the 0.7*VCC $V_{IH(MIN)}$ has a bit of a safety margin built in. This means that a lot of the time a 3.3 volt device will drive a 5 volt device just fine. This is especially true if the 3.3 volt rail is a little hot and the 5 volt rail a little low. This is a most unfortunate situation.

Systems designed without proper noise margins may work fine at room temperature on an engineer's workbench. Once these poorly designed systems hit mass production and are shipped to customers, invariably problems result. Sometimes the design flaws show up at temperature extremes, or when parts from specific batch of ICs from a particular manufacturer are used.

When interfacing 5-volt logic to 3.3-volt devices, such as the Rabbit 3000, be careful to select a compatible 5-volt logic family. The HCT, VHCT and HST devices are ideally suited for this situation.

5.2.2 Protecting Digital Inputs

Digital devices are susceptible to damage from all manner of electrical stresses. Protecting these devices is both art and science.

Interfacing to the External World

The science comes from our ability to model circuits and methodically test our designs. The art comes from the necessity to make sound trade-offs so that our designs are affordable yet well suited to their intended market. Fortunately, an ever-growing array of protective devices is available.

Gas discharge tubes (GDTs), also called spark gap suppressors, are found in telecommunications equipment. These devices are constructed by precisely placing two or three electrodes in a sealed glass chamber filled with specifically selected gasses.

The GDT is placed between a protected line and ground. Under normal line conditions, the GDT looks like an open circuit. When a transient event pushes the voltage between the two electrodes above a spark-over threshold, the gas ionizes and conducts. The GDT diverts the potentially destructive transient energy from the protected electronics to ground.

Once the gas inside the GDT is ionized, it only takes a relatively low voltage, called the holdover or glow voltage, to keep the device in conduction. This feature precludes most GDTs from AC power protection.

For example, consider a GDT with a 700-volt spark-over voltage and a 50-volt holdover voltage that is placed on a 110 VAC line. Assume a transient event causes spark-over to occur. Once the fault clears, the normal AC mains voltage would hold the GDT in conduction. The 110 VAC nominal line voltage is higher than the 50-volt holdover voltage. The GDT would cause a short circuit and would self-destruct.

There are GDTs specifically designed for AC operation that will cease to conduct at the AC zero crossing. Most GDTs do not stop conducting quickly enough and are not intended for AC power line protection, but rather for telco and other lower voltage protection.

GDTs are capable of repeatedly shunting thousands of amps for short periods of time. These devices generally cost $1.50–$5.00. Photo 1 shows several GDTs. These devices are fairly large compared to the components found on a microprocessor's PCB. The right-most GDT in Figure 5.1 is an SMT (Surface Mount Technology) device.

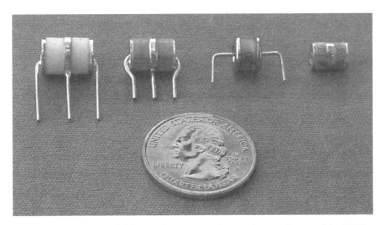

Figure 5.1: GDTs are available with two or three electrodes and in PTH or SMT.

Chapter 5

If a system has long external sensor leads, a GDT is good insurance against transient voltage induced by near lightning strikes. GDTs are the slowest of the transient suppression devices we will examine. GDTs are best suited if transient events are expected to last milliseconds or longer.

GDTs operate best when they are used with another form of protection. This cascaded arrangement is called "coordinated protection." The GDT is generally placed nearest the transient event and is considered the primary protective element.

Figure 5.2 shows a GDT combined with a metal oxide varistor (MOV) in a coordinated protection scheme. The designer should select a MOV that has a lower clamping voltage than the GDT spark over voltage. The resistors limit current and will help to dissipate the transient energy as heat.

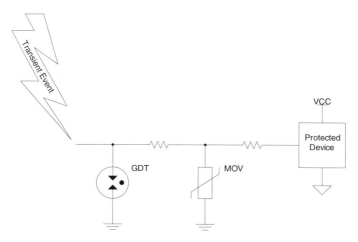

Figure 5.2: Coordinated protection provides multiple layers of incrementally faster protection.

MOVs are faster devices than GDTs. They can be purchased in a variety of physical sizes. The larger the MOV, the more energy it can dissipate before suffering permanent damage. Surface mount (SMT) MOVs are small devices and have limited ability to dissipate energy. The pin-through-hole (PTH) devices can be quite large and can dissipate much larger amounts of heat.

MOVs are formed by mashing together tiny bits of zinc oxide until they form a shape to which two electrodes can be soldered. Secret sauce ingredients are also added, most consisting of other metal oxides.

Each boundary between Zinc oxide particles acts as a little zener diode. The massive combination of random particles statistically acts like one big back-to-back zener diode.

The breakdown voltage for MOVs is less accurate than that for zener diodes. The response time of MOVs is usually slower than zener diodes. A MOV's primary advantage over a zener is the ability to dissipate more power than a zener diode.

MOVs are available with breakdown voltages in the 10's to 100's of volts.

Many engineers believe that MOVs, like fuses, are sacrificial devices. A MOV is expected to splatter its guts all over the PCB while valiantly protecting the electronics. This is flawed thinking. MOVs fully recover after a transient event occurs. This assumes that the power dissipated in the MOV was within the MOV's specified safe operating area (SOA).

It is true that the breakdown voltage of a MOV may change a few percent during the first several clamping episodes.

For a MOV to provide long-term protection, the system designer must select a large enough MOV to handle the anticipated currents. Also, the MOV's initial specified breakdown voltage must be high enough that if the breakdown voltage should decrease 10%, it will not will drop low enough to fall in the operating voltage range of the protected signal.

Because MOVs have a relatively fast response time and are capable of dissipating large amounts of energy for short periods, they often find application in protecting AC power lines. Figure 5.3 shows a common AC protection scheme.

When a transient over-voltage condition occurs, the MOVs will clamp the high voltage spike to ground. If the fault is sustained, the sustained high current through the fuses will cause one or both fuses to open. The MOVs protect against momentary transients. In the event of sustained over voltages, the combination of MOVs and fuses protects both the "Protected Device" and the MOVs from damage. The only sacrificial protective elements in Figure 5.3 are the fuses.

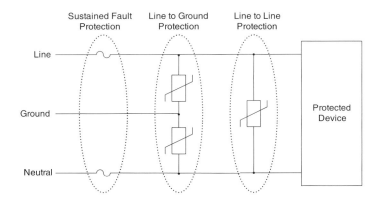

Figure 5.3: While useful for protecting digital lines, MOVs are often found across AC lines.

Chapter 5

A device often confused with a MOV is a proper TVS (transient voltage suppressor). The TVS is a semiconductor device that can be modeled as two back-to-back zener diodes. Figure 5.4 shows the schematic symbol for several protective devices.

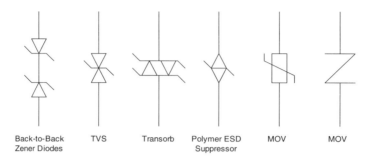

Figure 5.4: The schematic symbol for a TVS, Transorb or polymer ESD Suppressor derives from the schematic for two back-to-back zener diodes.

A TVS is the fastest of the clamping devices. They are also the least capable of carrying large currents for extended periods of time.

Marketing departments are forever trying to differentiate their product from the competition. Years ago, General Semiconductor (now part of Vishay) coined the word "Transorb" to distinguish their TVS from the competition. Figure 5.4 shows the symbol used for a Transorb.

TVS devices are best suited in protecting against electro-static discharge (ESD). These devices are often found as a secondary or even tertiary protective devices in coordinated protection networks.

There are other protective devices. Cooper Bussmann has a device they have dubbed "Polymer ESD Suppressors with SurgX® Technology." They are small SMT devices, and can carry only a few tens of amps. As technology advances, the number of options open to the system designer for circuit protection increases.

One of the simplest tools available to the designer is the diode. These can be used to great effect as an ESD protection devices. Many IC's have ESD protection diodes on their I/O pins. A simplified model of on-chip ESD protection is shown in Figure 5.5a.

The internal protection diodes in Figure 5.5a can be problematic in mixed rail systems. For example, consider the circuit in Figure 5.5b, where a CMOS device is powered from a 3.3-volt rail, and the device's input is driven from a 5-volt rail. The CMOS's VDD side diode will enter conduction under forward bias. Unless a there is a device to limit current, the high-side diode will be damaged. Most likely a significant portion of the device will also be collaterally damaged.

Some device families are "5-volt tolerant." A common way to implement this is to remove the high-side internal ESD protection diode. Figure 5.5c shows a 5-volt tolerant CMOS device.

To protect the input in a mixed rail system, an external diode will need to be added between the highest rail and the IC's input. Schottky diodes are often used because of their fast switching times. Figure 5.5c shows a configuration suitable for protecting a mixed 3.3 volt and 5.0 volt system.

Interfacing to the External World

The resistor in Figure 5.5c limits the current into either diode. Coupled with the parasitic capacitance of the schottky and the CMOS device's input capacitance, the resistor forms an RC low-pass filter. This will slow down high-speed transient events allowing the diodes extra time to enter conduction.

Plain, old-fashioned carbon-composition resistors are the best type of resistors for this application.

Metal film resistors have patterns etched into their film to trim the resistance to the desired value. ESD has a tendency to jump the insulative gaps in the metal film. During a transient event, this reduces the effective resistance of the resistor. Furthermore, if ionized or carbonized paths form, the resistor's value will be altered permanently.

Surface mount resistors have an added disadvantage over their larger PTH brethren. Under conditions of high current, "hot-spots" will form in SMT resistors. This is due to non-uniform current densities in the resistive film. These hot spots can permanently alter the resistor's value. Ohm's law tells us that if we have voltages in the thousands or tens-of-thousands of volts, we will have high currents during ESD events.

A coordinated network can be constructed to offer protection beyond that of Figure 5.5c. Figure 5.5d shows a two-stage network. The primary protection is a TVS working against the impedance generated by a ferrite bead. The two diodes and resistor provide the secondary protection.

These solutions are fairly expensive in terms of component

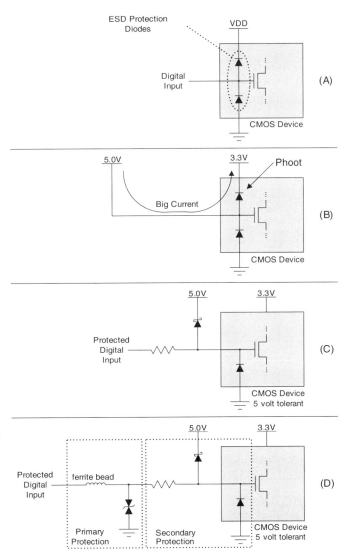

Figure 5.5: On-board ESD protection diodes offer protection, but can also complicate the design of multivoltage systems.

Chapter 5

count, board space and assembly cost. If protection must be added to a digital input and cost is an overriding factor, a simple RC network coupled with the internal protection diodes can provide a reasonable amount of protection.

The biggest problem with an RC network as a front line of defense against high-speed high-energy transients is the capacitor's parasitic effective series resistance (ESR) and effective series inductance (ESL). The ESR allows undesired high voltages to develop on the protected node. The ESL reduces the response time of the capacitor.

The good news is that the internal protection diodes can usually handle the leading edge of a transient event. This gives the RC network time to act as a filter.

Some integrated circuit manufacturers integrate high-end ESD suppression into their devices. For example, RS-232 and RS-485 transceivers are available with protection that guarantees the device can withstand repeated ±15 KV ESD hits. Analog switch manufacturers now offer devices with similar levels of protection.

The protected RS-232 and RS-485 transceivers are plug-in replacements for older, non-protected transceivers. The protected devices are not significantly more expensive than their unprotected counter parts. In this day and age of CE marks and emphasis on building robust devices, there is little reason not to use an ESD protected transceiver.

The best way to evaluate the level of protection any of these circuit topologies provide is through testing with an ESD generator. Schaffner EMC Inc, has excellent ESD simulators, also called ESD guns. These tools allow an engineer to zap a circuit under test with up to ± 21 kilovolts of simulated ESD.

Testing a few sample circuits on a bench isn't a particularly large sample set. But design is an exercise in trade-offs and risk management. For most products, ESD testing of a handful of model samples is sufficient.

Testing will also unmask hidden problems that will not show up in mathematical models or design equations. For example, consider the sample PCB layout shown in Figure 5.6. ESD will arc between the vias and damage the microprocessor. The best ESD protection can be made useless by sloppy PCB routing. Auto-routers should never be allowed to route protected networks. Unprotected signals should be kept well away from traces that go off board.

During ESD testing, a valuable technique is to darken the lab and while injecting ESD, look at the circuit board and visually check for arcing. Problems such as those shown in Figure 5.6 will become apparent.

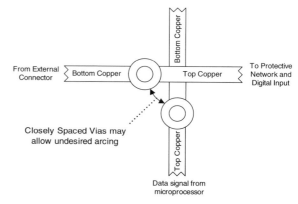

Figure 5.6: Even the best ESD protected input may be defeated by bad PCB layout.

Interfacing to the External World

5.2.3 Expanding Digital Inputs

One challenge faced by designers is how to add I/O to a processor. Greg Young, while working as a design engineer at Z-World, once said, "Every I/O pin has two struggling to get out." There are numerous techniques for expanding inputs, shown in Figure 5.7. Each has advantages and disadvantages.

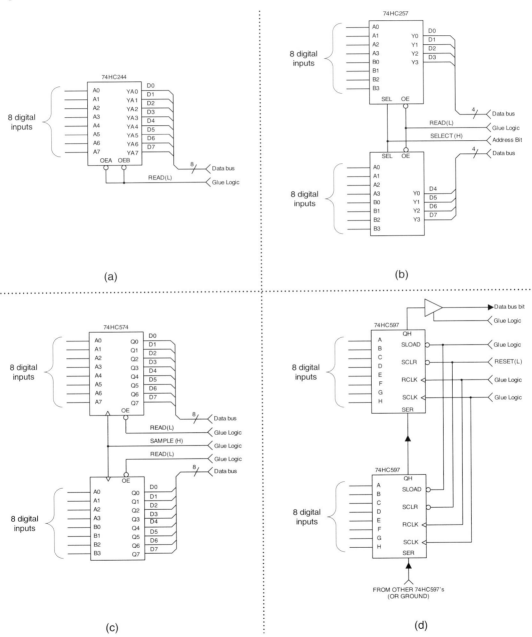

Figure 5.7: There are many ways to add digital inputs.

Chapter 5

All of the circuits shown in Figure 5.7 refer to HC logic devices. The HC logic family devices operate well from 3.3-volt or 5-volt rails. An abundance of logical functions are available.

An engineer faced with the task of building an interface between a 5-volt external system and a 3.3-volt processor core will have to consider noise margins. For example, the circuits in Figure 5.7 will work well if the HC devices are powered from 5 volts as long as there exists an HCT or HST buffer between the 3.3-volt core and the HC device. In most cases this means that HCT should be used for the glue logic.

In other cases, if the engineer can locate HCT or VHCT parts with equivalent logic functions to the HC parts shown, it may be preferable to replace the HC parts with another logic family.

From this point on, we will assume that we are not interfacing to 5-volt external logic and the HC logic parts are driven from a 3.3-volt rail. This will allow us to focus on the issues of capacitive loading, simultaneous sampling and general interfacing logic techniques.

As we look through these interfacing examples, we should consider that the underlying logical concepts are more important than the particular device implementations shown. The specific devices shown have been quite useful in designs, but there are always newer logic families and alternate devices available.

For example, the 74HC244 shown in Figure 5.7a is an example of an octal buffer. There are other parts available that perform the same function, but with different pin outs. For example, the 74HC245 is a bi-directional buffer that is often used in place of the 74HC244. The 74HC245 has a shorter propagation delay than the 74HC244 and may already be a line item on a design's bill of material. Of course the designer must decide which way to hard wire a 74HC245's DIR pin to ensure the device operates as a buffer in the correct direction.

Figure 5.7a is one of the most often seen methods for adding digital inputs to a processor's data bus. When a processor wants to read the digital inputs, the 74HC244's output-enable is asserted and data flows through the 74HC244 and onto the data bus. The biggest disadvantage of the 74HC244 is the 20pF worst-case capacitance of a tri-stated output. If more than a couple of these are added to a data bus, the capacitive load on the CPU's data bus may become intolerable.

The Rabbit 3000 offers a helping hand to the designer that wants to plop down a fistful of 74HC244s on the data bus. The Rabbit's Auxiliary Data Bus was added to the processor to minimize the capacitive loading on the high-speed memory bus.

If many 74HC244's are used to expand the I/O in a Rabbit-based design, the designer should consider using the Auxiliary Data Bus.

Figure 5.7b shows a technique that uses multiplexers to implement additional inputs. The 74HC257's SEL signal determines if the A0..3 or B0..3 inputs are presented to the Y0..3 outputs. Each 74HC257 tri-stated output capacitance is 15pF worst-case. Since each output actually corresponds to two inputs, the total data bus loading is 7.5pF per input. This compares favorably to the 20pF per input of the 74HC244 solution.

Sometimes a system is required to simultaneously sample more inputs that the data bus has bits. In this case, a latch can be employed. Figure 5.7c shows how two 74HC574's can be used to simultaneously capture 16-bits of data. The 74HC574's tri-state capacitance is 15pF per pin.

Interfacing to the External World

The 74HC574's sister chip, the 74HC374, has the same functionality but a different pinout. Depending on the PCB routing, one or the other IC's will be preferable. The 74HC574 has all of the inputs on one side of the chip, the outputs are on the other side. Most of the time, routing a PCB will be easiest with the 74HC574.

When minimizing capacitive loading on the bus is paramount, the scheme shown in Figure 5.6d should be considered. The serial shift chain only uses a single bit from the data bus. The only capacitive load on the bus is that presented by the tri-state buffer.

Each 74HC597 contains 16 flip-flops comprising two 8-bit registers. One register is the input latch. This is loaded by a rising edge on RCLK. The second register is the shift register. Data is moved from the input latch to the shift chain by asserting SLOAD(L) (active low).

Once the data is in the shift register rising edges on SCLK cause the data to be shifted out through QH. Each time a shift occurs, the "A" bit of the shift register is loaded from the SER input. 74HC597's can, for all practical purposes, be cascaded indefinitely.

SCLR(L) clears all of the registers in the shift-chain. This can be connected to the system's RESET(L) signal if the designer wants the shift-chain to be initialized with all zeros. Many designs will just tie SCLR(L) HIGH (inactive) and save the trouble of routing the trace. The shift chain will be loaded from the input latches on each read.

The concepts demonstrated in Figure 5.7 can be mixed and matched to suit the application. For example, eight shift chains from Figure 5.7d could use a single 74HC244 from Figure 5.7a as the tri-state output buffer.

Now, we can hear all the recent graduates leafing through these very pages and wondering aloud, "Where are the FPGAs and CPLDs? That's what real-engineers use to implement digital logic. Right?"

Unless the designer has some reason to use CPLDs or FPGAs beyond expanding digital I/O, they are a bad idea. They are expensive, require programming, are power hungry and worst of all, are often single source components (meaning they are available from only one manufacturer).

About the only advantage an FPGA or CPLD has over discrete HC or VHC components is board space. That's only true if the number of I/O pins is fairly high. If only 8 inputs need to be added, the board area-per-input ($mm^2 \cdot input^{-1}$) is tough to beat for a single 16-pin SSOP. Further marginalizing the FPGA/CPLD density advantage is the fact that most external I/O signals will require ESD protection networks. Since a large board area will be needed, using a high pin-density TSOP doesn't practically buy anything.

Surprisingly, it's hard to beat the simplicity, price and power consumption of HC and VCH logic parts for expanding digital inputs.

5.2.4 Expanding Digital Outputs

Expanding a system's digital output count is similar to expanding the digital inputs. It boils down to adding flip-flops that retain and present (to other devices) values written by the processor. The same issues of capacitive data bus loading exist for output logic as did for input logic. Additionally, issues of initialization, current drive capability and tri-state ability must be considered.

Figure 5.8 shows four schemes for implementing digital outputs. Each circuit has subtleties that will make it suited to some applications but not to others.

Figure 5.8a shows a simple 8-bit latch (74HC574). This is the same device used in Figure 5.7c to implement digital inputs. One troubling issue is that of initialization. The 74HC574 doesn't have a RESET pin.

The design shown in Figure 5.8a works around the lack of a RESET pin on the 74HC574. Upon system reset, 74HC574's outputs are tri-stated by the external flip-flop. This allows the external resistors to fix the port's state. The advantage of this method is that some of the outputs may be tied low and some high. This allows greater flexibility over a latch that uses a RESET (or CLR) pin to fix all of the outputs low upon initialization. Of course, the resistors will not be able to source or sink as much current as the 74HC574's outputs.

The added expense of the external flip-flop and resistors may be undesirable. This is especially true if an "all zero" initialization state is required. The circuit in Figure 5.8b shows how a 74HC273 can be used to implement a digital output with an all zero initialization. The CLR(L) pin can simply be wired to the system reset.

Both the circuits in Figures 5.8a and 5.8b suffer from the problem of placing a fairly high capacitive load on the data-bus per digital output. The figure in 5.8c shows how to use a 74HC259 bit addressable latch with global CLR(L) to reduce capacitive bus loading.

The 74HC259 has one annoying feature. The LOAD(L) signal that writes data into the flip-flops is level sensitive. This places a requirement on the processor bus to hold the data bit (DO in Figure 5.8c) valid after the LOAD(L) is brought high.

Some processors do not expect this hold time requirement. I/O devices are expected to capture their data when the WRITE single is first asserted. The processors have setup and hold time around the leading edge of the WRITE signal assertion. When the WRITE signal is de-asserted, some processor datasheets state a zero minimum hold time. The Rabbit has configurable hold times.

Figure 5.8d shows how to use the 74HC594 to implement a series of digital outputs. The 74HC594 does not offer an output tri-state. A similar chip, the 74HC595 swaps the output latch clear (RCLR(L)) for an output enable. Using a 74HC595 and a series of pull-up and pull-down resisters, as in Figure 5.8a, allows individual outputs to be initialized to HIGH or LOW states as the application may require.

There are other schemes for adding digital I/O to a processor. CPLDs and FPGAs are currently "all the rage." For reasons described earlier, they should be used only when absolutely required.

Regardless of the technique used to implement the digital output, there is a question of protecting the output from an over-current condition. Most of the CMOS parts presented have a maximum output current rated at 20 mA. If an output is inadvertently shorted to a power rail, damage can occur.

In systems with multiple PCBs, cable harnesses carry digital outputs between boards. If the outputs are used purely for logic functions, then a resistor placed in series with the output is well tolerated. Low input-current requirements ensure that the drop across a series resistor

Interfacing to the External World

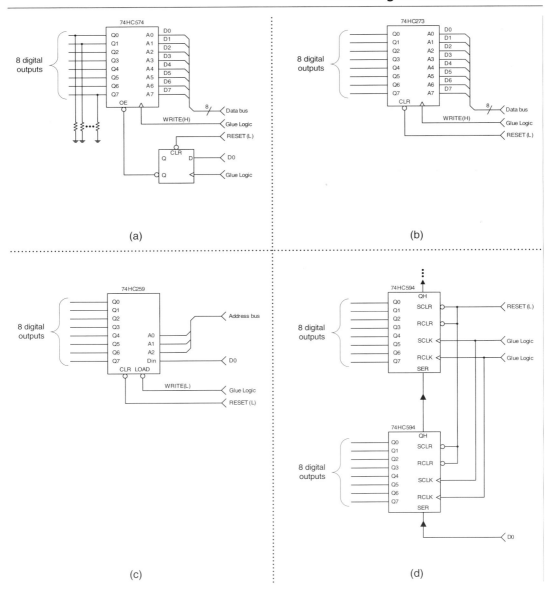

Figure 5.8: A variety of latches and shift registers allows the designer to make trade-offs to meet the application's requirements.

will be minimal. Under a fault condition, such as when a cable harness is incorrectly connected, the series resistor limits the current into or out of the digital output thereby protecting the driver IC.

ESD protection techniques are based on the same principles discussed earlier for digital inputs. GDTs, MOVs, and TVSs are all good options for diverting potentially damaging transient energy away from a digital output pin.

5.3 High Current Outputs

Digital outputs are fine and dandy, but embedded systems usually need to control actuators with digital ICs. The limited current available from a CMOS output is seldom enough to drive much beyond an LED.

The usual suspects for implementing high current outputs are bipolar junction transistors (BJTs), Darlington pairs, MOSFETs, electromechanical relays and solid-state relays (SSRs). We will have a brief look at each of these tools and examine the strengths and pitfalls of each.

5.3.1 BJT-based Drivers

Bipolar Junction Transistors (BJTs) are one of the most cost effective ways to implement a "high current driver." Discrete transistors are available in PCB mountable packages from the rice-sized SOT-23 to strawberry-sized TO-3.

When a BJT is saturated, $V_{ce(sat)}$ will be finite and nonzero. For small signal transistors switching small collector currents, 100 mV is a good estimate for $V_{ce(sat)}$. As the collector current goes up, so will $V_{ce(sat)}$.

The power dissipated in a transistor due to the collector current will be $V_{ce(sat)} * I_c$. The base current will also contribute $V_{be} * I_b$ watts to the total power dissipated.

Smaller devices can shed less heat than larger packages. Transistors like the MMBT2222A in a SOT-23 can only dissipate about 350 mW @ 25°C. The PN2222A in a pin-through-hole TO-92 is rated for 650 mW @ 25°C.

Transistors, such as the Zetex FMMT625 have a combination of low $V_{ce(sat)}$ and high current-transfer-ratio (also called beta in saturation, β_{SAT}, or $h_{FE(SAT)}$). This combination minimizes power dissipation.

Figure 5.9a shows the simplest single transistor low-side driver, also called a sinking driver. After power dissipation, the biggest issue to consider in this topology is the base current required to keep the transistor in saturation.

Over temperature, a ten is a conservative value for $h_{FE(SAT)}$. Some higher end devices, such as the FMMT625 have an $h_{FE(SAT)}$ twice that. This is a far cry from the value of 100 that many engineers use.

The reason for this discrepancy is that transistors biased in the active region have a much higher current gain than transistors in saturation. While a β of 100 might be a conservative parameter for amplifier design, it is ten times too high for most transistors operating in saturation.

The practical implication is that the base drive for a single transistor driver may be a burden on the CMOS output connected to the BJT. For example, if the high-current driver in Figure 5.9a is expected to switch a 500 mA load, a base current of 50 mA is required.

CMOS devices specify a maximum I_{cc} that may be pulled from VCC or sunk into GND. The 74HC574, which is one of the more robust CMOS parts, has an $I_{cc(max)}$ of 70 mA.

If all eight outputs are used then only (70 mA / 8 outputs) 8.75 mA per output (on average) is available. That leaves no safety margin for the 74HC574.

Interfacing to the External World

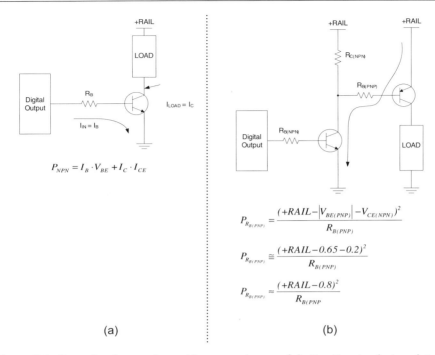

Figure 5.9: Even simple transistor drivers require careful attention to design details.

Most CMOS devices are less capable than the 74HC574. The 74HC259 has an $I_{cc(max)}$ of only 50 mA. Distributed over eight outputs, only 6.25 mA per output (on average) is available.

Building drivers capable of sourcing or sinking many amperes will usually require multiple stages of current amplification.

Figure 5.9b shows a two-stage driver. Heating in the PNP's base resistor is a key design consideration.

The voltage imposed across the PNP's base resistor is $+RAIL - |V_{BE(PNP)}| - V_{CE(SAT)NPN}$. This approximates to $(+RAIL - 0.8)$. For high rail voltages the V^2/R heating in the PNP's base resistor can be very high. For every doubling of voltage imposed across the resistor, the power dissipated quadruples.

For example, if the circuit shown in Figure 5.9b is designed to allow a 50 mA PNP base current and +RAIL is 30 volts, then $R_{B(PNP)}$ will have to be approximately 600 ohms. The power dissipated in $R_{B(PNP)}$ will be,

$$POWER = I \cdot V = I^2 R = \frac{V^2}{R}$$

$$P_{RB(PNP)} = 0.050 \cdot (30 - 0.8) \approx 0.050^2 \cdot 600 \approx \frac{(30-0.8)^2}{600} \approx 1.5\ watts$$

Chapter 5

If eight channels are designed into a device, then (8 * 1.5 watts) 12 watts of heat will have to be removed from the circuit. This may require ventilation and possibly a fan.

The least expensive transistors cost only pennies, but the assembly cost dwarfs the component cost. A good rule of thumb to use for insertion cost is $0.12 per part. This can be used for both PTH and SMT parts. Devices that require a heat sink, like a TO-3 or TO-220, will have additional charges.

Figure 5.9a requires two parts. The resistor and transistor may only cost a couple of pennies, but the insertion cost will be 2 × $0.12 = $0.24. The circuit in Figure 5.9b has five components for a total of $0.12 × 5 = $0.60 in assembly costs. For systems that require many I/O pins, this will quickly become expensive. For applications requiring many high-current drivers, there are ICs available containing eight channels of Darlington drivers.

Figures 5.10a and 5.10b show the ULN2803 and UDN2985 respectively. The ULN2803 is a sinking driver and the UDN2985 is a sourcing driver. Both of these devices may be driven from CMOS outputs.

The ULN2803 can sink up to 500 mA per pin, although the device is limited by the total power the DIP package can dissipate. This means the entire device can sink about 500 mA split up between the eight channels.

The UDN2985 can source around 250 mA. The maximum rail voltage is 30 volts. Like the ULN2803, the UDN2985 is limited by the package's heat shedding ability.

Both devices have integral fly-back suppression diodes. When driving electromechanical relays, the fly-back suppression diodes protect the transistors from the back EMF generated by the relay coil.

One disadvantage of these integrated drivers is the Darlington's inability to pull the output to the rail. In a Darlington, the output transistor is never driven fully into saturation. A Darlington can only pull an output within 1.2–2.5 volts of the rail.

The UDN2985's output stage is not a proper Darlington configuration, but is most accurately referred to as a compound PNP output stage. As with the Darlington configuration, the drive transistor is never saturated and subsequently the output stage will only drive within a volt or two of the rail.

For example, let's say a ULN2803 is going to drive a relay with a 5-volt coil and the top of the relay is connected to a 5-volt rail. The Darlington can only develop about a 3.5-volt drop across the relay coil. The remaining 1.5 volts will be across the Darlington.

Many 5-volt relays have a pick-up voltage (the voltage at which the relay is guaranteed to operate) that is higher than 3.5 volts. This means the ULN2803 may not be capable of driving the 5-volt relay with only a 5-volt rail.

Another place prepackaged Darlington drivers fall short is when people use them to drive digital inputs. The Darlington's voltage drop can eat up ALL of CMOS device's noise margin, especially when the CMOS device is being driven from a 3.3-volt or 2.7-volt rail.

Interfacing to the External World

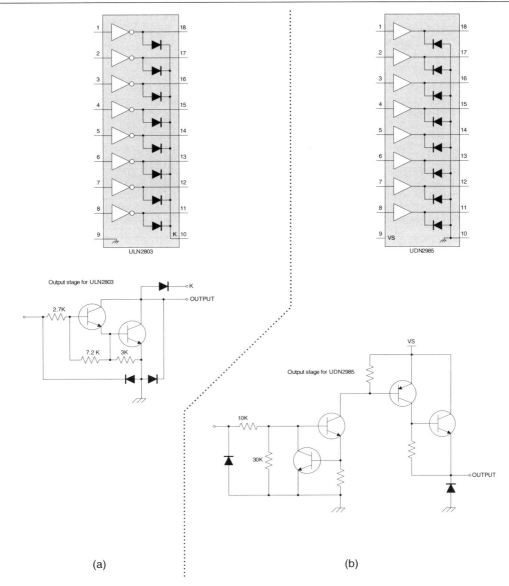

Figure 5.10: Prepackaged drivers simplify a designer's life.

5.3.2 MOSFETs

Metal Oxide Semiconductor Field Effect Transistors (MOSFETs) have many advantages over BJTs. Their low $R_{DS(ON)}$ allows MOSFETs to switch much higher currents than BJTs. They can be paralleled to share current. MOSFETs have a theoretically zero sustained drive current.

Disadvantages include maximum V_{GS} restrictions—usually around ±20 volts. MOSFETs are notoriously sensitive to ESD damage. Through careful design, these issues can be managed.

Chapter 5

There are hundreds of MOSFETs on the market. Table 5.1 highlights several inexpensive devices that cover a range of performance. The devices in Table 5.1 can be driven with a $|V_{GS}|$ of 5 volts, allowing 5-volt logic to turn them on and off.

Table 5.1: These readily available SMT MOSFETs are good choices for new designs.

Manufacturer	P/N	Polarity	Package	$V_{DS(MAX)}$ (volts)	$I_{D(MAX)}$ (Amps)	$V_{GS(MAX)}$ (volts)
Fairchild Semiconductor	FDN5618P	P-Channel	SOT-23	−60	−1.25	±20
Fairchild Semiconductor	FDN5630	N-Channel	SOT-23	60	1.7	±20
Fairchild Semiconductor	NDS332P	P-Channel	SOT-23	−20	−1.0	±8
Fairchild Semiconductor	NDS335N	N-Channel	SOT-23	20	1.7	20
Fairchild Semiconductor	NDT2955	P-Channel	SOT-223	−60	−2.5	±20
International Rectifier	IRFL014	N-Channel	SOT-223	60	2.7	±20
International Rectifier	IRF4905S	P-Channel	D²PAK	−55	−74.0	±20
International Rectifier	IRF1010NS	N-Channel	D²PAK	55	75.0	±20

The last two lines in Table 5.1 indicate very high drain currents are possible. The conditions under which these currents could be obtained are highly unlikely to be realized in a practical design. The case temperatures must be maintained at 25°C to get the −74 or +75 amp drain currents listed by the manufacturer. Perhaps with some sort of spray cooling or chilled heatsink technology this might be possible. Most designs are lucky to have circulating air in a vented box. Under these conditions, using the IRF4905S or IRF1010NS to switch a few amps is reasonable, provided that sufficient V_{GS} is available.

The circuit shown in Figure 5.11 uses an IRF4905S to switch a 30-volt source into a load. A zener diode is used to clamp the maximum V_{GS} to less than the 20 volt maximum allowed by the datasheet.

Unlike BJTs, there are no gate resistors associated with MOSFETs that carry a significant current. This means less power is dissipated driving a MOSFET than driving a BJT—see Figure 5.9b.

A neat trick that can be used to save a few pennies in a design, such as that shown in Figure 5.11, is to use a reverse biased base-emitter junction in place of the zener diode. The Zetex FMMT491 can be used for both the current sink NPN shown in Figure 5.11 and in place of the zener diode. The base-emitter junction breaks down around 10 volts. Since the IRF4905S turns on pretty hard with a V_{GS} of −10 volts, this technique will allow the elimination of a line item from the BOM without compromising the circuit's performance.

Some caution is advised with this sort of hack. The Zetex FMMT491 indicates the emitter base junction will breakdown at 5 volts with 100 uA. Through experimentation, one can see that with 2 mA, the FMMT491's V_{EB} is about 10 volts. Hitting the FMMT491 with a can of freeze-mist and a blast from a hot-air gun will show the FMMT491 will work pretty well over a broad range of temperatures.

Interfacing to the External World

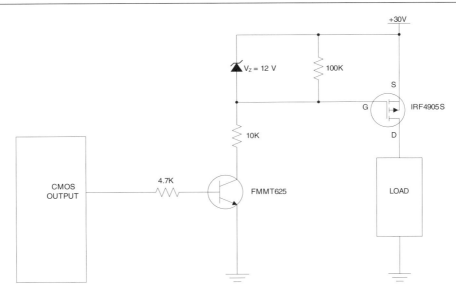

Figure 5.11: A zener diode is a simple way to limit V_{GS} to a reasonable level.

As a "general rule," engineers frown upon relying on a device's undocumented characteristics. Engineers are also the ones responsible for making trade-offs. Sometimes the careful application of common sense coupled with a little experimentation will yield an equitable trade-off that runs counter to the "general rules." Sometimes engineering is as much art as science.

As with BJT circuits, IC manufactures offer devices that have multiple MOSFET drivers in a single package. One such family of devices is the Power Logic available from Texas Instruments and STMicroelectronics.

The Power Logic devices are available in common logic functions, such as latches (TPIC6259, STPIC6A259 and TPIC6B273) and shift registers (TPIC6595). The output stage is an open drain N-channel MOSFET. There are integrated fly-back suppression diodes for driving inductive loads.

Depending on the number of Power Logic channels sinking current in a device, and the ambient temperature, each channel can continuously sink 150 mA to 500 mA. Some devices allow peak currents up to 1.5 amperes. The outputs are usually rated for 45 to 50 volts.

Unlike the Darlington high-current drivers, the Power Logic devices will pull their outputs very close to ground. Additionally, FETs share current well. Multiple channels of Power Logic outputs can be paralleled to obtain higher current drive capacity than is available from only one channel.

Texas Instruments offers the devices in DIP packages. STMicroelectronics has datasheets for SMT versions of some devices.

With the advent of inexpensive, robust and beefy MOSFETs, a designer has options that didn't exist with BJTs. In particular, higher load currents can be switched with lower drive currents.

Chapter 5

5.3.3 Electromechanical Relays

When an application calls for a switch with low contact resistance, an electromechanical relay is sometimes the best choice. Like all electronic components, relays have evolved rapidly in the last few decades.

Relay manufacturers measure reliability in minimum expected operations. Today it's common to see tens-of-millions of expected operations before contact failure.

Relays are available in a SMT and PTH packages for circuits that must switch milliamps to a few amps. Milk carton sized relays, called contactors, are also available for applications that need to switch hundreds of amps.

Small relays are often used to drive larger contactors. In these situations, the small relay is called a pilot relay. Since driving a contactor is simply a matter of driving a pilot relay, we'll examine smaller relays in a bit more detail.

The Omron G6B series PTH relay is notable for having contacts rated at 5 amps up to 250 VAC. The Omron contacts can switch a maximum 150-watt (1250 VA) resistive load. This is only impressive when one considers the device's volume is only about 2.4 cubic centimeters.

SMT relays such as the TQ-SMD series from NAIS are also available. These small DPDT relays occupy less than a cubic centimeter. The contacts are rated to switch a 60-watt (or 62.5 VA) resistive load with a maximum current of 2 amps.

Relays, like all switches, have a finite contact resistance. This is best measured under maximum current load.

The practical implication of contact resistance is heat. When the contacts carry current, the contact resistance causes Joule heating proportional to I^2R.

The Omron G6B has a 30 milliohm contact resistance. The smaller TQ-SMD contacts have more than twice the contact resistance of the Omron (75 milliohm).

The Omron G6B relays achieve such outstanding contact ratings by having BIG contacts. Opening a G6B with a hacksaw will reveal the disproportionate large contacts for the volume of the relay. To get the relay to perform with large contacts and a relatively small coil, Omron has incorporated a "helper" magnet in the G6B series.

The helper magnet effectively polarizes the relay. When the coil is energized with the proper polarity, the helper magnet is attracted to the coil and the relay's armature is actuated. When the coil is de-energized, the contacts move back into their original position. A reverse energized coil will actually repulse the armature and the contacts will not actuate.

Regardless whether the relay has a helper magnet or not, the driver circuit must be capable of handling the coil's fly-back voltage.

Placing a diode across the relay coil is the most common method of clamping the fly-back voltage to a level that will be tolerated by transistor drivers. Packaged drivers, such as the ULN2803 and UDN2985, shown in Figure 5.10, contain a fly-back suppression diode. Figure 5.12 shows an example of a discrete low-side BJT driver sinking current for a relay with an accompanying fly-back suppression diode.

Interfacing to the External World

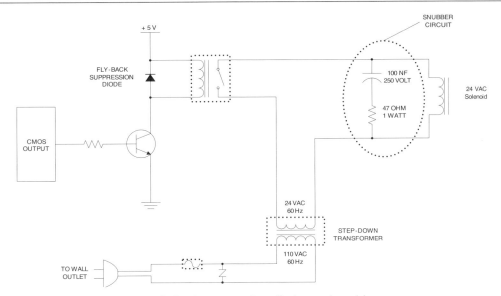

Figure 5.12: Fly-back suppression diodes and snubbers are both tools to manage the energy stored in inductive elements.

Relays are often used to drive motors, contactors, solenoids and other magnetic devices. When driving inductive loads, a designer must take into account the energy that will be stored in the load's magnetic field. If the load is driven with a DC source, a fly-back suppression diode can be used.

If AC is used to drive the load, a simple diode will not work. A simple RC network, called a snubber network, can be used quite effectively to dissipate stored energy. Figure 5.12 shows an example of a snubber network.

When the relay is opened, the snubber and load form a damped LCR circuit. Since both the inductor and capacitor are theoretically purely reactive devices, they will not dissipate any energy. The energy stored in the load will be dissipated as heat in the snubber's resistor.

In practice, all parasitic resistances will dissipate energy as heat. The effective series resistance of the capacitor will limit the (ripple) currents the capacitor can handle. The DC resistance of the inductor will generate heating (copper losses) in the inductor, as will core heating (iron) loses.

The consequences for not using a mechanism, like a snubber, to burn off undesired energy can be as innocuous as a little arcing on the relay contacts or as serious as arcing that reaches back into relay's coil driver. Small relays can offer several thousand volts of isolation between the contacts and coil circuit. Inductive fly-back can often exceed this level. In either case, undesired radiated emissions from the arcing will occur.

Figure 5.12 shows a snubber placed across an inductive load. This arrangement can reduce noise significantly. The snubber provides a conductive path for the current that the inductor's stored energy will produce. Inductive flyback voltages will be significantly reduced when the relay opens.

Chapter 5

In practical systems, the length of wire between the switch and the inductive load is often sufficient to produce a bit of back EMF across the relay. This will contribute to arcing in the switch contacts. Arcing generates unwanted EMI and if severe will cause deterioration of the relay contacts.

A common practice is to place the snubber directly across the switch contacts as shown in Figure 5.13. This minimizes arcing and the unwanted associated emissions. Another advantage to this configuration is that the relay is often located on a PCB. This means it is fairly easy to add a snubber as the resistor and capacitor are intended to be mounted on a PCB.

A snubber configuration as shown in Figure 5.12 is sometimes overly costly because the solenoid may not have any PCB to which to mount a snubber. This means either the device must be "fly wired" and glued or a printed circuit board must be added. Either way, labor and material costs increase.

As a practical matter, if the wire lengths between the switch and the inductive load are long, the area enclosed by the wire loop can become quite large. Depending on the frequencies involved, this can make for a pretty good radiator. In some cases, a snubber may need to be placed across both the switch and the load to bring emission levels down to an acceptable level.

The decision to place a snubber across the switch depends upon the geometry of the system, the currents being switched and the magnitude of the inductive load. One other factor is the construction of the switch.

Mechanical contacts arc. Semiconductor based switches do not. A snubber that is used to reduce arcing and the associated broadband radiated emissions may not be required if the mechanical relay is replaced with a solid-state relay.

On the other hand, if the size of the inductive load is unknown, or the length of wire between the switch and the load is long, a snubber near the switch is good insurance against high flyback voltages damaging the switch.

Whenever there is a snubber placed across a switch, leakage must be considered. The impedance of the snubber's capacitor is given by,

$$C_{IMPEDANCE}(\omega) = \frac{1}{C \cdot s} = \frac{1}{C \cdot j\omega} \Rightarrow (\ )C_{IMPEDANCE}(f) = \frac{1}{C \cdot 2\pi f j} \Rightarrow |C_{IMPEDANCE}(f)| = \frac{1}{C \cdot 2\pi f}$$

ω is radian frequency; $s = j\omega$; f is frequency in Hz; $j = \sqrt{-1}$

In Figure 5.13, when the relay contacts are open, there is still a conductive path for the 60 Hz AC across the contacts. The impedance is,

$$Contact_{(OFF)IMPEDANCE}(60Hz) = \frac{1}{100 \cdot 10^{-9} \cdot 2 \cdot \pi \cdot 60} \angle -90° + 47\angle 0° \approx 26.5k\Omega\angle -90°$$

In the 24 VAC system of Figure 5.13, 26.5k ohms allows about a milliamp of leakage current. This is not sufficient to keep the solenoid energized.

Interfacing to the External World

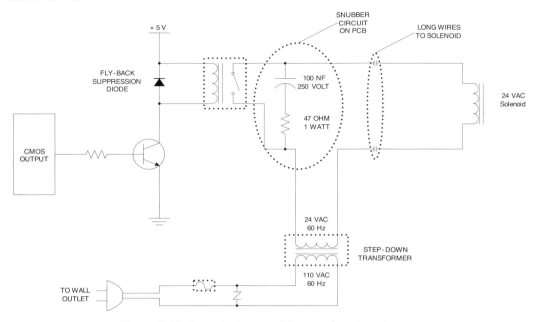

Figure 5.13: In a pinch, a snubber can be placed across the relay contacts instead of the inductive load.

In some cases, a snubber across a relay's contacts can be problematic. If the relay is switching a high voltage, the snubber can allow enough leakage current to give an incautious human a good jolt. Depending on the system this can be handled with safety interlocks and strict lock-out/tag-out procedures. In other systems, it may simply be unacceptable to allow any leakage current across the relay contacts. These are trade-offs the designer must make.

Another situation to look out for is when a snubber allows enough leakage current to keep the magnetic load actuated when it should be off. The case of a pilot relay driving a contactor is a classic case.

Contactors (and relays) have a pick-up current and a hold-current. The pick-up current is the amount of current required to develop a sufficient magnetic field to actuate the contactor. The hold-current is required to hold the contactor in an actuated position once the contactor has been actuated.

The pick-up current is generally much higher than the hold current. Some relay manufacturers specify a pick-up voltage and a hold-voltage. By using the impedance of the relay or contactor's coil, one can calculate the currents in question.

The hold-current will often be specified as a maximum value. This means that even less current will be required to hold the device in an actuated state than the specified hold-current.

An undesirable situation will occur if a pilot relay's snubber allows a leakage current higher than the **minimum** hold-current for the contactor. In essence, the pilot relay will turn on the contactor, but the leakage current will never allow the contactor to fully de-energize.

Chapter 5

If the system designer is suffering particularly bad karma, the snubber's leakage current will not be enough to keep most contactors energized. Symptoms of an inappropriately designed snubber may not show up until systems are deployed in the field. Besides being somewhat embarrassing, this type of bug can be hard to find and expensive to fix.

Snubbers are invaluable tools for managing energy stored in inductive loads. The designer will have to trade off component values to achieve the responsiveness, costs, size and leakage currents required for each specific system.

For small inductive loads (fractional horse power loads), values for the capacitor are often found between 10 nF and 100 nF. It's common to over specify the working voltage for the capacitor. In systems like those shown in Figure 5.13 high voltage transients from the AC mains will be stepped down and coupled to the capacitor.

Another condition to be aware of is the failure of the galvanic isolation between the primary and secondary of the step down transformer. If this occurs, the ideal situation would be for the fuse to open, see Figure 5.13. In another case, the solenoid may be damaged. If the snubber's capacitor is over specified, as is the case in Figure 5.13, under no circumstances should the controller board be damaged. Worse case, a technician may have to replace a fuse, the step-down transformer and the solenoid. None of this requires repairing a PCB.

Resistor values tend to range from 10 ohms to 100 ohms. As a matter of good practice, the resistor should be specified to be flameproof and, if size and cost permit, over specified by a watt or two.

When making these trade-offs, the best advice is "test early and test often." Better to spend a little more money early on in a design to find out that a design trade-off will push emissions above allowable limits than to get a design set in concrete only to find the same information out later. A good test lab will be willing to work with a designer throughout the design process—not just at the end of a design.

Another useful tool is a circuit simulator, like SPICE. Simulators are notorious for being so strict and finicky that getting a simulation to complete is nearly impossible. Some other simulators are so over-simplified that getting data the designer can trust is difficult. Like all technologies, circuit simulators have improved over time.

Linear Technologies (www.linear.com) offers a free SPICE implementation called Switcher-CAD III. This tool has an easy to use graphical front end, while allowing access to the SPICE netlist. SwitcherCAD III is a third generation SPICE engine and is optimized for simulating switching power supplies. The tool works wonderfully for the general purpose simulation.

On the CD-ROM packaged with this book, we have provided simulations for the circuits shown in Figures 5.12 and 5.13. A quick visit to www.linear.com will let the reader download and install the most current version of SwitcherCAD III.

5.3.4 Solid-State Relays—SSRs

Solid-state relays are simply optoisolated silicon switches. These devices are available in a wide range of form factors and power ratings. Currently, the "contact" side of SSRs can be

Interfacing to the External World

found with BJTs, SCRs, TRIACs, and MOSFETs. Some devices even contain circuitry to delay the "contact" from turning on until the AC voltage across the "contact" is at a zero crossing.

SSRs, when correctly sized, offer an advantage in mean time between failure (MTFB) over electromechanical devices. This comes from the fact that SSRs have no moving parts. SSRs, like all electrical components, have their own trade-offs.

SSRs are more expensive than electromechanical devices, and tend to dissipate more heat than their electromechanical counterparts. When specifying an SSR, the designer must pay careful attention to the device's derating curve.

For example, the SSRs in the Crydom MCX family are sold as 5 amp relays. The derating curve shows that at ambient temperatures above 30°C the device is to be derated by 120 mA/°C. This means if the SSR is intended to operate in a 60°C ambient environment, the device is only capable of carrying about 3 amps.

This heavy temperature derating can cause a designer to have to specify larger SSRs and thereby incur even more cost. Electromechanical relays also have temperature derating curves, but SSRs generally have the more onerous restriction.

In addition to MTBF, SSRs offer an advantage over electromechanical devices on the drive side. Driving an SSR is simply driving an LED. SSRs often have a fairly wide input range. The Crydom MCX series has input options spanning 3–15 VDC to 90–140 VAC(rms).

SSR input currents are usually a few milliamps to a few tens of milliamps. The MCX datasheet lists typical input currents from 5 mA to 15 mA depending on input configuration.

An SSR's driver circuit does not have to contend with inductive fly-back. However, switching inductive loads is still something to which the system designer must pay careful attention. When switching DC into inductive loads, designers can use a fly-back suppression diode to manage back-EMF. Circuits switching AC into inductive loads will benefit from the same simple snubber circuit shown in Figures 5.12 and 5.13.

SSRs, when properly applied, have outstanding life expectancies. They may require substantial heat sinking or air circulation. SSRs are usually more expensive than their electromechanical cousins. SSRs are wonderful innovations, but they are not the best solution for every situation.

5.4 CPLDs and FPGAs

Complex programmable logic devices (CPLDs) and field programmable gate arrays (FPGAs) have been a boon for system designers. The high gate density available allows complex functions to be implemented in a configurable device. Programmable logic devices are available in ever-faster speeds grades.

FPGAs are available as SRAM based devices that must be programmed on boot. An 8-pin boot EEPROM is the usual method. There are some antifuse-based FPGAs that are the equivalent of OTP devices.

Chapter 5

CPLDs are generally flash-based devices. Once programmed, they retain their configuration even during power-down.

These devices can be used to best effect by placing them in the path of high-speed data to do filtering or other high-speed manipulation. Figure 5.14 shows how FPGAs can be placed in a video path to do image filtering and other video related tasks.

CPLDs and FPGAs are architecturally different creatures, but for our purposes the distinction is academic. Generally we can assume a CPLD will be smaller, cheaper and faster than FPGAs.

Figure 5.14 shows a block diagram of a reasonably sophisticated single CCD camera. The system provides a processed analog output as well as a web connection. The web connection in this system is primarily used to configure the image processing. The system also allows a frame to be captured from either the raw RGB image path or from the processed RGB image path. The captured image can be pulled into the Rabbit-based RCM3700 for analysis or sent over the Ethernet connection.

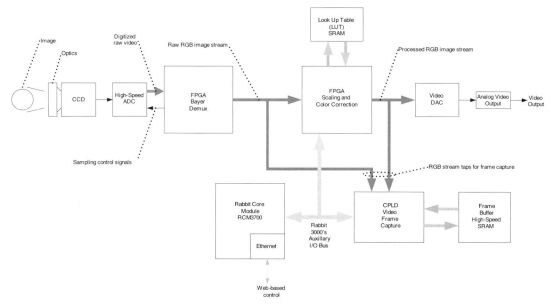

Figure 5.14: A digital video camera is an application that benefits from the versatility and high speed of FPGAs.

Color cameras that use a single CCD have a color filter (or mask) over the CCD that makes some pixels sensitive to red light, some to green and some to blue. The most common of these masks is called a Bayer color filter. When reading a Bayer encoded image, a system must decode (or demultiplex) the image and form single digital pixels each with a red, green and blue component by weighting several adjacent color selective pixels from the sensor. In Figure 5.14, the first FPGA in the video path is dedicated to this function.

Once the raw video image has been converted to a digital steam of RGB information, the camera places a second FPGA in the signal path to apply additional processing to the image.

Interfacing to the External World

This may include scaling the image, digitally zooming on a portion of the image, or doing color correction.

The RCM3700 can communicate with the image processing FPGA via the Rabbit's Auxiliary I/O bus. This is an 8-bit bus and is comparatively slow. The Rabbit isn't moving real time video data across the bus, just configuration data for the FPGA.

The look-up-table (LUT) connected to the second FPGA is a pretty common feature found in image processors. The RCM3700 can load the LUT through the image processing FPGA. Once the LUT is loaded, the FPGA can use the data in the LUT as part of the filtering algorithm.

Figure 5.14 shows a CPLD acting as a high-speed data acquisition controller and memory interface. The CPU can command the CPLD to start a high-speed capture of a single frame from either the raw or processed video streams. Once an image frame is collected in the frame buffer SRAM, the CPU uses the CPLD as glue logic to access the contents of the frame buffer.

The Rabbit 3000A has two new block move instructions (LSDDR and LSIDR) that allow the CPU to move data from repeated reads from a single IO port to multiple RAM locations. These block move instructions could be used to move a video image efficiently from the frame buffer into the RCM3700's internal SRAM.

If a system has need of a CPLD or FPGA, extra I/O pins can often be used to expand digital I/O lines. Since the expensive programmable logic already exists in the system, the extra I/O pins are essentially free.

For example, if the camera needed PTZ (pan / tilt / zoom) controls or a local man-machine interface (MMI), extra pins from the CPLD or image processing FPGA can serve these functions.

5.5 Analog Interfacing—An Overview

So far, we have looked at expanding a controller's capacity for turning devices ON and OFF and sensing whether a device is ON or OFF. Most of the world is analog. Temperature, color, strain, sound intensity, velocity, pressure and innumerable other environmental quantities are analog. Our discussion will not be complete unless we look at pulling analog signals into a digital environment.

5.5.1 ADCs

An analog to digital converter (ADC) is a device that maps an analog signal, usually a voltage, to a digital code. These devices can be implemented in a variety of ways to optimize for certain characteristics. For example delta-sigma ($\Delta\Sigma$) converters, are generally slow, but have very high resolution and moderate cost. The $\Delta\Sigma$ converters are found in many biophysical sensing instruments.

Flash converters are the fastest converters available, but are expensive and have relatively low resolutions (6 to 12 bits). These types of converters are found in video applications, communications systems and digitizing oscilloscopes.

Successive approximation (SAR) converters use a binary search algorithm and a single comparator to accomplish a conversion. They strike a middle ground between $\Delta\Sigma$ converters and flash

Chapter 5

converters. SAR converters are available between 8 and 16-bits with reasonably high sample rates at moderate prices. SAR converters are the workhorses found in many embedded systems.

ADC's have nonidealities. Most ADC datasheets do a fairly good job of characterizing the device performance. Sometimes a datasheet may give "typical" performance parameters or only characteristics at 25°C. In these situations, further testing and a few phone calls to the manufacturer will often yield a more concrete characterization.

An ADC is just one component in a chain of components that comprise a data acquisition channel (DAQ channel). Unless the ADC selected is a terrible device, the errors associated with the ADC will probably be swamped by errors associated with other components in the DAQ channel.

5.5.2 Project 1: Characterizing an Analog Channel

The ADC's reference voltage (or current), signal conditioning op-amps, gain resistors, filter capacitors, noisy power supplies, parasitic noise coupling from poor PCB layout and even raw sensor errors will introduce noise and errors into any DAQ channel. Characterizing an ADC is less important than characterizing an overall DAQ channel.

There are a few types of analysis that can be used to characterize a DAQ channel. These roughly divide into two groups. DC characterization is concerned with identifying offset errors, gain errors and noise when a DC signal is applied to the DAQ channel. AC characterization is concerned with looking at these errors as various frequencies are applied to the DAQ.

There are many papers that discuss academically rigorous methods for characterizing DAQ channels. In some systems an exhaustive characterization may be justified. However, most practicing engineers will get a long way with simple DC characterization. In this section, we are going to work through a DC noise analysis of one of the Rabbit Semiconductor's RCM3400 DAQ channels.

Linux-based tools will be used to do the data collection and analysis. These tools are well developed and, best of all, are free. Linux is an ideal platform for laboratory PCs collecting, logging and analyzing data.

Some level of noise is present in any analog circuit, dependent on the physical properties of the ICs, noise coupled in through the physical layout of the printed circuit board, ripple present on the DC power supplies, and so forth. By applying a known DC signal to our DAQ channel, we can sample our ADC readings and learn a great deal about the quality of the system.

For this section, we will use equipment setup as in Figure 5.15.

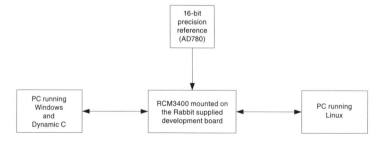

Figure 5.15: A simple setup allows a quick characterization of the RCM3400's analog channels.

Interfacing to the External World

Our analysis will yield a quantitative set of numbers characterizing the offset and noise performance of the DAQ channel. From the acquired dataset, we will also generate a histogram as a qualitative aid for the engineer to see how the noise is distributed.

The characterization will proceed as follows.

- The RCM3400 will acquire measurements of the precision reference
- The RCM3400 will transmit the sampled data to the Linux PC
- Linux-based software will analyze the data statistically and display it graphically

The data collection phase will complete entirely before the sampled data is transmitted to the Linux PC. This reduces the chance that any activity on the serial ports will be coupled as noise into the ADC channels. Furthermore, sampling the ADC channel as quickly as possible minimizes the likelihood of thermal drift and offset changes smearing the data in our histogram. We are trying to characterize the RCM3400's best-case performance.

Sampling the Precision Reference with Dynamic C

We must decide how many ADC samples we need in order to be statistically significant. Arbitrarily, we pick 32,768 (32K) samples. This number is not so large that it will take too long to acquire the data, and it doesn't seem so small as to be insignificant. Picking a "statistically meaningful" sample size for a project like this is a bit of an art.

For our example, we are using a battery-powered, precision 16-bit voltage reference that produces 2.4995 ± 0.0005 volts with noise less than 1 part in 65,536 (16-bits). The RCM3400 ADC will accept voltages up to 20V, so we are well within the physical limits of the Rabbit's hardware.

Let's consider the storage required for our sampled data. The RCM3400 API includes a function named `anaIn()` that returns an int when it reads from the ADC. The ADC itself returns a value between 0 and 2047, and the driver expands that to include two special (negative) values indicating abnormal conditions. While the valid ADC values fit into 11 bits, it is convenient to treat them as 16-bit integers for our purpose. The RCM3400 has ample RAM available to store the extra information.

If we lacked sufficient RAM, we could use another, more compact format to store our samples. With no loss in fidelity, we could pack eight 11-bit samples into 11 bytes, a savings of 5 bytes over using 16-bit integers. Alternatively, if we could accept some loss in fidelity, we could store the difference between successive samples instead of the samples themselves. If we assume that successive samples are likely to be nearly equal, the difference between them should be close to zero. Storing an 8-bit difference between samples would require half of the storage space used to store 16-bit integers. A limitation of this approach is that if the sampled signal changes too much between successive samples, the difference will not fit into 8 bits and thus we lose fidelity.

Although the RCM 3400 has sufficient RAM available to store 32K integers, declaring an array like this:

```
int storage[32768];
```

yields the compiler error message: "Array dimension is too large; the size of a dimension cannot exceed 32768." Even if we decide that 32767 samples would be sufficient, we cannot

Chapter 5

proceed naively because (a) if the storage array is declared as a global variable, we get the compiler error "Out of variable data space." and (b) if the storage array is declared as an 'auto' variable, and thus stored on the runtime stack, our program fails at runtime because we've corrupted the stack with our data array.

Fortunately, other methods to use memory on the Rabbit exist and are easy. Dynamic C allows us to allocate large memory blocks from extended memory at runtime using the xalloc() function call. Additionally, Dynamic C lets us read and write to integers stored in extended memory using the xgetint() and xsetint() functions. Now that we have considered the storage requirements, we can press ahead with understanding how to acquire the samples.

The RCM3400 library includes functions to configure the DAQ channel's gain and read the ADC. Sample code included with Dynamic C shows how these functions are used. We started with the sample code's function sample_ad() and then modified it to suit our purposes.

The enclosed CD contains the complete program, NoiseCheck.c, that was used to sample the RCM3400's ADC, store the samples and report them via the serial port.

The code begins by defining several macros: GAINSET is used to configure the programmable ADC gain in the function sample_ad(); NUMSAMPLES is the number of ADC samples we will take; DINBUFSIZE and DOUTBUFSIZE configure the size of the receive and transmit buffers used by the serial drivers.

The function sample_ad() takes one parameter—an integer that represents which ADC channel we wish to sample from. The ADC has eight channels, numbered 0 through 7. The original sample_ad() function provided with Dynamic C had a second parameter—the number of samples to take and returned the average of the samples. Using the average of a number of samples is a common data acquisition technique to help mitigate the impact noise on the DAQ channel. Since our goal is to determine the noise that might be present, we do not want to average ADC samples. Thus, we removed the second parameter for our program.

Transmitting the Sampled Data to the Linux PC

The collected samples will be transferred from the RCM3400 to the Linux PC via RS-232. The RCM3400 has several serial channels available, and most PCs are equipped with at least one RS-232 port.

Z-World provides a convenient library of functions in Dynamic C for handling the serial ports. The functions that we will use are serDopen(), serDwrFlush(), serDrdFlush(), serDclose(), serDputs(), and serDwrUsed(). The prefix "serD" is used in the function names to indicate that the function affects serial channel D. Similar functions exist for all serial channels A through F. The open and close functions are housekeeping functions to configure the serial port. wrFlush() and rdFlush() are used to ensure that the transmit and receive buffers are cleared before we use them. puts() is used to queue a string in the output buffer for asynchronous transmission. wrUsed() returns the number of bytes that are currently queued for transmission. Behind the scenes, the serial library handles the asynchronous transmission and reception of serial data by installing interrupt handlers to deal with the various buffers and UART registers involved. This library makes using Rabbit UARTs for RS-232 convenient. More information on serial I/O is presented in Chapter 7.

Interfacing to the External World

Under Linux we have even fewer functions to worry about. Historically, Unix and its descendents (like Linux) treat hardware devices as files. To refer to a hardware device, the programmer need only refer to a special file name that corresponds to that device, using the same software functions as one would for any other file (e.g., open(), close(), read(), and write()). Additionally, we can use the usual shell commands for file manipulation to send and receive data from the serial ports under Unix/Linux.

It's quite convenient to write small programs that each solve a small piece of the complete a task. In this case, there will be one small program (GetData.sh on the enclosed CD) that reads data from the serial port and saves it to a data file. There will be a second program (Histogram.pl on the enclosed CD) that reads the data file and produces a statistical analysis of the sample data and saves a histogram of the data to a second file. A third program (gnuplot) will read the histogram data to produce a graph of the data.

We must choose a data format that both the RCM3400 and the Linux PC can support. Since there's no need for high-speed data transfer, we will use ASCII text to send the sample data. Because the data analysis will happen on the Linux PC, we might want to send each sample as a hexadecimal number, in a format easily handled by the usual Unix tools. This leads to the simple format below.

```
<start of data><carriage return / linefeed>
<sample 1 value><comma><carriage return / linefeed>
...
<sample 32768 value><comma><carriage return / linefeed>
<end of data><carriage return / linefeed>
```

Each `<sample>` will look like: 0xXXXX where X is a digit 0..9 or letter a..f and the start and end of data messages are ASCII text.

Strictly speaking, the `<comma>` characters in the message stream are redundant since each line will be ended by a carriage return and line feed. However, they could be useful if the data file ends up being imported into a spreadsheet program. The start and end of message strings would normally be well defined, except that the data extraction tools on Unix are generally very forgiving about such things.

In NoiseCheck.c, we used the string "`Start of data transfer`" to indicate the start of the message and the string "`End of data transfer`" for the end of message. Both humans and Linux can easily parse the resulting data file.

Linux Data Capture Program Listing

The enclosed CD contains the file getdata.sh, also shown in Listing 5.1. This is the bash shell script that was used to record the received ADC sample data on the Linux PC.

Linux is generally configured with online manuals for their system programs. For more information regarding the programs mentioned, please refer to the online "man" pages. For example, to get more information on "`stty`" at a Linux command prompt just type `man stty`.

The script begins by defining two variables, COMPORT and FILEBASE, which hold the filename of the serial device and the base filename of the received data files, respectively. Next, the script uses the STTY program to set the receive and transmit baud rates for the serial device. The default settings for number of data bits, stop bits, and parity mode are acceptable and are not changed here.

> Unix serial ports were historically used to connect the computer to teletypewriters, and STTY reflects this heritage by having a plethora of options that would be used only for that purpose. In general, when using the serial ports on a Unix machine for data communication, these options should be disabled.

The script continues by generating a unique file name where the received data will be stored. The DATE command is called to return the Julian date (the day of the year) followed by the time in 24-hour format. This data string is appended to the value of the FILEBASE variable to generate a file name.

Finally, a terse line explanation is given to the user and the heart of the script commences. The CAT program is used to concatenate the contents of a list of files and print the resulting data stream on the STDOUT device, which is by default the user's console. We invoke CAT giving the name of the serial device as the file that we want to send to STDOUT. We then pipe (|) that data stream to the TEE command, which copies its input to a file as well to the STDOUT data stream. The result of this command is that the contents of the serial port are both copied to the file named by FILENAME as well as echoed to the screen for the user to watch. This program ends when the user types control-C.

If the user didn't want to see the received serial data, the user could change the last line to read:

    cat $COMPORT > $FILE

Also, if the user didn't want to write to a new file each time this script was run, but instead append new data to the end of the current file, the last line could be changed to:

    cat $COMPORT | tee -a $FILEBASE

Both of the previous changes changes would be written as:

    cat $COMPORT >> $FILEBASE

If the user didn't want to have the current shell process to wait for a control-C, then the last command should be placed in the background via the & shell operator as follows:

    cat $COMPORT >> $FILEBASE &

Note that even if the & is used to "background" the script, the script can continue to echo data to the console. This feature is not very useful, and probably demonstrates more flexibility than ease of use on the part of Unix and derivative operating systems.

Since power outages are common, it's reasonable to want to make a Linux PC begin logging data upon boot without user intervention. Unix provides many ways for the user to have programs invoked automatically, some based on a schedule via the AT and CRON programs, some based on the current state of the operating system via the INIT program.

Interfacing to the External World

> The simplest of the mechanisms is a shell script that the system runs after all other boot-up functions have completed. This script is located in different places on different flavors of Unix, but on many systems the script is named /etc/rc.local. One need only add the commands in Listing 5.1 to the end of the /etc/rc.local script, or the rc.local script could invoke a separate shell script that contains the commands from Listing 5.2. In this case, the user should definitely background the CAT command and not echo data to the STDOUT data stream.

Listing 5.1: Bash shell script used to Receive and Record the Sample Data

```
#!/bin/sh
#
# acquire a batch of data from the serial port
#
# Kelly Hall, 2003

COMPORT=/dev/ttyS1
FILEBASE=data.log

# set up the com port
stty -F $COMPORT ispeed 19200 ospeed 19200

# get the time/date
DATE=`date +%j-%T`
FILE=$FILEBASE"-"$DATE

# tell the user what's happening
echo "logging data from $COMPORT to $FILE"
echo "press control-C to exit"

cat $COMPORT | tee $FILE
```

Analyzing the Data Graphically and Numerically

Once the ADC sample data has been transferred to the Linux PC, we want to analyze the data for trends. One form of qualitative graphical analysis is a histogram. This is a plot of each data value read versus the number of times that value occurred in the data set. Visual inspection of the histogram will quickly tell us if the noise "looks" right. For example, multiple humps, also called modes, are indicative of nonrandom (coupled) noise. If the histogram is a single bell-shaped curve, then we can look to a more quantitative analysis that assumes the noise is gaussian.

Chapter 5

Other analyses we will perform include calculating the mean of the data set, the standard deviation of the data, and the calculated noise in the ADC channel represented as both voltages and bits of error. This type of analysis is quantitative.

Since the RCM3400 ADC accepts input voltages in the maximum range of 0 V through 20 V, and returns a code based on the input voltage that ranges from 0 to 2047 (11 bits), each ADC code represents approximately (20 V–0 V) / 2048 = 9.8 mV.

Additionally, the ADC has a gain control that rescales the input signal: with a Gain of 2, the input signal ranges from 0 V to 10 V, so each ADC code would represent 4.9 mV. Gains of 4, 5, 8, 10, 16 and 20 are available, as well as a gain of 1, which corresponds to the 0 V–20 V range, discussed above. As the gain increases, the range of input voltage decreases accordingly.

Our experimental precision reference based on the AD780 generates a constant 2.500 V signal. We can use gains of 1, 2, 4, or 5 to measure the signal. If the gain was set any higher, the maximum input voltage would be lower than 2.500 V and thus our input signal would be out of range.

Once we know the gain setting of the ADC, we can convert between input voltages and ideal ADC codes by multiplying or dividing the size of each code by the appropriate amount. For example, with a gain of 1, a 3.3 V input should produce the ADC code of (3.3 V / 9.8 mV per code) = 337. Similarly, an ADC code of 1250 should correspond to an input voltage of (1250 * 9.8 mV per code) = 12.25 V.

The first part of our data analysis will compute the average returned ADC code for the input voltage. This will likely differ somewhat from the expected ADC code that we can calculate above using the gain and resolution of the converter. The difference between the expected value and the mean is called an "offset." We can express offset in ADC codes or in Volts.

The standard deviation (σ) of the data is a measure of how spread out the data is from the mean. σ is conveniently mathematically equal to the RMS noise of the ADC channel. We can measure this noise in ADC codes, or in Volts. The base-2 logarithm of the standard deviation is the RMS noise in bits, and if we subtract that from the advertised resolution of the ADC we can obtain the effective resolution of the channel in bits.

The overall peak-to-peak noise in the sample set is taken to be $\pm 3 * \sigma$ (in ADC codes). Some texts take $\pm 3.3\sigma$ to be the peak-to-peak noise. We can convert this into bits of peak-to-peak noise via the base-2 logarithm of $3 * \sigma$, and thus we compute the "noise free resolution" of the ADC channel by subtracting the bits of noise from the advertised resolution of the ADC.

The quantitative characterization techniques presented here are useful in comparing one system to another as long as the engineer consistently applies the same functions to data from different systems.

Linux Data Reduction Program Listing

Listing 5.2 shows the Perl script used to calculate the histogram of the ADC sample data, as well as the various statistical data we are looking for to characterize the noise on the ADC channel.

Perl comes standard with most Linux distributions. It is also available for free download for both Linux and Windows® from www.ActiveState.com.

Interfacing to the External World

The script begins by loading the Perl Module 'Statistics::Lite'. This module is available from the Comprehensive Perl Archive Network at http://www.cpan.org/. This module provides simple access to standard deviation, mean, and variance functions over a data set.

The script continues by declaring a hash table to store a count of how many times we have seen each ADC code, and by declaring a list to hold the complete data set. Next, two constants are declared to hold the desired input voltage and the number of bits of resolution of the ADC.

Next, the script pulls the name of the data file and the maximum input range (in volts) from the command line. This allows us to reuse the script on different data files and with different gain settings.

Next, the script gets to work by resetting a counter of the number of data lines read, and proceeds to open the input file. The script loops through each line of the input file using the <> operator. For each line of input, we try to match the input line to a regular expression of the form:

$0xXXXX,^ where: $ means the beginning of the line
0x are literal characters
X means a nonspace character (letter or digit)
, is a literal character
^ means the end of the line

The exact regular expression in the `if` statement is slightly different—it includes optional whitespace following the comma, and if successful it saves off the four X characters into a special variable. If the if statement succeeds, then the matched part of the expression is used as the ADC code, and we save that into the variable $new_data. $new_data is a string that begins with "0x...." and while that's a legal hash table key for Perl, it's more convenient to convert that string to an integer before we use it as a key. We use the Perl function `oct()` to do this and save the integer into the variable named $num. Then, $num is used as a key into the hash table, and we update the count stored at that location in the table. Then, we insert the $new_data into our complete list of all data read stored in @data. Finally, we increment the number of lines read in the variable $count. When we've read all the data in the input file, the while loop terminates and we close the input file.

The script generates the histogram and writes it to a new file. The histogram is placed in a file with a name generated from the name of the file passed on the command line with "-histogram" appended. We initialize a new variable that will hold the sum of all the data values we read. Then we loop through the hash table, in order of the keys. For each key, we print out the key itself, some whitespace, and the count from the hash table. We also update our running sum. When all the keys have been processed, we close the output file and print out the $sum we've calculated. It should be the same as the number of lines of input data we read above.

Next the script computes some statistics. The `Statistics::Lite` module generates all of the data that it can, and it saves its results in the hash named %results.

The last thing the script does is print out the results.

Chapter 5

Listing 5.2: Perl Script used for Data Analysis

```perl
#!/usr/bin/perl

# some handy stats functions
use Statistics::Lite qw(:all);

# we need a hash to store the codes we see
my %codes;
my @data;

# constants for this analysis
$inputvoltage = 2.500;
$ADCbits = 11;

# get the file name from the command line
$fname = $ARGV[0] or die "usage: $0 fname maxRange";
$maxRange = $ARGV[1] or die "usage: $0 fname maxRange";

open DATA, $fname or die „can't open $fname";
$count = 0;
while(<DATA>) {
  # we just want lines of the form $0xXXXX,^
  if( m/^(0x\w\w\w\w),\s*$/ ) { # if the regexp matches, save off the data
      $new_data = $1;              # update the bin
      $codes{$new_data} += 1;
      $num = oct $new_data;        # convert to an integer
      push @data, $num;            # add to our raw data array
      $count++;                    # update our count
   }
}
close DATA;

# open the output file
$fname = $fname . „-histogram";
open OFILE, „>$fname" or die „can't open output file $fname";
$sum = 0;
foreach $code (sort keys %codes) {
  print OFILE oct($code) . „ " . $codes{$code} . „\n";
  $sum += $codes{$code};
}
close OFILE;

# now do the analysis
%results = statshash @data;

printf „\nAssuming an %3d bit converter with 0 to %3dV input range\n", $ADCbits, $maxRange;
printf „Assuming a %6.3f volt precision reference\n\n", $inputvoltage;
$voltspercode = $maxRange / (1<<$ADCbits);
printf "1 Code = %7.3g Volts\n\n", $voltspercode;
```

(Listing 5.2 continued on next page)

Interfacing to the External World

Listing 5.2: Perl Script used for Data Analysis (continued)

```
printf "Measured Mean (codes)   =   %9.4f\n", $results{mean};

printf "Expected Mean (codes)   =   %9.4f\n", ($inputvoltage /
$voltspercode);
$offset = $inputvoltage - $voltspercode * $results{mean};
printf "offset (volts)          = %7.3f Volts\n\n", $offset;

$noiseRMSvolts = $results{stddev}*$voltspercode;
$noiseRMScodes = $results{stddev};
printf "RMSnoise(ADC codes)         = %7.3f\n", $noiseRMScodes;
printf "RMSnoise(volts)             = %8.4f\n\n", $noiseRMSvolts;
printf "Noise(pk-to-pk)(ADC codes)  = %7.3f\n", 3.3*$noiseRMScodes;
printf "Noise(pk-to-pk)(volts)      = %8.4f\n", 3.3*$noiseRMSvolts;
```

Linux Histogram Visualization

Gnuplot was used to plot and save the histogram graphics. Gnuplot is highly adept at plotting data from files such as these. Invoke gnuplot as follows:

```
$ gnuplot
```

This will return a report similar too:

```
            G N U P L O T
            Version 3.7 patchlevel 3
            last modified Thu Dec 12 13:00:00 GMT 2002
            System: Linux 2.4.20-20.9smp

            Copyright(C) 1986 - 1993, 1998 - 2002
            Thomas Williams, Colin Kelley and many others

            Type `help` to access the on-line reference manual
            The gnuplot FAQ is available from
            http://www.gnuplot.info/gnuplot-faq.html

       Send comments and requests for help to <info-
       gnuplot@dartmouth.edu>
            Send bugs, suggestions and mods to <bug-gnuplot@dartmouth.edu>


    Terminal type set to 'unknown'
    gnuplot>
```

At the gnuplot> prompt, just type the command

```
Plot "HISTOGRAMFILE" with boxes
```

Chapter 5

Where HISTOGRAMFILE is the name of the file containing the histogram data. For example,

```
plot "data.log-352-15:44:06-histogram" with boxes
```

Gnuplot can generate EPS (encapsulated postscript files) suitable for importing into word processors:

```
gnuplot> set terminal postscript eps
gnuplot> set output "histogram.eps"
gnuplot> plot "data.log-352-15:44:06-histogram" with boxes
gnuplot> set output
gnuplot> exit
```

For more information on the versatile gnuplot program, please refer to the gnuplot man page or the gnuplot homepage `http://www.gnuplot.org/`.

Sample ADC Noise Quantification and Visualization

Here we put together the work in the preceding sections and characterize the first DAQ channel on the RCM3400 development board.

Since our voltage reference was 2.5 V we had the option of using ADC gain settings of 1, 2, 4 and 5. We ran the analysis for all gain settings.

The histograms showed that the data reported from the ADC was tightly clumped around the mean value. In each histogram, only two or three bins had data in them. An ideal ADC in a noiseless system would only report data in one bin. In real life systems, a histogram with only two or three bins is excellent. Figure 5.16 shows a representative histogram from our experiments. The reported ADC codes are on the horizontal axis. The number of occurrences is shown on the vertical axis.

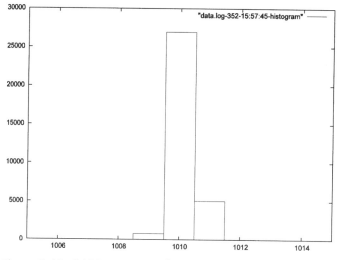

Figure 5.16: GAIN = 4, 0–5 volt input range, 2.5 volt stimulus.

Interfacing to the External World

An example of the textual output from our Perl script is shown here,

```
Assuming an  11 bit converter with 0 to   20V input range
Assuming a   2.500 volt precision reference

1 Code = 0.00977 Volts

Measured Mean (codes)    =    251.9695
Expected Mean (codes)    =    256.0000
offset (volts)           =    0.039 Volts

RMSnoise(ADC codes)      =    0.172
RMSnoise(volts)          =    0.0017

Noise(pk-to-pk)(ADC codes) =  0.568
Noise(pk-to-pk)(volts)     =  0.0055
```

The results of our four experiments are summarized in Table 5.2.

Table 5.2: The RCM3400 performed admirably over all of the ranges measured.

	Gain of 1 **0-20 volt range**	**Gain of 2** **0-19 volt range**	**Gain of 4** **0-5 volt range**	**Gain of 5** **0-4 volt range**
Volts per code	9.77 mV	4.88 mV	2.44 mV	1.95 mV
Measured offset	39 mV	36 mV	34 mV	32 mV
Measured RMS noise in codes	0.172	0.409	0.403	0.675
Measured RMS noise in volts	1.7 mV	2.4 mV	1.0 mV	1.3 mV
Measured pk-pk noise in codes	0.568	1.617	1.329	2.227
Measured pk-pk noise in volts	5.5 mV	7.9 mV	3.2 mV	4.3 mV

If the RMS noise on a channel is greater than 1 bit, we can compute the effective resolution of the channel from,

$$RESOLUTION_{EFFECTIVE} = ADC\ bits - LOG_2\left(Noise_{RMS}\ in\ codes\right)$$

The RCM3400 has RMS noise levels so low as to render the computation of effective resolution meaningless. The logarithm becomes negative for RMS noise levels less than one code. The conclusion we can draw from our experiments is that the effective resolution of the RCM3400 is a full 11 bits for gains of 1, 2, 4 and 5.

If the peak-to-peak noise on a channel is greater than 1 bit, we can compute the noise free resolution of the channel from:

$$RESOLUTION_{NOISE\ FREE} = ADC\ bits - LOG_2\left(Noise_{PK-PK}\ in\ codes\right)$$

In our experiment, we see that the worst case peak to peak noise on the RCM3400 channel 1 occurs with a gain of 5 and is 2.227 codes. From the equation above, we compute,

$$RESOLUTION_{NOISE\ FREE\ Gain\ =\ 5} = 11 - LOG_2(2.227) = 11 - \frac{\ln(2.227)}{\ln(2)} = 9.84\ bits$$

Our analysis of the RCM3400's DAQ channel showed admirable performance. For this example, we only examined one of the eight single-ended (or four differential) DAQ channels on the RCM3400. A careful engineer will characterize all DAQ channels used in a system.

The DC analysis techniques presented here will allow an engineer to get a good feel for how accurately the ADC is reporting sensor data under ideal conditions. This is the best performance that the engineer can expect from the system without calibration or averaging data samples.

5.6 Conclusion

Companies are rolling more and more features into silicon, making the system designer's job easier. However, the system designer must still exercise caution when devising an interface between real world sensors and a processor. Issues of ESD, bus loading and power consumption still exist and must be handled by the system engineer.

The Rabbit 3000 has simplified the bus loading issue by providing an auxiliary I/O bus allowing peripheral devices to be added without unduly loading the high-speed memory bus. The core module designs address issues of memory interfacing, battery backup, system reset, power supervision and, on some cores, analog interfacing.

Some core modules from Rabbit Semiconductor have provided low noise DAQ channels. For example, the RCM3400 will drop right into many applications needing one to eight solid 11-bit resolution DAQ channels.

CHAPTER 6

Introduction to Rabbit Assembly Language

Assembly language has long been a favorite of programmers, for many reasons. Looking back into history, some of the early computers (such as the MITS Altair 8080) could only be programmed in assembly language. Even some industrial machines, which were considered "powerful" for their period, had primitive programming technologies by present day standards. For example, in 1974, the Raytheon RDS 500 was originally designed to track missile trajectories but had to be programmed one assembly instruction at a time, using bit switches on the front panel. Soon after personal computers began to get popular, compilers for higher-level languages began to appear for them, and programmers could program these machines in BASIC, C, FORTRAN, Pascal, and even Forth.

Programmers found that certain things were best done with assembly language. In some cases, high-level languages did not give programmers the level of control they desired to manipulate machine internals, while in other cases, given the CPU clock speeds of that time, certain things happened too slowly unless they were done in assembly language. Even higher level languages such as C and "simple" languages such as BASIC allowed programmers to manipulate bits and bytes and embed assembly language code between higher-level statements. Moreover, the limited amount of memory available in such systems further necessitated the need for efficient code, thus forcing programmers to use assembly language.

Although much has changed from those days, programmers working with microcontrollers still conclude that some things are best done with assembly language. Although CPU clock speeds have drastically increased in the last three decades, most programmers will agree that the most efficient code, in the context of execution time and memory usage, is still written with assembly language. Wherever things have to be done with tight timing constraints, programmers will explore whether they should resort to assembly language. For example, using an 8-bit microcontroller to implement modem connection sequences, a programmer can usually only rely on assembly language code to generate and detect tones in the precise time windows needed.

This chapter starts with an overview of the Rabbit instruction set. Subsequent sections cover some useful concepts—passing parameters between assembly language and C, and coding in a mixed C / Assembly environment. The chapter concludes with a number of projects, highlighting low-level control of on-chip and off-chip peripherals, as well as the ability of assembly language to make things happen in critical time windows.

Chapter 6

6.1 Introduction to the Rabbit 3000 Instruction Set

Rabbit assembly instructions are used to:

- Load data into specific registers or memory locations. These instructions are used to load data from a source memory location or register into a register (usually the accumulator), perform some operations, and transfer the results back into a register or memory location.

- Exchange contents of certain registers. When operations affect the contents of certain registers, and the programmer wishes to save the original contents of the registers, the programmer can use these exchange instructions to leave the original registers alone, work an alternate set of registers, and once the work is completed, the original register contents maybe restored. When the registers are "exchanged out," (also called "swapped out") the contents of the original registers will be unaffected by the previous operations.

- Push data on and off the stack. While keeping track of subroutines' return addresses and their parameters, the stack is a handy place to store data temporarily[1]. Various instructions are provided to "push" data on the stack and "pop" it back into a given register. The programmer must be careful to remove as much data from the stack as was pushed onto the stack, otherwise a stack overflow condition can develop.

- Perform arithmetic and Boolean operations. Almost all of the 8-bit arithmetic operations involve the accumulator, while 16-bit operations involve 8-bit register pairs. The Rabbit even has a multiply instruction!

- Test and manipulate individual bits. These instructions come in especially handy for I/O operations when the code needs to know if a bit got set by an external input or when the program needs to turn an external device on or off by manipulating a port bit. Additionally, bits of internal registers and memory locations can be tested. For example, bits in the status register can be tested to determine if the result of a prior arithmetic operation required a carry or borrow.

- Copy entire blocks of memory. Some of the Rabbit's instructions allow the programmer to copy an entire block of memory in as few as four instructions.

- Jump to alternate sections of code. Assembly programs can make branching decisions, usually by testing results of operations. Various instructions allow programmers to test for certain Boolean conditions and make branching decisions accordingly.

Although the Rabbit instruction set is derived from the Z80 instructions, quite a few new instructions have been added to the Rabbit 3000 microprocessor.

The assembly language instructions and the register designations are NOT case sensitive. However, there are times when it is advisable to use specific case in order to make the code easier to read. For example, the HL register pair can be easily misinterpreted by the reader if it is in lower case: hl. Some type fonts do not do well differentiating between a lower case l and the digit 1.

---

[1] This is the mechanism used to store "auto" variables in Dynamic C as well as most other C compilers.

Introduction to Rabbit Assembly Language

Rabbit assembly instructions generally consist of an opcode followed by zero or more operands. The opcode is the "instruction" while the operands are data. Operands can take many different forms. For example, they may consist of a 16-bit address or be a single bit. Table 6.1 shows how we represent operands and result of operations on CPU flags. The following instruction descriptions use the operand abbreviations from this table.

Table 6.1: Operands used in the Rabbit instruction set.

Operand	Meaning
b	Bit select: 000 = bit 0, 001 = bit 1, 010 = bit 2, 011 = bit 3, 100 = bit 4, 101 = bit 5, 110 = bit 6, 111 = bit 7
cc	Condition code select: 00 = NZ, 01 = Z, 10 = NC, 11 = C
d	7-bit (signed) displacement. Expressed in two's complement.
dd	Word register select destination: 00 = BC, 01 = DE, 10 = HL, 11 = SP
dd'	Word register select alternate: 00 = BC', 01 = DE', 10 = HL'
e	8-bit (signed) displacement added to PC.
f	Condition code select: 000 = NZ (non zero), 001 = Z (zero), 010 = NC (non carry), 011 = C (carry), 100 = LZ[2] (logical zero), 101 = LO[3] (logical one), 110 = P (sign plus), 111 = M (sign minus)
m	MSB of a 16-bit constant.
mn	16-bit constant.
n	8-bit constant or LSB of a 16-bit constant.
r, g	Byte register select: 000 = B, 001 = C, 010 = D, 011 = E, 100 = H, 101 = L, 111 = A
ss	Word register select (source): 00 = BC, 01 = DE, 10 = HL, 11 = SP
v	Restart address select: 010 = 0020h, 011 = 0030h, 100 = 0040h, 101 = 0050h, 111 = 0070h
xx	Word register select: 00 = BC, 01 = DE, 10 = IX, 11 = SP
yy	Word register select: 00 = BC, 01 = DE, 10 = IY, 11 = SP
zz	Word register select: 00 = BC, 01 = DE, 10 = HL, 11 = AF

---

[2] Logical zero if all four of the most significant bits of the result are 0.
[3] Logical one if any of the four most significant bits of the result are 1.

Chapter 6

Rabbit assembly instructions are divided into the following groups:

- Load and Store
    - Load Immediate Data
    - Load & Store to Immediate Address
    - 8-bit Indexed Load and Store
    - 16-bit Indexed Load and Store
    - Register to Register Moves
- Exchange Instructions
- Stack Manipulation Instructions
- Arithmetic and Logical Operations
    - 8-bit Arithmetic and Logical Operations
    - 16-bit Arithmetic and Logical Operations
- 8-bit Bit Set, Reset and Test
- 8-bit Increment and Decrement
- 8-bit Fast Accumulator Operations
- 8-bit Shifts and Rotates
- Instruction Prefixes
- Block Move Instructions
- Control Instructions—Jumps and Calls
- Miscellaneous Instructions

To those familiar with assembly programming, several groups will look familiar. Most processors have "load and store" and "arithmetic and logical" instructions. Some groups will look familiar to Z-80 enthusiasts such as the group containing "exchange instructions." However, the Rabbit also has a number of unique instructions such as those found in the "block move group" and "instruction prefix" group.

The tables presented in this text use a format and nomenclature consistent with the Rabbit 2000/3000 Microprocessor Instruction Reference Manual. Each instruction description shows entries in a table with the following headings.

Instruction	Clk	A	I	S	Z	V	C	Operation

The **Instruction** column contains the instruction mnemonic and opcode format.

The **Clk** column indicates the number of machine cycles required for the instruction to execute.

The **A** column indicates what effect the ALTD prefix instruction has on the instruction. The following table shows the key for the "A" column.

Symbol	Description
F	ALTD selects alternate flags
R	ALTD selects the alternate destination register
SP	ALTD operation is a special case

Introduction to Rabbit Assembly Language

The **I** column indicates what effect the IOI and IOE prefix instructions have on the instruction. The following table shows the key for the "I" column.

Symbol	Description
S	IOI and IOE affect source
D	IOI and IOE affect destination

The **S**, **Z**, **V**, and **C** columns correspond to the Sign, Zero, Overflow and Carry flags. These are found in the Rabbit's "Flags" register (sometimes called the Status or Status Flags register). The Overflow flag is sometimes referred to as the Logical/Overflow or **LV** flag in the Rabbit processor documentation. The following table shows the key for the symbols used in the flags column.

Symbol	Description
*	Flag affected
-	Flag unaffected
0	Flag is cleared
1	Flag is set
V	Arithmetic Overflow is stored
L	Logical Result is stored

6.1.1 Load Immediate Data

Instructions that belong to this addressing mode load a constant into the destination register or register pair. This is called the "immediate" addressing mode because the constant to be loaded immediately follows the opcode for the load instruction. Table 6.2 lists instructions in the "immediate" group.

Table 6.2: Load immediate data.

Instruction	Clk	A	I	S	Z	V	C	Operation
LD IX,mn	8			-	-	-	-	IX = mn
LD IY,mn	8			-	-	-	-	IY = mn
LD dd,mn	6	r		-	-	-	-	dd = mn
LD r,n	4	r		-	-	-	-	r = n

The following instructions illustrate how immediate data is loaded in 8 and 16-bit registers:

```
ld a, 5           ; register a gets a value of 5 (decimal)
ld ix, 0x1234     ; the ix register gets a value of 1234 (hex)
```

If a programmer programs a loop that always executes a fixed number of times, a register can be set up as the loop counter and the register can be initialized the first time using an immediate instruction. The following instructions illustrate how this is done:

```
#define COUNTER   240            ; counter constant (decimal)
    ld    BC, COUNTER            ; load counter value
```

6.1.2 Load and Store to Immediate Address

In this addressing mode, one of the operands is a register, while the other operand is fetched from memory. Depending on the instruction, a 16-bit address is used to point to source or destination data. This instruction set has the term "immediate address" in it because the 16-bit address immediately follows the opcode. Table 6.3 lists instructions in this group.

Table 6.3: Load and store to immediate address.

Instruction	clk	A	I	S	Z	V	C	Operation
LD (mn),A	10	d	-	-	-	-	-	(mn) = A
LD A,(mn)	9	r	s	-	-	-	-	A = (mn)
LD (mn),HL	13		d	-	-	-	-	(mn) = L; (mn+1) = H
LD (mn),IX	15		d	-	-	-	-	(mn) = IXL; (mn+1) = IXH
LD (mn),IY	15		d	-	-	-	-	(mn) = IYL; (mn+1) = IYH
LD (mn),ss	15		d	-	-	-	-	(mn) = ssL; (mn+1) = ssH
LD HL,(mn)	11	r	s	-	-	-	-	L = (mn); H = (mn+1)
LD IX,(mn)	13		s	-	-	-	-	IXL = (mn); IXH = (mn+1)
LD IY,(mn)	13		s	-	-	-	-	IYL = (mn); IYH = (mn+1)
LD dd,(mn)	13	r	s	-	-	-	-	ddL = (mn); ddH = (mn+1)

The above instructions are useful for performing simple pointer-based operations. The 16-bit pointer can store or retrieve one or two bytes to or from memory (or I/O, if the I/O is memory mapped). For example, if a program has to use a number of variables, the variables can be stored in memory locations and the results of operations can be stored in the variables.

The following instructions illustrate some examples of this instruction group:

```
        #define BUFFER_SIZE 64       ; spce to allocate for buffer
        char    bytestoread          ; number of bytes to read
        int     bytecounter          ; number of bytes that have been read

        ld      a, (bytestoread)     ; find out how many bytes to read

        ld      (bytecounter), hl    ; update byte counter

        ld      ix,(sp+2+BUFFER_SIZE)       ; ix=buffer
```

6.1.3 8-bit Indexed Load and Store

In their simplest form, these instructions come in handy for performing pointer-based operations. For instance, if a program has to add a set of values stored in RAM, the programmer can set up a register pair to point to the starting location of the table in RAM, read the values one by one, and keep adding them to a destination register.

Introduction to Rabbit Assembly Language

In a more complex application of indexed addressing, the contents of an index register are added to a displacement to compute the address of the operand. This is what the "IX+d" and "IY+d" instructions below accomplish, where "d" is the 8-bit displacement. C language programmers can think of the "index with displacement" model as a structure where the index register points to the beginning of the structure and the displacement is used to point to elements within the structure. Table 6.4 lists 8-bit indexed instructions:

Table 6.4: 8-bit indexed load and store.

Instruction	clk	A	I	S	Z	V	C	Operation
LD A,(BC)	6	r	s	-	-	-	-	A = (BC)
LD A,(DE)	6	r	s	-	-	-	-	A = (DE)
LD (BC),A	7		d	-	-	-	-	(BC) = A
LD (DE),A	7		d	-	-	-	-	(DE) = A
LD (HL),n	7		d	-	-	-	-	(HL) = n
LD (HL),r	6		d	-	-	-	-	(HL) = r = B, C, D, E, H, L, A
LD r,(HL)	5	r	s	-	-	-	-	r = (HL)
LD (IX+d),n	11		d	-	-	-	-	(IX+d) = n
LD (IX+d),r	10		d	-	-	-	-	(IX+d) = r
LD r,(IX+d)	9	r	s	-	-	-	-	r = (IX+d)
LD (IY+d),n	11		d	-	-	-	-	(IY+d) = n
LD (IY+d),r	10		d	-	-	-	-	(Iy+d) = r
LD r,(IY+d)	9	r	s	-	-	-	-	r = (IY+d)

A simple example of indexed addressing would be a program that adds a series of contiguous integers in RAM. The programmer can set up a 16-bit register pair to point to the start of the table, and then load each integer into the accumulator. The program can add individual integers to a 32-bit result value in RAM.

The following instructions illustrate 8-bit indexed load and stores:

```
    ld          a,(iy+8)            ;
    ld          (ix+8),b            ;
```

6.1.4 16-bit Indexed Load and Store

The 16-bit loads and stores are similar to their counterparts described in Section 6.1.3, except that in this case the source or destination is a 16-bit register pair or two contiguous locations in memory.

Because these instructions require fetching addresses from memory, adding a displacement, and storing the contents to a destination register pair, the instructions require a relatively large number of clock cycles to execute. Table 6.5 lists 16-bit indexed instructions:

Table 6.5: 16-bit indexed load and store.

Instruction	clk	A	I	S	Z	V	C	Operation
LD (HL+d),HL	13		d	-	-	-	-	(HL+d) = L; (HL+d+1) = H
LD HL,(HL+d)	11	r	s	-	-	-	-	L = (HL+d); H = (HL+d+1)
LD (SP+n),HL	11			-	-	-	-	(SP+n) = L; (SP+n+1) = H
LD (SP+n),IX	13			-	-	-	-	(SP+n) = IXL; (SP+n+1) = IXH
LD (SP+n),IY	13			-	-	-	-	(SP+n) = IYL; (SP+n+1) = IYH
LD HL,(SP+n)	9	r		-	-	-	-	L = (SP+n); H = (SP+n+1)
LD IX,(SP+n)	11			-	-	-	-	IXL = (SP+n); IXH = (SP+n+1)
LD IY,(SP+n)	11			-	-	-	-	IYL = (SP+n); IYH = (SP+n+1)
LD (IX+d),HL	11		d	-	-	-	-	(IX+d) = L; (IX+d+1) = H
LD HL,(IX+d)	9	r	s	-	-	-	-	L = (IX+d); H = (IX+d+1)
LD (IY+d),HL	13		d	-	-	-	-	(IY+d) = L; (IY+d+1) = H
LD HL,(IY+d)	11	r	s	-	-	-	-	L = (IY+d); H = (IY+d+1)

The following instructions utilize 16-bit indexed load and stores:

```
    ld          hl,(ix+4)           ;
    ld          (ix+2),hl           ;
```

6.1.5 Register to Register Moves

This addressing mode is also referred to as "register addressing," since all the operands are the CPU registers. As the name implies, these instructions copy contents of one register into another—any of the 8-bit registers can be moved into any other 8-bit register. This is necessary because most of the math and logical operations are performed in an accumulator (register A or register pair HL). As shown in Table 6.6, certain 16-bit registers can also perform a move into other 16-bit registers.

Because register to register moves happen within the CPU and do not require operands to be fetched from memory, the instructions take few clock cycles.

Introduction to Rabbit Assembly Language

Table 6.6: Register to register moves.

Instruction	clk	A	I	S	Z	V	C	Operation
LD r,g	2	R		-	-	-	-	r = g (r, g any of B, C, D, E, H, L, A)
LD A,EIR	4	fr		*	*	-	-	A = EIR
LD A,IIR	4	fr		*	*	-	-	A = IIR
LD A,XPC	4	R		-	-	-	-	A = MMU
LD EIR,A	4			-	-	-	-	EIR = A
LD IIR,A	4			-	-	-	-	IIR = A
LD XPC,A	4			-	-	-	-	XPC = A
LD HL,IX	4	R		-	-	-	-	HL = IX
LD HL,IY	4	R		-	-	-	-	HL = IY
LD IX,HL	4			-	-	-	-	IX = HL
LD IY,HL	4			-	-	-	-	IY = HL
LD SP,HL	2			-	-	-	-	SP = HL
LD SP,IX	4			-	-	-	-	SP = IX
LD SP,IY	4			-	-	-	-	SP = IY
LD dd',BC	4			-	-	-	-	dd' = BC (dd': BC', DE', HL')
LD dd',DE	4			-	-	-	-	dd' = DE (dd': BC', DE', HL')

For example, consider the IIR register: it points to an interrupt vector table specific to internally generated interrupts. When a programmer wishes to use internally generated interrupts, the IIR register cannot be immediately loaded with a value—such an opcode does not exist. Instead, the immediate value can be loaded in the accumulator, and the accumulator can be copied into the IIR.

6.1.6 Exchange Instructions

The CPU contains an "alternate register set," where register pairs AF, HL, BC, and DE contain their alternates, AF", HL', BC', and DE', respectively. Under normal circumstances, the programmer does not use the alternate register pairs, since Dynamic C uses them for its own purposes.

As the name indicated, the "Exchange" instructions swap contents of 16-bit registers with special alternate registers. For instance, the "EX DE',HL" instruction swaps the contents of HL with those of alternate register pair DE'. There are two special cases:

- The "EXX" instruction does three swaps with a single instruction: BC, DE and HL.
- The "EX AF, AF'" instruction treats the accumulator and flags register as a register pair and swaps them out with their alternate registers.
- The "ALTD" prefix, covered later in Section 6.1.14, allows instruction to directly access the alternate registers, without exchanging all the registers

Chapter 6

Table 6.7 lists various exchange instructions.

Table 6.7: Exchange instructions.

Instruction	clk	A	I	S	Z	V	C	Operation
EX (SP),HL	15	r		-	-	-	-	H <-> (SP+1); L <-> (SP)
EX (SP),IX	15			-	-	-	-	IXH <-> (SP+1); IXL <-> (SP)
EX (SP),IY	15			-	-	-	-	IYH <-> (SP+1); IYL <-> (SP)
EX AF,AF'	2			-	-	-	-	AF <-> AF'
EX DE',HL	2		s	-	-	-	-	if (!ALTD) then DE' <-> HL else DE' <-> HL'
EX DE',HL'	4		s	-	-	-	-	DE' <-> HL'
EX DE,HL	2		s	-	-	-	-	if (!ALTD) then DE <-> HL else DE <-> HL'
EX DE,HL'	4		s	-	-	-	-	DE <-> HL'
EXX	2			-	-	-	-	BC <-> BC'; DE <-> DE'; HL <-> HL'

Figure 6.1 illustrates how the register pairs are swapped:

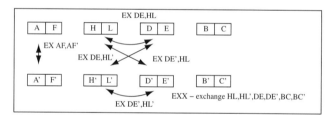

Figure 6.1: Exchange Instructions

6.1.7 Stack Manipulation Instructions

It is assumed the reader knows what a stack is and how it works. Table 6.8 lists the stack manipulation instructions. The first instruction in this group (ADD SP, d) adds a one-byte displacement to the stack. This is a quick way for the programmer to move the stack pointer forward by up to 255 bytes, without changing the contents of the stack. The displacement is always positive. The main reason for the instruction would be for reserving a block of memory on the stack.

The other instructions in this group either push 16-bit register pairs on the stack, or pop them off.

These instructions allow the programmer to store (push) data and variables on the stack, and then read them back (pop them) when required. One has to be careful in dealing with the stack when working with subroutines or interrupts, since these elements use the stack to remember where to return after they are done. For instance, if a main program has to pass

Introduction to Rabbit Assembly Language

data to a subroutine through the stack, the calling program can push the data on the stack. When the subroutine is called, it cannot simply read the data using the current stack pointer, since the subroutine call has altered the stack contents and the stack pointer—the stack pointer now has to be adjusted in order to pop the relevant data correctly, and then adjusted again to point to the return address of the calling program.

In Dynamic C it is the responsibility of the CALLING routine to restore the stack pointer once the called routine has returned. The called routine accesses the passed values using SP relative indexed addressing.

Table 6.8: Stack manipulation instructions.

Instruction	clk	A	I	S	Z	V	C	Operation
ADD SP,d	4	f		-	-	-	*	SP = SP + d — d=0 to 255
POP IP	7			-	-	-	-	IP = (SP); SP = SP+1
POP IX	9			-	-	-	-	IXL = (SP); IXH = (SP+1); SP = SP+2
POP IY	9			-	-	-	-	IYL = (SP); IYH = (SP+1); SP = SP+2
POP zz	7	r		-	-	-	-	zzL = (SP); zzH = (SP+1); SP=SP+2 — zz= BC,DE,HL,AF
PUSH IP	9			-	-	-	-	(SP-1) = IP; SP = SP-1
PUSH IX	12			-	-	-	-	(SP-1) = IXH; (SP-2) = IXL; SP = SP-2
PUSH IY	12			-	-	-	-	(SP-1) = IYH; (SP-2) = IYL; SP = SP-2
PUSH zz	10			-	-	-	-	(SP-1) = zzH; (SP-2) = zzL; SP=SP-2 — zz= BC,DE,HL,AF

6.1.8 8-bit Arithmetic and Logical Operations

The Rabbit microprocessor performs the typical arithmetic and logical operations found in other 8-bit processors: adds, subtracts, complements, ANDs, ORs, XORs. There are no multiply or divide instructions in this instruction group, although the 16-bit instruction group allows for a multiply involving 16-bit register pairs.

There are special add and subtract instructions that also take the carry bit into account.

These instructions work using the accumulator and an operand that can be an immediate value, another 8-bit register, or a memory location pointed to by the HL, IX or IY register pairs.

Chapter 6

Table 6.9 lists 8-bit arithmetic and logical operations.

Table 6.9: 8-bit arithmetic and logical operations.

Instruction	clk	A	I	S	Z	V	C[4]	Operation
ADC A,(HL)	5	fr	s	*	*	V	*	A = A + (HL) + CF
ADC A,(IX+d)	9	fr	s	*	*	V	*	A = A + (IX+d) + CF
ADC A,(IY+d)	9	fr	s	*	*	V	*	A = A + (IY+d) + CF
ADC A,n	4	fr		*	*	V	*	A = A + n + CF
ADC A,r	2	fr		*	*	V	*	A = A + r + CF
ADD A,(HL)	5	fr	s	*	*	V	*	A = A + (HL)
ADD A,(IX+d)	9	fr	s	*	*	V	*	A = A + (IX+d)
ADD A,(IY+d)	9	fr	s	*	*	V	*	A = A + (IY+d)
ADD A,n	4	fr		*	*	V	*	A = A + n
ADD A,r	2	fr		*	*	V	*	A = A + r
AND (HL)	5	fr	s	*	*	L	0	A = A & (HL)
AND (IX+d)	9	fr	s	*	*	L	0	A = A & (IX+d)
AND (IY+d)	9	fr	s	*	*	L	0	A = A & (IY+d)
AND n	4	fr		*	*	L	0	A = A & n
AND r	2	fr		*	*	L	0	A = A & r
CP (HL)	5	f	s	*	*	V	*	A - (HL)
CP (IX+d)	9	f	s	*	*	V	*	A - (IX+d)
CP (IY+d)	9	f	s	*	*	V	*	A - (IY+d)
CP n	4	f		*	*	V	*	A - n
CP r	2	f		*	*	V	*	A - r
OR (HL)	5	fr	s	*	*	L	0	A = A | (HL)
OR (IX+d)	9	fr	s	*	*	L	0	A = A | (IX+d)
OR (IY+d)	9	fr	s	*	*	L	0	A = A | (IY+d)
OR n	4	fr		*	*	L	0	A = A | n
OR r	2	fr		*	*	L	0	A = A | r
SBC (IX+d)	9	fr	s	*	*	V	*	A = A - (IX+d) - CY
SBC (IY+d)	9	fr	s	*	*	V	*	A = A - (IY+d) - CY
SBC A,(HL)	5	fr	s	*	*	V	*	A = A - (HL) - CY
SBC A,n	4	fr		*	*	V	*	A = A-n-CY (cout if (r-CY)>A)
SBC A,r	2	fr		*	*	V	*	A = A-r-CY (cout if (r-CY)>A)
SUB (HL)	5	fr	s	*	*	V	*	A = A - (HL)
SUB (IX+d)	9	fr	s	*	*	V	*	A = A - (IX+d)
SUB (IY+d)	9	fr	s	*	*	V	*	A = A - (IY+d)
SUB n	4	fr		*	*	V	*	A = A - n
SUB r	2	fr		*	*	V	*	A = A - r
XOR (HL)	5	fr	s	*	*	L	0	A = [A & ~(HL)] | [~A & (HL)]
XOR (IX+d)	9	fr	s	*	*	L	0	A = [A & ~(IX+d)] | [~A & (IX+d)]
XOR (IY+d)	9	fr	s	*	*	L	0	A = [A & ~(IY+d)] | [~A & (IY+d)]
XOR n	4	fr		*	*	L	0	A = [A & ~n] | [~A & n]
XOR r	2	fr		*	*	L	0	A = [A & ~r] | [~A & r]

[4] SBC and CP instruction output inverted carry. C is set if A<B if the operation or virtual operation is (A-B). Carry is cleared if A>=B. SUB outputs carry in opposite sense from SBC and CP.

Introduction to Rabbit Assembly Language

The following instructions illustrate 8-bit arithmetic and logical operations:

```
#define   BIT_MASK       0xA5
cp  a,b              ;
and BIT_MASK         ;
cp  0x01             ;
xor 0x80             ;
```

6.1.9 16-bit Arithmetic and Logical Operations

These instructions, shown in Table 6.10, are similar to their 8-bit counterparts except for one key difference: because the Rabbit CPU only has 8-bit registers, the 16-bit math instructions use register **pairs** as operands.

There is a multiply instruction that uses BC and DE as operands and stores the result in HL and BC. There are various increment, decrement, and rotate instructions that involve register pairs.

Table 6.10: 16-bit arithmetic and logical operations.

Instruction	clk	A	I	S	Z	V	C	Operation
ADC HL,ss	4	fr		*	*	V	*	HL = HL + ss + CF — ss=BC, DE, HL, SP
ADD HL,ss	2	fr		-	-	-	*	HL = HL + ss
ADD IX,xx	4	f		-	-	-	*	IX = IX + xx — xx=BC, DE, IX, SP
ADD IY,yy	4	f		-	-	-	*	IY = IY + yy — yy=BC, DE, IY, SP
ADD SP,d	4	f		-	-	-	*	SP = SP + d — d=0 to 255
AND HL,DE	2	fr		*	*	L	0	HL = HL & DE
AND IX,DE	4	f		*	*	L	0	IX = IX & DE
AND IY,DE	4	f		*	*	L	0	IY = IY & DE
BOOL HL	2	fr		*	*	0	0	if (HL != 0) HL = 1, set flags to match HL
BOOL IX	4	f		*	*	0	0	if (IX != 0) IX = 1
BOOL IY	4	f		*	*	0	0	if (IY != 0) IY = 1
DEC IX	4			-	-	-	-	IX = IX – 1
DEC IY	4			-	-	-	-	IY = IY – 1
DEC ss	2	r		-	-	-	-	ss = ss - 1 (ss= BC, DE, HL, SP)
INC IX	4			-	-	-	-	IX = IX + 1
INC IY	4			-	-	-	-	IY = IY + 1
INC ss	2	r		-	-	-	-	ss = ss + 1 (ss= BC, DE, HL, SP)
MUL	12			-	-	-	-	HL:BC = BC * DE, signed 32 bit result. DE unchanged
OR HL,DE	2	fr		*	*	L	0	HL = HL | DE — bitwise or
OR IX,DE	4	f		*	*	L	0	IX = IX | DE
OR IY,DE	4	f		*	*	L	0	IY = IY | DE
RL DE	2	fr		*	*	L	*	{CY,DE} = {DE,CY} — left shift with CF
RR DE	2	fr		*	*	L	*	{DE,CY} = {CY,DE}
RR HL	2	fr		*	*	L	*	{HL,CY} = {CY,HL}
RR IX	4	f		*	*	L	*	{IX,CY} = {CY,IX}
RR IY	4	f		*	*	L	*	{IY,CY} = {CY,IY}
SBC HL,ss	4	fr		*	*	V	*	HL=HL-ss-CY (cout if (ss-CY)>hl)

6.1.10 8-bit Bit Set, Reset and Test

The bit manipulation instructions are used to set or reset bits, as well as test the status of bits. Consider an 8-bit parallel port that uses certain bits for outputs; in order to change the status of just one bit, the contents of the port can be read, one bit changed using these instructions, and the contents can be written back to the port. This sequence is often referred to as a read-modify-write operation.

Instructions in this group, shown in Table 6.11, allow bits to be manipulated in certain 8-bit registers, or memory locations pointed to by HL, IX and IY register pairs.

Table 6.11: 8-bit bit set, reset and test.

Instruction	clk	A	I	S	Z	V	C	Operation
BIT b,(HL)	7	f	s	-	*	-	-	(HL) & bit
BIT b,(IX+d))	10	f	s	-	*	-	-	(IX+d) & bit
BIT b,(IY+d))	10	f	s	-	*	-	-	(IY+d) & bit
BIT b,r	4	f		-	*	-	-	r & bit
RES b,(HL)	10		d	-	-	-	-	(HL) = (HL) & ~bit
RES b,(IX+d)	13		d	-	-	-	-	(IX+d) = (IX+d) & ~bit
RES b,(IY+d)	13		d	-	-	-	-	(IY+d) = (IY+d) & ~bit
RES b,r	4	r		-	-	-	-	r = r & ~bit
SET b,(HL)	10		b	-	-	-	-	(HL) = (HL) | bit
SET b,(IX+d)	13		b	-	-	-	-	(IX+d) = (IX+d) | bit
SET b,(IY+d)	13		b	-	-	-	-	(IY+d) = (IY+d) | bit
SET b,r	4	r		-	-	-	-	r = r | bit

In the above table, "bit" is a value between 0 and 7, where 7 is the most significant bit.

The following instructions illustrate various bit set, reset and test operations:

```
#define ON_BIT   4       ; bit to turn motor on
set ON_BIT, a            ; turn motor on

bit 7,(hl) ; test busy bit
res 7,(hl) ; clear busy bit
```

6.1.11 8-bit Increment and Decrement

These instructions operate on locations pointed to by HL, IX or IY register pairs, as well as directly on the CPU's 8-bit registers. An increment or decrement is always quicker than an "add one to" or "subtract one from," respectively, because no immediate data has to be loaded. Unlike the 16-bit INC/DEC instructions, their 8-bit counterparts do affect the status flags. The 8-bit increment and decrement instructions are shown in Table 6.12.

Introduction to Rabbit Assembly Language

Table 6.12: 8-bit increment and decrement instructions.

Instruction	clk	A	I	S	Z	V	C	Operation
DEC (HL)	8	f	b	*	*	V	-	(HL) = (HL) - 1
DEC (IX+d)	12	f	b	*	*	V	-	(IX+d) = (IX+d) -1
DEC (IY+d)	12	f	b	*	*	V	-	(IY+d) = (IY+d) -1
DEC r	2	fr		*	*	V	-	r = r – 1
INC (HL)	8	f	b	*	*	V	-	(HL) = (HL) + 1
INC (IX+d)	12	f	b	*	*	V	-	(IX+d) = (IX+d) + 1
INC (IY+d)	12	f	b	*	*	V	-	(IY+d) = (IY+d) + 1
INC r	2	fr		*	*	V	-	r = r + 1

6.1.12 8-bit Fast Accumulator Operations

These instructions are considered "fast" because they do not load immediate data, or use pointers to point to operands—everything happens right in the accumulator. Table 6.13 lists instructions in this group.

Table 6.13: 8-bit fast accumulator operations.

Instruction	clk	A	I	S	Z	V	C	Operation
CPL	2	r	-	-	-	-	-	A = ~A
NEG	4	fr		*	*	V	*	A = 0 – A
RLA	2	fr		-	-	-	*	{CY,A} = {A,CY}
RLCA	2	fr		-	-	-	*	A = {A[6,0],A[7]}; CY = A[7]
RRA	2	fr		-	-	-	*	{A,CY} = {CY,A}
RRCA	2	fr		-	-	-	*	A = {A[0],A[7,1]}; CY = A[0]

6.1.13 8-bit Shifts and Rotates

These instructions allow bits to be shifted or rotated in and out of a register or a memory location. Before we proceed, it is important to have a distinction between shifts and rotates:

- A "shift" happens when bits in a register or memory location are shifted one place, left or right. The bit shifted out goes to the carry flag, while a 0 gets shifted in. There is a special case (SRA) where the most significant bit remains unchanged.

- A "rotate" is different from a shift because bits are shifted out through the carry flag, and then shifted back in (see Figure 6.2). There are special cases about which bit gets shifted in, the carry bit or the bit being shifted out.

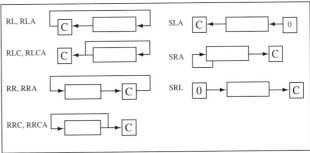

Figure 6.2: 8-bit shifts and rotates.

Chapter 6

There are many reasons a programmer may want to shift bits:

- A "left shift" has the same affect as a "multiply-by-two" while a "right shift" performs a "divide-by-two." This is likely the quickest way to achieve these results. A program may call for receiving a byte, reversing the order, and transmitting out the result. Once a byte has been received and verified for correctness, it is very easy to shift its bits out through the carry flag, and rotate those bits into a separate register or memory location. This would effectively place bits in reverse order into the destination. In the case of division by two, there is a loss of resolution, but when performing integer math, a reduced precision is inevitable.

- Programmers sometimes use parallel port pins to transmit serial data, especially in cases where the serial ports are already being used. When transmitting data serially, a programmer will typically store the byte to be transmitted in a parallel register, and will shift each bit that has to be transmitted. Similarly, on the receiving end, each received bit will be shifted in so that a complete byte can be put together. Using port pins as serial ports in this manner is often referred to as "bit-banging" a serial port.

Table 6.14 lists instructions in the "shift and rotate" group.

Table 6.14: 8-bit shifts and rotates.

Instruction	clk	A	I	S	Z	V	C	Operation
RL (HL)	10	F	b	*	*	L	*	{CY,(HL)} = {(HL),CY}
RL (IX+d)	13	F	b	*	*	L	*	{CY,(IX+d)} = {(IX+d),CY}
RL (IY+d)	13	F	b	*	*	L	*	{CY,(IY+d)} = {(IY+d),CY}
RL r	4	fr		*	*	L	*	{CY,r} = {r,CY}
RLC (HL)	10	F	b	*	*	L	*	(HL) = {(HL)[6,0],(HL)[7]}; CY = (HL)[7]
RLC (IX+d)	13	F	b	*	*	L	*	(IX+d) = {(IX+d)[6,0], (IX+d)[7]}; CY = (IX+d)[7]
RLC (IY+d)	13	F	b	*	*	L	*	(IY+d) = {(IY+d)[6,0], (IY+d)[7]}; CY = (IY+d)[7]
RLC r	4	fr		*	*	L	*	r = {r[6,0],r[7]}; CY = r[7]
RR (HL)	10	F	b	*	*	L	*	{(HL),CY} = {CY,(HL)}
RR (IX+d)	13	F	b	*	*	L	*	{(IX+d),CY} = {CY,(IX+d)}
RR (IY+d)	13	F	b	*	*	L	*	{(IY+d),CY} = {CY,(IY+d)}
RR r	4	fr		*	*	L	*	{r,CY} = {CY,r}
RRC (HL)	10	F	b	*	*	L	*	(HL) = {(HL)[0],(HL)[7,1]}; CY = (HL)[0]
RRC (IX+d)	13	F	b	*	*	L	*	(IX+d) = {(IX+d)[0], (IX+d)[7,1]}; CY = (IX+d)[0]
RRC (IY+d)	13	F	b	*	*	L	*	(IY+d) = {(IY+d)[0],(IY+d)[7,1]}; CY = (IY+d)[0]
RRC r	4	fr		*	*	L	*	r = {r[0],r[7,1]}; CY = r[0]
SLA (HL)	10	F	b	*	*	L	*	(HL) = {(HL)[6,0],0}; CY =(HL)[7]
SLA (IX+d)	13	F	b	*	*	L	*	(IX+d) = {(IX+d)[6,0],0}; CY = (IX+d)[7]
SLA (IY+d)	13	F	b	*	*	L	*	(IY+d) = {(IY+d)[6,0],0}; CY = (IY+d)[7]
SLA r	4	fr		*	*	L	*	r = {r[6,0],0}; CY = r[7]

Introduction to Rabbit Assembly Language

Table 6.14: 8-bit shifts and rotates (continued).

Instruction	clk	A	I	S	Z	V	C	Operation
SRA (HL)	10	F	b	*	*	L	*	(HL) = {(HL)[7],(HL)[7,1]}; CY = (HL)[0]
SRA (IX+d)	13	F	b	*	*	L	*	(IX+d) = {(IX+d)[7], (IX+d)[7,1]}; CY = (IX+d)[0]
SRA (IY+d)	13	F	b	*	*	L	*	(IY+d) = {(IY+d)[7], (IY+d)[7,1]}; CY = (IY+d)[0]
SRA r	4	fr		*	*	L	*	r = {r[7],r[7,1]}; CY = r[0]
SRL (HL)	10	F	b	*	*	L	*	(HL) = {0,(HL)[7,1]}; CY = (HL)[0]
SRL (IX+d)	13	F	b	*	*	L	*	(IX+d) = {0,(IX+d)[7,1]}; CY = (IX+d)[0]
SRL (IY+d)	13	F	b	*	*	L	*	(IY+d) = {0,(IY+d)[7,1]}; CY = (IY+d)[0]
SRL r	4	fr		*	*	L	*	r = {0,r[7,1]}; CY = r[0]

6.1.14 Instruction Prefixes

The Rabbit has two I/O spaces: internal I/O registers and external I/O registers. Instruction prefixes IOI and IOE can be used to generate code that accesses internal or external I/O registers instead of accessing memory. With these prefixes, any 16-bit memory address is decoded as an internal or external I/O address:

- The **IOI** instruction prefix causes the following instruction to access an internal I/O port instead of memory. Since the internal I/O peripherals occupy the first 256 bytes of the internal I/O space, the upper byte of the 16-bit memory address is ignored and the lower byte is used to access the internal I/O port.

The following instruction provides as an example of the IOI prefix. The use of shadow registers is defined in Section 6.3.7.

```
    xor     a               ; port e bit 1..7 inputs, 0 output
    ld      (PEFRShadow),a  ; update shadow register first
ioi ld      (PEFR),a        ; set up function register

    ld      a,0x01
    ld      (PEDDRShadow),a ; update shadow register first
ioi ld      (PEDDR),a       ; set up data direction register
```

- The **IOE** instruction prefix causes the following instruction to access an external I/O port instead of memory. Since external I/O peripherals can be mapped within a 64K space, the full 16-bit address is used to access external I/O ports. By default, writes are inhibited for external I/O operations and fifteen wait states are added for I/O accesses.

The following instruction provides as an example of the ALTD prefix:

```
    #define EXT_ADDRESS 0xFC00 ; external address

        ld      hl, EXT_ADDRESS ; set up pointer
    ioe ld      (hl),a          ; put data value externally
```

Chapter 6

- The **ALTD** prefix causes the immediately following instruction to affect the alternate flags, or use the alternate registers for the destination of the data, or both. Using this instruction allows the programmer to access the alternate register set or alternate flags directly, without the need to exchange all the registers. See Section 6.1.6 for more details on the alternate register set.

The following instruction provides as an example of the ALTD prefix:

```
altd    ex     de,hl        ; de = index (was in hl')
```

Instruction prefixes are shown in Table 6.15.

Table 6.15: Instruction prefixes.

Instruction	clk	A	I	S	Z	V	C	Operation
ALTD	2			-	-	-	-	alternate register destination for next Instruction
IOE	2			-	-	-	-	I/O external prefix
IOI	2			-	-	-	-	I/O internal prefix

6.1.15 Block Move Instructions

These instructions, listed in Table 6.16, are used to rapidly copy (or move) blocks of data from one part of memory to another. They work in the following manner:

- BC is set up as the counter—number of bytes to copy
- HL is set up to point to the source block
- DE is set up to point to the destination block

The instructions differ from each other in the following manner:

LDD:
- Set up DE to point to destination address
- Set up HL to point to source address
- One byte of data is copied from (HL) to (DE)
- both HL and DE are decremented

LDDR:
- Set up DE to point to highest address of destination block
- Set up HL to point to highest address of source block
- A block of data is copied from (HL) to (DE); block size is specified by BC
- both HL and DE are decremented during the operation

LDI:
- Set up DE to point to destination address
- Set up HL to point to source address
- One byte of data is copied from (HL) to (DE)
- both HL and DE are incremented

Introduction to Rabbit Assembly Language

LDIR:
- Set up DE to point to lowest address of destination block
- Set up HL to point to lowest address of source block
- A block of data is copied from (HL) to (DE); block size is specified by BC
- both HL and DE are incremented during the operation

Table 6.16: Block move instructions.

Instruction	clk	A	I	S	Z	V	C	Operation
LDD	10	D	-	-	*	-		(DE) = (HL); BC = BC-1; DE = DE-1; HL = HL-1
LDDR	6+7i	D	-	-	*	-		if {BC != 0} repeat:
LDI	10	D	-	-	*	-		(DE) = (HL); BC = BC-1; DE = DE+1; HL = HL+1
LDIR	6+7i	D	-	-	*	-		if {BC != 0} repeat:

6.1.16 Control Instructions—Jumps and Calls

Instructions in this group affect program flow, and can be divided into the following sections:

- Unconditional Jumps: These instructions start execution from another point in code. As Table 6.17A indicates, these instructions start program execution from a fixed address (PC=mn), or from an address pointed to the HL, IX or IY register pair. The LJP instruction is special because it allows a jump to be made to a computed address in XMEM.

Table 6.17A: Unconditional jump instructions.

Instruction	Clk	A	I	S	Z	V	C	Operation
JP mn	7			-	-	-	-	PC = mn
JP (HL)	4			-	-	-	-	PC = HL
JP (IX)	6			-	-	-	-	PC = IX
JP (IY)	6			-	-	-	-	PC = IY
JR e	5			-	-	-	-	PC = PC + e (if e==0 next seq instruction is executed)
LJP xpc,mn	10			-	-	-	-	XPC=xpc; PC = mn

- Conditional Jumps: These instructions, shown in Table 6.17B, cause a jump if one of the given conditions is met.

Table 6.17B: Conditional jump instructions.

Instruction	Clk	A	I	S	Z	V	C	Operation
JP f,mn	7			-	-	-	-	if {f} PC = mn
JR cc,e	5			-	-	-	-	if {cc} PC = PC + e

Chapter 6

- Subroutine Calls can be made to a 16-bit address using the CALL instruction, or to a computed address in XMEM, using the LCALL instruction, as shown in Table 6.17C.

Table 6.17C: Subroutine calls.

Instruction	Clk	A	I	S	Z	V	C	Operation
CALL mn	12			-	-	-	-	(SP-1) = PCH; (SP-2) = PCL; PC = mn; SP = SP-2
LCALL xpc,mn	19			-	-	-	-	(SP-1) = XPC; (SP-2) = PCH; (SP-3) = PCL; XPC=xpc; PC = mn; SP = (SP-3)

- Returns from subroutines can be made unconditionally, using the RET instruction, or utilize one of several flag conditions with the RET f instruction. LRET is used to return from subroutines stored in XMEM space. These instructions are shown in Table 6.17D.

Table 6.17D: Return instructions.

Instruction	Clk	A	I	S	Z	V	C	Operation
RET	8			-	-	-	-	PCL = (SP); PCH = (SP+1); SP = SP+2
RET f	8/2		-		-	-	-	If {f} PCL = (SP); PCH = (SP+1); SP = SP+2
LRET	13			-	-	-	-	PCL = (SP); PCH = (SP+1); XPC = (SP+2); SP = SP+3

Interrupt Service Routines (ISRs) present a special case:

- Unlike other processors, the Rabbit uses a "Return" (RET) instruction rather than a "Return from Interrupt" (RETI) instruction to return from an interrupt. The RETI instruction is shown in Table 6.17E and does not seem to serve any useful purpose.

In most cases, an interrupt service routine should be terminated as follows:

```
        ipres           ; restore the interrupt priority
        ret             ; return to interrupted code
```

A programmer does not usually have to do anything special to use this instruction; just putting a "RET" at the end of the interrupt service routine is sufficient. Chapter 7 provides many examples of this instruction.

Table 6.17E: Return from interrupt instruction.

Instruction	clk	A	I	S	Z	V	C	Operation
RETI	12			-	-	-	-	IP = (SP); PCL = (SP+1); PCH = (SP+2); SP = SP+3

Introduction to Rabbit Assembly Language

- Decrement and Jump if Non-Zero, shown in Table 6.17F. This instruction is especially handy for implementing loops. Any block of code preceding this instruction can be repeated, provided that it meets the following conditions:
  - The block of code can execute a maximum of 256 times (if B=0 at the start of the count).
  - The entire code block that has to be repeated must fit within this distance limitation. This is because the instruction performs a relative jump to within a 256-byte distance from the current instruction,

Table 6.17F: Decrement and jump instruction.

Instruction	Clk	A	I	S	Z	V	C	Operation
DJNZ j	5	R		-	-	-	-	B = B-1; if {B != 0} PC = PC + j

- Reset (RST) instruction. This instruction, shown in Table 6.17G, pushes the current Program Counter onto the stack and then starts program execution from vector v in the interrupt table. Storing the current program counter on the stack tells the interrupt where to resume operation after servicing the interrupt.

As shown in the table, RST 10, RST 18, RST 20, RST 28, and RST 38 are available. The RST 0x28 instruction is special because it transfers program execution to the Dynamic C debug kernel. This is the only reset instruction used by Dynamic C.

Table 6.17G: Reset instruction.

Instruction	Clk	A	I	S	Z	V	C	Operation
RST v	10			-	-	-	-	(SP-1) = PCH; (SP-2) = PCL; SP = SP - 2; PC = {R,v} v = 10,18,20,28,38 only

Chapter 6

6.1.17 Miscellaneous Instructions

The following instructions do not fit in the above groups; a brief description of each instruction is shown in Table 6.18:

Table 6.18: Miscellaneous instruction.

Instruction	clk	A	I	S	Z	V	C	Operation
CCF	2	f		-	-	-	*	CF = ~CF
IPSET 0	4			-	-	-	-	IP = {IP[5:0], 00}
IPSET 1	4			-	-	-	-	IP = {IP[5:0], 01}
IPSET 2	4			-	-	-	-	IP = {IP[5:0], 10}
IPSET 3	4			-	-	-	-	IP = {IP[5:0], 11}
IPRES	4			-	-	-	-	IP = {IP[1:0], IP[7:2]}
LD A,EIR	4	fr		*	*	-	-	A = EIR
LD A,IIR	4	fr		*	*	-	-	A = IIR
LD A,XPC	4	r		-	-	-	-	A = MMU
LD EIR,A	4			-	-	-	-	EIR = A
LD IIR,A	4			-	-	-	-	IIR = A
LD XPC,A	4			-	-	-	-	XPC = A
NOP	2			-	-	-	-	No Operation
POP IP	7			-	-	-	-	IP = (SP); SP = SP+1
PUSH IP	9			-	-	-	-	(SP-1) = IP; SP = SP-1
SCF	2	F		-	-	-	1	CF = 1

6.2 Some Unique Rabbit Instructions

Although the Rabbit 3000 instruction set is derived from Z-180 instructions, there are some significant differences. The Rabbit 3000 uses some special instructions in areas of I/O and memory access; these instructions do not exist for the Z-180. Moreover, the Rabbit 3000 has dropped some Z-180 instructions; programmers can refer to the Rabbit 3000 microprocessor user manual for details on these instructions.

6.2.1 Instructions That Make Some Operations More Efficient

- The DJNZ instruction, covered in Table 6.17F, is useful for repeatedly executing a block of code.
- The LDD, LDDR, LDI and LDIR instructions, covered in Table 6.16, are useful for moving a block of memory.
- The "LCALL xpc,mn" instruction, shown in Table 6.17C, allows a jump to a computed address in XMEM.
- The 16-bit multiply instruction, shown in Table 6.10, is useful for generating a signed 32-bit multiply out of the BC and DE registers.

6.2.2 Instructions That Bypass the Memory Management Unit

Chapter 2 covered the Memory Management Unit (MMU) in some detail. Instructions in Table 6.19 bypass the MMU and access the entire physical memory space directly. While the instructions use 16-bit register pairs, they target a 20-bit address space. The four most significant bits of the 20-bit address are derived from the four least significant bits of the accumulator (bits 3 though 0).

Table 6.19: Instructions that bypass the MMU.

Instruction	clk	A	I	S	Z	V	C	Operation
LDP (HL),HL	12	-	-	-	-	-	-	(HL) = L; (HL+1) = H. (Adr[19:16] = A[3:0])
LDP (IX),HL	12	-	-	-	-	-	-	(IX) = L; (IX+1) = H. (Adr[19:16] = A[3:0])
LDP (IY),HL	12	-	-	-	-	-	-	(IY) = L; (IY+1) = H. (Adr[19:16] = A[3:0])
LDP HL,(HL)	10	-	-	-	-	-	-	L = (HL); H = (HL+1). (Adr[19:16] = A[3:0])
LDP HL,(IX)	10	-	-	-	-	-	-	L = (IX); H = (IX+1). (Adr[19:16] = A[3:0])
LDP HL,(IY)	10	-	-	-	-	-	-	L = (IY); H = (IY+1). (Adr[19:16] = A[3:0])
LDP (mn),HL	15	-	-	-	-	-	-	(mn) = L; (mn+1) = H. (Adr[19:16] = A[3:0])
LDP (mn),IX	15	-	-	-	-	-	-	(mn) = IXL; (mn+1) = IXH. (Adr[19:16] = A[3:0])
LDP (mn),IY	15	-	-	-	-	-	-	(mn) = IYL; (mn+1) = IYH. (Adr[19:16] = A[3:0])
LDP HL,(mn)	13	-	-	-	-	-	-	L = (mn); H = (mn+1). (Adr[19:16] = A[3:0])
LDP IX,(mn)	13	-	-	-	-	-	-	IXL = (mn); IXH = (mn+1). (Adr[19:16] = A[3:0])
LDP IY,(mn)	13	-	-	-	-	-	-	IYL = (mn); IYH = (mn+1). (Adr[19:16] = A[3:0])

Chapter 6

6.3 Starting to Code Assembly with Dynamic C

Dynamic C provides a lot of freedom to the assembly language programmer. There are various ways of incorporating assembly language in Dynamic C:

- Assembly language statements can be embedded in C functions
- Entire functions can be written in assembly language
- C statements may be embedded in assembly code
- C-language variables may be accessed by the assembly code
- Parameters may be passed between C-language code and assembly code. This is described in Section 6.4.

6.3.1 Inline Assembly Code

Dynamic C allows the use of assembly language anywhere in the code; the programmer simply has to put "#asm" before and "#endasm" after the block of assembly language code that needs to be assembled. Program 6.1 illustrates this concept with a simple example:

Program 6.1: Inline assembly.

```
   printf ("\nWarning Light on\n");      // operating in C

   #asm
      ld      a, (PGDRShadow)            ; get contents of shadow register
      set     6, a                       ; turn LED on
      ld      (PGDRShadow), a            ; update shadow register
      ioi ld  (PGDR), a                  ; write data to port g
   #endasm

   get_status();                                          // back to C
```

6.3.2 "C" Wrappers

A block of assembly code can be encapsulated within a "C" wrapper. This is typically done when a "C" function is defined in order to contain assembly code. Parameters to the "C" function can be passed into assembly code using pointers. Function initPort in Program 6.11 provides an example of a C wrapper encapsulating assembly code.

A C function containing embedded assembly code may use a C return statement to return a value.

6.3.3 Standalone Assembly

Programmers can write entire functions in assembly and benefit from the speed of machine language. Assembly functions can be declared just as C functions and can receive and return parameters. As an example, Program 6.2 shows how an assembly function is used for fast table lookup and the value is returned to the C calling program.

Introduction to Rabbit Assembly Language

Program 6.2: Code fragment showing standalone assembly function.

```
int CRC_lookup (int value);

main()
{
   int i,j;
   i=1;
   j= CRC_lookup(i);
}

#asm
CRC_lookup::
   ...
   ld hl,a          // The return value is put in HL
   ret              // just before the function returns
#endasm
```

A few things to note about standalone assembly functions:

- A function prototype needs to be declared before the function declaration. Ideally, the function prototype is declared towards the beginning of the program. Defining a function prototype is important so that functions that have not been compiled may be called, and that the compiler can do typechecking on the function's parameters.
- Note the use of double colons in declaring function CRC_lookup—this is Dynamic C's standard method for calling standalone assembly routines from C. The assembly function is identified by the double colon "::" and the double colon construction declares that the label is of global scope.
- CPU registers used by the assembly program do not have to be saved upon entering the function or restored upon exit. However, as we will see in Chapter 7, if the standalone assembly function is an Interrupt Service Routine (ISR), this rule is reversed and CPU registers used by the ISR have to be saved upon entry and restored upon exit.

Chapter 6

6.3.4 Embedding C Statements in Assembly

Dynamic C allows embedding C statements in assembly. The C statements have to be preceded by a "c" indicating that they are to be compiled and not assembled. The following code fragment uses embedded C code to perform some math in some assembly code:

Program 6.3: Code fragment showing embedded C statements in assembly.

```
    int result (int x);

    #define DIVIDEND 57
    #define DIVISOR 19

    int temp1, temp2, temp3;

    //////////////////////////////////////////////////////////
    // int result (int x)
    // the assembly subroutine returns x + (DIVIDEND / DIVISOR)
    //////////////////////////////////////////////////////////
    #asm debug root
    result::
       // HL should contain parameter x

              ex de,hl                  ; store x in DE

              push    de                ; save needed registers

              c temp1 = DIVIDEND / DIVISOR;
              c temp2 = (DIVIDEND * 4) / (DIVISOR - 3);
              c temp3 = temp1 + temp2;

              pop     de                ; restore needed registers

              ld      a, (temp3)        ; load global variable

              ld h, 0                   ; add x and temp3
              ld l, a
              add hl, de

              ret                       ; return result in HL and DE
    #endasm

    main()
    {
    int x,y,z;

       for (x=0; x<1000; x++)
         {
              z = result(x);
              printf ("X = %d, Z = %d\n", x,z);
         } // loop

    } // main
```

Introduction to Rabbit Assembly Language

Note that register pair DE is saved before invoking the inline C statements. Since compiling these statements will change register contents, it is key to save important registers before invoking C statements, and restore them before continuing with assembly code.

The assembly statement `ld a, (temp3)` allows us to load a C variable in a CPU register. This concept is covered in Section 6.4.5.

6.3.5 Assembly Calling a C Program

Assembly code can call a C function, but it's a little more work than C calling an assembly function. The calling program has to perform the following housekeeping tasks before the call:

- Since the called C function may change register contents, the assembly calling program should save important registers prior to the call and restore them after the call. This can easily be done by pusing them on the stack. To be safe, the assembly program should always assume that the C function will change register contents.

- If the C function returns a `struct`, the calling program must reserve space on the stack for the returned structure.

- If the assembly calling program needs to pass parameters to the C code, it has to push the parameters on the stack in the right order.

- If the first argument is a pointer, an `int`, `unsigned int` or `char`, it should be loaded into HL. If it is a `long`, `unsigned long` or `float`, it should be loaded into the BC:DE combination.

- Finally, the calling program has to use the call instruction to call the C program.

Once the C function is done and code execution returns to the assembly calling program, the calling program has to perform the following additional housekeeping tasks:

- The program has to recover the stack space that it allocated to arguments. This is done by popping the variable off the stack, two bytes at a time.

- If the C function returns a `struct`, the calling program must recover the returned structure from the stack.

- If the calling program saved any registers by pushing them on the stack, it should restore them by popping them off the stack.

- Depending on the data type of the returned parameter, the calling program should retrieve the returned value from HL (if an `int`, `unsigned int` or `char`) or BC:DE (if a `long`, `unsigned long` or `float`).

Chapter 6

The following example illustrates a segment of assembly code calling a C program:

Program 6.4: Code segment showing assembly calling a C program.

```
/****************************************************************
    param3.c

    Description
    ===========
    This program demonstrates an assembly program calling a C
    function.

    The program passes parameters from assembly to C.

    Instructions
    ============
    1.  Compile and run this program.
    2.  Watch the results in the stdio window
****************************************************************/

int print_regs (int reg_HL, int reg_BC, int reg_DE, int reg_AF);

/////////////////////////////////////////////////////////////
// int print_regs (int x)
// prints x and returns x+1
/////////////////////////////////////////////////////////////

int print_regs (int reg_HL, int reg_BC, int reg_DE, int reg_AF)
{
    printf("HL= %4x BC= %4x DE= %4x AF= %4x\n",
            reg_HL, reg_BC, reg_DE, reg_AF);
}

main()
{
#asm debug
amain::

    ld hl, 10                   ; count down from this number

loop:
    push hl                     ; save critical registers


    ld BC, 0x2222
    ld DE, 0x3333
    ld A, 0x44

    push AF                     ; set up parameters
    push DE                     ; on the stack
    push BC                     ; for the C function
```
(Program 6.4 continued on next page)

Introduction to Rabbit Assembly Language

Program 6.4: Code segment showing assembly calling a C program (continued).

```
        push HL

        call print_regs         ; call C function to print value in HL

        pop HL                  ; clean up the stack
        pop BC
        pop DE
        pop AF

        pop hl                  ; restore critical registers

        dec hl                  ; count down
        ld   a, h               ; test if we're done
        or   l

        jr nz, loop
end:
        jp end                  ; infinite loop

#endasm

} // main
```

The `main()` part of the program is written in assembly, and calls the `print_regs` function to print the contents of registers passed in as parameters. The order of the parameters pushed on the stack has to correspond to the input parameters of the called function—the last parameter is pushed first and the first parameter pushed last.

The program uses HL as a counter and pushes and pops HL before and after calling the C subroutines, so that the HL register pair isn't affected by the C code. The program gives very different results if the HL register pair isn't saved and restored as shown.

After the call to `print_regs`, the assembly program pops the parameters off the stack, to get the stack back in the condition before the call. An alternative method is to directly manipulate the stack pointer.

6.3.6 Accessing I/O Ports in Assembly and Dynamic C

As we have covered so far, the processor is able to access registers and memory, do some math, perform some binary logic, and make decisions based on certain conditions. In order to be truly useful, the processor has to be able to interface with external elements such as:

- LEDs and LCDs
- Keypads and switches
- Buzzers
- Relays and solenoids
- analog-to-digital and digital-to-analog converters

Chapter 6

Such external elements interface between real-world data and digital signals the microprocessor understands best. The Rabbit microprocessor uses I/O ports to interface with such external devices. Some of these peripherals can be connected using parallel ports; others can use serial ports, and some peripherals use ports with specialized protocols such as SPI, I2C and 1-wire.

Since some of the projects in this book will make use of I/O ports, it is important to first understand how to set up these ports and make use of them (Chapter 7 provides more examples of using these ports). The basic idea of using the Rabbit's I/O ports is the following:

- Decide which I/O port needs to be used. This depends on which functions a particular port is able to serve and what alternate functions the programmer will have to give up in using a port. For instance, while Port C can be used for parallel I/O, the same port pins are used alternatively as four asynchronous serial ports, A though D.
- Initialize the port properly. This will include setting up the Direction, Function, Control and Data Registers for a given port. Here are a few examples:
  - Port B Data Direction Register (PBDDR) sets up bit direction for Port B, selecting which bits are to be used as inputs and which are to be used as outputs. Program 6.16A provides an example of using this port.
  - Port C Function Register (PCFR) sets up Port C for bitwise I/O or for supporting serial ports A through D.
  - Port E Function Register (PEFR) sets up Port E for bitwise I/O or for supporting the external I/O feature. In addition, Port E Control Register (PECR) controls clocking of the upper and lower nibble of the final output register of the port.
  - In addition, the Data Register for each port is used to write to or read from its corresponding port.

6.3.7 Using Shadow Registers for Port I/O

Shadow registers play an important role in Rabbit programming—they help the programmer keep track of what's written to write-only registers so that the programmer can have the means to "read" the contents of a write-only register.

Except for the port data registers PADR through PGDR, most of the remaining I/O-related registers are write-only. There is a shadow register associated with each of the write-only registers. In order to make them easy to remember, each shadow register is named after its corresponding write-only register. For instance, the shadow register for the Port E Function Register (PEFR) is called PEFRShadow. As long as the programmer updates the corresponding shadow register for a write-only register, the shadow register can be read back to find the register's contents.

In general, Zworld recommends that the shadow register be updated BEFORE the corresponding I/O register.

As an example, the following code fragment is part of an initialization routine for Port G for Program 6.5:

Introduction to Rabbit Assembly Language

Program 6.5: Port initialization and use of shadow registers.

```
#define DS1 6           //port G bit 6
#define DS2 7           //port G bit 7

/***************************************************************

This routine Initializes Port G to work with an LED on the proto board.
Input parameters: none
Output parameters: none
Register affected: A, plus port registers

***************************************************************/
#asm root
initPort::
// clear all bits for pclk/2
         ld      a,0x00
         ld      (PGCRShadow),a
   ioi   ld      (PGCR),a

// clear all bits for normal function
         ld      a,0x00
         ld      (PGFRShadow),a
   ioi   ld      (PGFR),a

         ld      (PGDCRShadow),a
// set bits 7,6 drive open drain
         or      0xC0

// clear bit 2 drive output
         res     2,a
         ld      (PGDCRShadow),a
   ioi   ld      (PGDCR),a

         ld      (PGDRShadow),a

// set bits 7,6 output high
         or      0xC0

// clear bit 2 output low
         res     2,a
         ld      (PGDRShadow),a
   ioi   ld      (PGDR),a

// set bits 7,6,2 to output, clear bits 5,4,3,1,0 to input
         ld      a,0xC4
         ld      (PGDDRShadow),a
   ioi   ld      (PGDDR),a

#endasm
```

Chapter 6

As shown in the above program, the initialization routine sets up port G in the following manner:

- PGCR: select a transfer clock of PCLK/2
- PGFR: select normal mode for the port (parallel and bitwise I/O)
- PGDCR: select normal mode for the required I/O pin (high and low) and not open drain
- PGDDR: set the appropriate port bits for input and output

Note that, in each case, the information written to a (write only) port register is also copied into its associated shadow register. Doing this keeps them synchronized and gives the programmer the ability to read the data back from the shadow register.

6.3.8 Using Library Functions for Port I/O

Although assembly code will run faster than compiled C code, programmers may choose to use C functions for performing port I/O. Dynamic C provides the following functions in this regard:

The following functions can be used to read internal I/O registers:

- `RdPortI(int PORT)`: returns PORT, high byte zero
- `BitRdPortI(int PORT, int bitcode)`: read bit position 0-7 from the port

The following functions can be used to write internal I/O registers:

- `WrPortI(int PORT, char *PORTShadow, int value)`: write value to a port and a shadow register
- `BitWrPortI(int PORT, char *PORTShadow, int value, int bitcode)`: set a port bit and update its associated shadow register as well

If a NULL pointer replaces the pointer to "PORTShadow" for the `WrPortI` function, the corresponding shadow register is not updated with the function call. A pointer to the shadow register is mandatory for `BitWrPortI()`.

The following functions can be used to access external I/O registers:

- `int RdPortE(int PORT)`: returns PORT, high byte zero
- `int BitRdPortE(int PORT, int bitcode)`: returns bit status
- `int WrPortE(int PORT, char *PORTShadow, int value)`: writes a byte to external port
- `int BitWrPortE(int PORT, char *PORTShadow, int value, int bitcode)`: writes a bit to external port

If a NULL pointer replaces the pointer to "PORTShadow" for the `WrPortE` function, the corresponding shadow register is not updated with the function call. A pointer to the shadow register is mandatory for `BitWrPortE()`.

Once a port is initialized, the following statement can be used to read the data register of an I/O port; this example reads the Port A Data Register (PADR):

```
k=RdPortI(PADR);
```

6.4 Passing Parameters Between C and Assembly

Once a program is divided into separate functions, it becomes useful for the functions to be able to call each other for their own needs. Moreover, parameters provide portability to a function, so that it can be used in more than one program. Dynamic C allows parameters to be passed between C and assembly functions, and this section will illustrate various combinations of parameter passing with examples.

Before moving further, it is important to know the Dynamic C data types and how much space is needed to represent each data type. Table 6.20 lists the Dynamic C data types:

Table 6.20: Dynamic C data types.

Data Type	Description
char	8-bit unsigned integer. Range: 0 to 255 (0xFF)
int	16-bit signed integer. Range: −32,768 to +32,767
unsigned int	16-bit unsigned integer. Range: 0 to +65,535
long	32-bit signed integer. Range: −2,147,483,648 to +2,147,483,647
unsigned long	32-bit unsigned integer. Range 0 to $2^{32} - 1$
float	32-bit IEEE floating-point value. The sign bit is 1 for negative values. The exponent has 8 bits, giving exponents from −127 to +128. The mantissa has 24 bits. Only the 23 least significant bits are stored; the high bit is 1 implicitly. (Rabbit controllers do not have floating-point hardware.) Range: 1.18×10^{-38} to 3.40×10^{38}
enum	Defines a list of named integer constants. The integer constants are signed and in the range: −32,768 to +32,767.

In examining the parameters passed through the stack for a function call, it will be important to keep in mind the above data types. Knowing the number of bytes for each parameter is key in determining how to fetch parameters from the stack.

6.4.1 "C" Passing a Single Parameter to Assembly

A C program calling an assembly function puts the one or two-byte parameter into HL, with register H containing the most significant byte. A four-byte parameter goes in BC:DE with register B containing the most significant byte. Program 6.6 illustrates how this is done: the calling program passes a single parameter "x" to the assembly routine `result`, which extracts the argument from the HL register pair.

Program 6.6: Passing a single parameter from C to assembly.

```
/*************************************************************
    Param1.c

    Description
    ===========
    This program demonstrates parameter passing
    between assembly and C.
```
(Program 6.6 continued on next page)

Chapter 6

Program 6.6: Passing a single parameter from C to assembly (continued).

```
        The program passes one parameters from C to assembly, and
        one result back.
        Instructions
        ============
        1.   Compile and run this program.
        2.   Watch the results in the stdio window
    ***************************************************************/


        int result (int x);

/////////////////////////////////////////////////////////////
// int result (int x)
// the assembly subroutine returns sum of x + 1
/////////////////////////////////////////////////////////////
#asm root
result::
    // HL contains parameter x

            inc hl                      ; x = x+1
            ret                         ; pass x back through HL
#endasm


main()
{
int x,z;

   for (x=0; x<10; x++)
      {
            z = result(x);
            printf ("X = %d, Result = %d\n", x,z);
      } //for

} //main
```

6.4.2 "C" Passing Multiple Parameters to Assembly

Multiple arguments are passed to assembly using the stack. The following rules apply here:

- If the first argument has one or two bytes (`int, unsigned int, char, pointer`), only the first argument is put into HL (with H containing the most significant byte).
- If the first argument has four bytes (`long, unsigned long, float`), the first argument is put into BC:DE (with register B containing the most significant byte)
- All arguments, including the first, are pushed on the stack. The first argument is pushed last.

As shown in Section 6.4.4, the assembly program returns values in the same manner: using HL for one or two bytes in return, or BC:DE for four-byte arguments.

Introduction to Rabbit Assembly Language

Program 6.7 demonstrates two integers that are passed from a C program to an assembly function:

Program 6.7: Passing multiple parameters to an assembly function.

```c
/***************************************************************
    Param2.c

    Description
    ===========
    This program demonstrates parameter passing
    between assembly and C.

    The program passes two parameters from C to assembly, and
    one result back.

    Instructions
    ============
    1.  Compile and run this program.
    2.  Watch the results in the stdio window
****************************************************************/

int result (int x, int y);

///////////////////////////////////////////////////////////
// int result (int x, int y)
// the assembly subroutine returns sum of x and y
///////////////////////////////////////////////////////////
#asm root
result::
    // (sp+2) should contain first parameter x
    // (sp+4) should contain second parameter y


    ld hl,(sp+4)                ; get y
    ex de,hl                    ; store in DE

    ld hl,(sp+2)                ; get x

    add hl, de                  ; return result in HL and DE

    ret
#endasm


main()
{
int x,y,z;
```

(Program 6.7 continued on next page)

Chapter 6

Program 6.7: Passing multiple parameters to an assembly function (continued).

```
while (1)
{

   for (x=0; x<10; x+=2)
     {
            for (y=1; y<11; y+=2)
            {
            z = result(x,y);
            printf ("X = %d, Y = %d, Z = %d\n", x,y,z);
            } // inner loop

     }  // outer loop

   printf("\n\n");

} // while

} // main
```

If the number of parameters is small, the assembly function can easily retrieve them by reading offsets from the stack pointer, knowing the size of the parameter variables.

In order to gain a better understanding of how parameters are passed, it will be useful to have a visual representation of the stack and parameters. Figure 6.3 is a simplified view[5] of how parameters are passed on the stack:

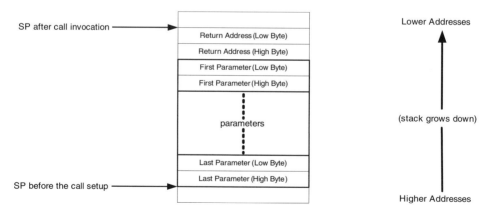

Figure 6.3: Passing parameters on the stack

---

[5] The figure does not show the IX register, which, if used as a frame reference point, is also pushed on the stack

Introduction to Rabbit Assembly Language

The following key concepts are at work here:

- The stack pointer decrements with each "push" operation.
- When a variable is stored on the stack, the most significant byte is pushed first.
- When a number of variables are passed to a function, the last parameter is pushed first and the first parameter is pushed last.

One can easily map the parameters from Program 6.7 to Figure 6.3 and see how the stack decrements as parameters are set up for the call, and what the stack looks like just after the program starts executing the `result` function.

6.4.3 "C" Passing a Pointer to Assembly

While it is easy to pass a single variable as a parameter, it becomes tedious to pass a large number of variables or an array. Except when single variables are passed, Dynamic C uses the stack to pass parameters, and, as the stack grows, it may become tedious for the programmer to efficiently manage the stack in such a case. It is often simpler just to pass a pointer as a parameter and let the function access variables in memory.

For example, the following function will add two 64-bit numbers together. The same program could be written in C, but it would be many times slower because C does not provide an add-with-carry operation (`adc`).

Program 6.8A: Passing a pointer to assembly.

```
    void eightadd( char *ch1, char *ch2 )
    {
    #asm
        ld    hl,(sp+ch2)         ; get source pointer
        ex    de,hl               ; save in register DE
        ld    hl,(sp+ch1)         ; get destination pointer
        ld    b,8                 ; number of bytes
        xor   a                   ; clear carry

    loop:
        ld    a,(de)              ; ch2 source byte
        adc   a,(hl)              ; add ch1 byte
        ld    (hl),a              ; store result to ch1 address
        inc   hl                  ; increment ch1 pointer
        inc   de                  ; increment ch2 pointer
        djnz  loop                ; do 8 bytes
        ; ch1 now points to 64 bit result
    #endasm
    }
```

Chapter 6

The following function accepts a pointer to a table and an index value into the table, and returns an entry from the table. The function becomes useful when it is quicker to do a table lookup than to compute a value. As an example, Program 6.8B looks up square roots and squares for integers between 0 and 15:

Program 6.8B: Passing a pointer to an assembly-based table lookup function.

```c
int t_lookup(char *table_pointer, int index);    // function prototype
char result;

// Table contains squares of integers 0 through 15
const char squaretable[] =
{0, 1, 4, 9, 16, 25, 36, 49, 64, 81, 100, 121, 144, 169, 196, 225};


// Table contains integer square roots of integers 0 through 15
const char roottable[] =
{0, 1, 1, 1, 2, 2, 2, 2, 2, 3, 3, 3, 3, 3, 3, 3};



main()
{
   char i;

   while (1)
   {
   for (i=0; i<16; i++)
   {
   printf ("Number: %d, Squared: %d\n",
              i, t_lookup(squaretable, i));
   }

   printf ("\n\n");

   for (i=0; i<16; i++)
   {
   printf ("Number: %d, Root: %d\n", i, t_lookup(roottable, i));
   }

   printf ("\n\n");
   } //while
} //main



#asm root
t_lookup::
// first input parameter is the pointer to the table
```

(Program 6.8B continued on next page)

Program 6.8B: Passing a pointer to an assembly-based table lookup function (continued).

```
// second input parameter is the index into the table

    EX      DE, HL          ; put table address into DE
    LD      HL,(SP+4)       ; get index into table

    ADD     HL, DE          ; table address of value
    LD      A, (HL)         ; get result from table

    LD      H,0             ; return value is put
    LD      L,A             ; in HL just before t_lookup() returns

    ret
#endasm
```

We will put this function to good use later in this chapter and in Chapter 7.

6.4.4 Assembly Returning Results to "C"

The following rules apply to an assembly function returning results to the C calling program:

- If the result is an int, unsigned int, char, or a pointer, return the result in HL (register H contains the most significant byte).
- If the result is a long, unsigned long, or float, return the result in BC:DE (register B contains the most significant byte).
- Therefore, before executing the "return" statement, the assembly program has to place the return results in the appropriate register pair.

6.4.5 Accessing "C" Variables in Assembly

Accessing static variables is simple, because the symbol evaluates to the address directly. The code segment in Program 6.9 shows how to read a C variable in assembly.

Program 6.9: Accessing "C" variables in assembly.

```
// compute value and write to Port A Data Register
value=x+y;

#asm

        ld a,(value)    ; value to write
ioi     ld (PADR),a     ; write value to PADR

#endasm
```

This method is not efficient, and can create issues when used with recursive code and multitasking.

Chapter 6

6.5 Project 1: Creating a Delay Routine

On one hand, microprocessors are designed to run as fast as possible, and on the other hand, programmers sometimes need to insert delays in their code. Detecting events within a window of time, generating precisely timed signals, following the manufacturer's timing recommendations in initializing a device, and deliberately slowing down portions of program execution for debugging are examples of activities that cannot be carried out without delays.

One of the authors started his career writing firmware for modems, when 2400-baud modems were the fastest in the world. Such modems were downward compatible; an originating (calling) modem would start out listening for a modem responding with a 2400-baud connection. If the originating modem did not find such a response, it would drop down to a 1200-baud connection, and so on, all the way down to a 300-baud connection.

Standards exist for modem connection sequences—on one side of the line, these standards tell originating modems when to listen for what type of response, and on the other side, the standards dictate what tones the answering modems have to generate and for how long. In addition, there are various timeouts when the connect sequence may momentarily get interrupted and the modems have to attempt reconnection—the timing windows for these are defined as well so that the modems are able to re-connect. Such a connection sequence is shown in Figure 6.4:

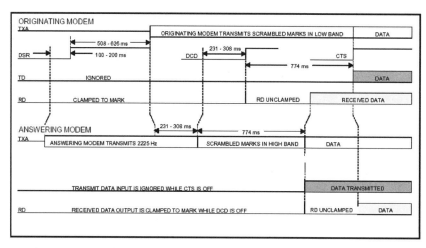

Figure 6.4: Modem connection sequence for the Bell 212 connection.

It is impossible for such connections to take place unless transmission and recognition windows are timed precisely—this is done with the use of delays.

Consider the X10 protocol that uses "carrier current transmission" to transfer information over household electrical wiring. The transmitting device waits for a zero crossing (when the 60 Hz AC wave crosses the zero-volt line) and then sends timed bursts of 125-KHz signals before the next zero crossing. The microprocessor can neither generate a 125-KHz signal, nor

Introduction to Rabbit Assembly Language

can it time the generation of the signals, unless it is able to wait for the right time to do the right thing. The ability to incorporate delay loops, when required, makes it possible for us to develop such an application.

6.5.1 Theory: What Types of Delays Can Be Created

The Rabbit programmer can easily create delays using three different techniques:

- Using assembly language or C statements to create delay loops
- Setting up timers and interrupts to create delays.
- Using Dynamic C's built-in delay functions.

This section will highlight only the loop-based method to illustrate how a programmer can create delays with assembly language. The other two methods are highlighted in Chapter 7.

6.5.2 Implementation: Creating Loop-based Delays in Assembly

One way to create a delay is to keep the microprocessor busy for some time. Knowing that a Rabbit microprocessor is operating at tens of megahertz, each instruction ends up taking a few microseconds to execute. Even if a typical instruction takes 4 clock cycles, the microprocessor has to execute a lot of instructions just to give us a millisecond of delay. This is where loops come in—the programmer can make the microprocessor execute instructions in a loop and decide how many times to execute the loop to give us the required delay.

There is a strong reason for using assembly language, and not a high-level language for incorporating loop-based delays. Knowing how many clock cycles each assembly instruction takes, the programmer can structure the loops to consume a precise numbers of clock cycles. This gives the programmer excellent control over the loops' granularity.

Another reason for using assembly and not a high-level language for implementing loop-based delays is that different language vendors may build there tools differently, so the same high-level code may generate varying amounts of delay, depending on the compiler / interpreter used, while assembly language routines will always assemble to generate the instructions expected by the programmer.

Program 6.10A illustrates a simple loop. Each assembly language instruction is followed by the number of clock cycles it will take to execute:

Program 6.10A: Code fragment for generating delay with a simple loop.

```
            LD      A,      0FFh        ; 4         loop counter
LOOP:       DEC     A                   ; 2
            JP      NZ,     LOOP        ; 7
```

Chapter 6

The total delay therefore depends on the total number of clock cycles executed by the loop, plus the 4 clock cycles it takes to set up the accumulator for the loop. When the total number of clock cycles to be executed is multiplied by the clock period of the microprocessor, we will get the total delay generated by the loop.

$T_D = T_c * \Sigma$ (clock cycles executed)

Where T_D = Total Delay and T_c = Clock Period

> **Note**: The Rabbit RCM 3200, operating at 29.4 MHz clock, gives us 0.034014 microseconds per clock period

We can break it down further to incorporate the time spent outside the loop, and the number of times the loop has to execute:

$T_D = \text{Delay}_{outside} + \text{Delay}_{loop}$

Where,

$\text{Delay}_{outside} = T_c * \Sigma$ (clock cycles executed outside the loop)

and,

$\text{Delay}_{loop} = T_c * \Sigma$ (clock cycles executed inside the loop) * number of times the loop is executed

Therefore,

$T_D = T_c * [\Sigma$ (clock cycles executed outside the loop) $+ \Sigma$ (clock cycles executed inside the loop) * number of times the loop is executed]

Looking at Program 6.10, we can calculate T_D as follows:

$T_D = 0.034014$ uS $* [4 + 9 * 255]$, which gives us about 78 microseconds.

Another factor in considering timing loops is the CPU clocks required to enter and return from the code if it is a subroutine or function.

See spreadsheet DELAY.XLS on the enclosed CD-ROM for these calculations.

A delay of 78 microseconds is not useful for most practical applications. Interfacing to peripherals may require delays in the order of milliseconds, while blinking lights or scrolling text across displays will need hundreds of milliseconds of delays. Program 6.10A clearly reaches its limits with the maximum value in the 8-bit loop counter.

Introduction to Rabbit Assembly Language

Through the use of 16-bit loop counters and nested loops, Program 6.10B is able to achieve much longer delays:

Program 6.10B: Code fragment for generating delay using nested loops.

```
            PUSH    AF              ; 10 save registers before starting
            PUSH    BC              ; 10
            PUSH    HL              ; 10

            LD      BC,     0FFh    ; 6 load outer loop constant
OUTER:      LD      HL,     00FFFh  ; 6 load inner loop constant
INNER:      DEC     HL              ; 2
            LD      A, H            ; 2
            OR      L               ; 2 repeat inner loop
            JP      NZ, INNER       ; 7 until HL is decremented to zero

            DEC     BC              ; 2 repeat outer loop
            LD      A, B            ; 2 until BC is
            OR      C               ; 2 decremented to zero
            JP      NZ, OUTER       ; 7

            POP     HL              ; 7 restore registers
            POP     BC              ; 7
            POP     AF              ; 7
```

The "housekeeping" or "loop initialization" section of this routine (first five instructions) takes 51 clock cycles to push (30 cycles) and pop (21 cycles) the register pairs used in the routine, plus another 6 cycles to load register pair BC. The total loop delay is calculated as follows:

Each iteration of the inner loop (INNER) takes 13 clock cycles (2+2+2+7). Six clock cycles are used to set up the loop by loading register pair HL. The outer loop (OUTER) takes 32 clock cycles (13 clock cycles for the INNER loop, plus 6 clock cycles to set up the inner loop, plus 13 clock cycles for the outer loop), plus 6 clock cycles to set up the OUTER loop by loading register pair BC.

Therefore, the overall delay is given by:

$T_D = M_c * (13 * N_i + 6) * (32 * N_o + 6)$, where M_c is the time per machine cycle, and N_i and N_o represent immediate values loaded into HL and BC, respectively.

Using a 29.4 MHz clock, we get M_c to be 0.034014 microseconds.

The spreadsheet DELAY.XLS works out these calculations, and Table 6.21 gives us the following register values for the required delays:

Table 6.21: Register values for Program 6.5B.

Required Delay	Value for HL	Value for BC
1 mS	0x0010	0x0080
20 mS	0x015F	0x0080
50 mS	0x01B9	0x00FF
100 mS	0x01B8	0x0200
250 mS	0x0226	0x0400
500 mS	0x2000	0x008A

The spreadsheet works with Microsoft Excel's "Goal Seek" function to find the values for the above table. The "Goal Seek" function requires the "Analysis ToolPak" to be loaded into Excel, and, if it is not loaded, the spreadsheet returns the #NAME? error.

The "Analysis ToolPak" is loaded with the following steps:

- Click "Add-Ins" on the Tools menu
- Select the "Analysis ToolPak" box In the "Add-Ins available" list
- Click OK.

The above instructions are current as of Microsoft™ Office releases 97 through 2003.

6.5.3 Final Thought

Because integer values are loaded into the register pairs, a certain amount of rounding off happens in order to arrive at the delay values. The rounding error may not be a significant issue for most applications, especially where delays are used for human interface devices such as flashing LEDs and debouncing switches. For other applications requiring precise delays, the keen programmer will find optimal values for loop counters that provide the lowest rounding errors.

6.6 Project 2: Blinking an LED

To blink an LED, one has to simply turn it on and off sequentially. However, just turning on and off an LED can produce a surprising number of effects. Consider an LED that is turned on for a time T_{ON} and off for a T_{OFF}. If T_{ON} is short and T_{OFF} is long, the LED emits brief bursts of light followed by a long period of darkness. In systems that need to provide a visual indication that they are operating, sometimes called a heartbeat, this arrangement is often used. Battery powered systems especially benefit from this arrangement as the LED only consumes power for a short period of time.

When a programmer thinks of "blinking an LED" the first thought that might occur would be to let T_{ON} equal T_{OFF}. This arrangement can be seen on many appliances and has a "look at me" factor. Although simple to do, this arrangement can become irritating over time.

Introduction to Rabbit Assembly Language

Another effect is to let T_{ON} be long and T_{OFF} be short. This can be used as a heartbeat, but consumes more power than the first example. This visually says "I'm on and working and am only briefly taking the time to blink this light because I'm quite busy just now, thank you very much."

The ratio of T_{ON} to total period is referred to as the device's duty cycle. Duty cycle is expressed as a percentage and is defined as,

$$DUTY\ CYCLE = \frac{T_{ON}}{T_{OFF} + T_{ON}} \cdot 100\%$$

If the frequency is high enough, the intensity of the LED as perceived by the human eye is proportional to the duty cycle. So at a low frequency such as ½ hertz, a ten percent duty cycle looks like a heart beat on a smoke-detector, while at 200 hertz the same ten percent duty cycle looks like a dim LED.

By adjusting frequency and duty-cycle the programmer can produce a remarkably wide range of visual effects. And it all boils down to simply turning the LED on and off at the right time.

The next section shows how to use assembly code to control an LED.

6.6.1 Implementation: blinking an LED in assembly

Program 6.11 (`flashled1.c`) illustrates a simple piece of code that uses the delay routine above to blink an LED. This program was written to run on the Rabbit 3200 prototyping board.

Program 6.11: Code fragment for flashing an LED.

```
/*************************************************************
    flashled1.c

    Description
    ===========
    This assembly program flashes the DS1 LED
    on the prototyping board.

<code deleted for brevity>

//////////////////////////////////////////////////////////
// DS1 led on protoboard is controlled by port G bit 6
// turns on if state = 0
// turns off if state = 1
//////////////////////////////////////////////////////////
#asm root
DS1led::
        push    af                          ; save registers we'll use

// if (state == ON) then turn LED on
        ld      a,l                         ; HL contains 0 for ON
        or      a,h
        jr      nz,turn_LED_off             ; turn LED off if HL not 0
```

(Program 6.11 continued on next page)

Chapter 6

Program 6.11: Code fragment for flashing an LED (continued).

```
        //use shadow register to keep other bit values
            ld      a,(PGDRShadow)
            res     DS1,a                       ; clear bit 6 only

            jp      DS1led_done                 ; ready to write to port
    turn_LED_off:
        //use shadow register to keep other bit values
            ld      a,(PGDRShadow)
            set     DS1,a                       ; set bit 6 only

    DS1led_done:
            ld      (PGDRShadow),a              ; update shadow register
    ioi     ld      (PGDR),a                    ; write data to port g

            pop     af                          ; restore saved registers
            ret
    #endasm

    //////////////////////////////////////////////////
    //////////////////////////////////////////////////
    void main()
    {
        initPort();                             // initialize port G only

        while (1)
        {
            DS1led(ON);

            Delay1mS(500);                      //on for 500 ms

            DS1led(OFF);

            Delay1mS(500);                      //off for 500 ms

        } // while

    } // main
```

The "`Delay1mS`" subroutine routine uses the delay algorithm covered in Section 6.5 to create a 1-millisecond delay, but with a twist: the subroutine runs the delay algorithm inside a loop so that an external parameter can be used to create delays longer than 1 millisecond. There is additional time spent in executing the "`for`" loop but the delay is considered negligible and therefore ignored in our calculation.

Since the DS1 LED is connected to Port G, the program starts off by initializing Port G in the `initPort()` routine. As covered in Section 6.4.5, the programmer has to follow a sequence in order to initialize the port. The `initPort()` routine is described in Section 6.4.6.

Introduction to Rabbit Assembly Language

The main section of the program then executes an endless loop and alternately calls the `DS1led` routine to turn the LED on and off, inserting a delay after turning the LED on or off.

Each time the `DS1led` routine is called, it checks the input parameter to see whether the LED needs to be turned on or off. It then loads the shadow register for the port data register and, depending on whether the bit corresponding to the LED needs to be set or reset, it modifies the port data register and the corresponding shadow register. This keeps the two registers synchronized and serves the required function of affecting the LED state.

A few additional elements should be noted here, that relate to a mixed C / assembly environment:

- The use of double colons ("::") declares the assembly labels.
- `Delay1mS` uses a C wrapper around an assembly function.
- The `DS1led` routine uses the HL register pair as a parameter to indicate the LED on / off status.
- Shadow registers are updated before their associated port registers.
- The "ioi" instruction prefix is used to access the CPU's internal registers.

6.6.2 Using #GLOBAL_INIT

When embedded programs start to run, the first thing they often do is to initialize various hardware devices and control registers. Dynamic C provides the "GLOBAL_INIT{}" function chain for this purpose. The following characteristics should be noted about GLOBAL_INIT{}:

- Code placed in the GLOBAL_INIT{} section is automatically run once when the program starts up.
- The function GLOBAL_INIT{} may be called explicitly at any time with the statement: _GLOBAL_INIT();. The programmer should be careful because invoking this function at other times will reset some critical Dynamic C system variables.
- Any number of #GLOBAL_INIT sections may be used in the code. The order in which they are called is indeterminate since it depends on the order in which they were compiled.

Since Program 6.11 uses the `initPort()` function to initialize various I/O ports, we can use the GLOBAL_INIT{} function as follows:

Program 6.12: Code fragment showing GLOBAL_INIT.

```
void main()
{

   #GLOBAL_INIT
   {
    initPort();                    // initialize port G only
   }

<code deleted for brevity>
```

Chapter 6

We will continue to use GLOBAL_INIT{} in other programs that follow. In later sections, we will use GLOBAL_INIT{} to also initialize variables and arrays.

6.6.3 Using Dynamic C to Create a Delay

Although this chapter is devoted to assembly language, it is prudent to mention that the programmer can use Dynamic C to create a delay. The "Delay1mS" routine in Program 6.11 can be replaced by C code as shown in Program 6.11:

Program 6.13A: Creating delay with Dynamic C.

```c
/***************************************************************

This routine creates a 1 millisecond delay.
Input parameters: number of milliseconds to delay.
Output parameters: none

***************************************************************/

void Delay1mS(int mSdelay)
{

    auto unsigned long t0;

    t0 = MS_TIMER;
    while (MS_TIMER < (t0 + mSdelay))
            /* do nothing */;

}
```

Dynamic C provides a number of system variables to keep time. Programmers can access these variables to find out elapsed time between events, such as how much time is taken to execute a section of code. Dynamic C automatically updates the following system variables:

- MS_TIMER: updated once every millisecond.
- SEC_TIMER: updated once every second.
- TICK_TIMER: updated 1024 times per second (the frequency of the periodic interrupt).

Since Dynamic C uses these variables for its own housekeeping, they should not be changed by the application program.

As shown in Program 6.13A, the programmer can utilize these system variables to create delay. However, in certain cases, the watchdog timer (covered in Chapter 7) may kick in and reset the system. Therefore, the watchdog should periodically be reset in the delay loop, as shown in Program 6.13B:

Introduction to Rabbit Assembly Language

Program 6.13B: Code segment showing delay loop with watchdog access.

```
/*************************************************************

This routine creates a one second delay.
Input parameters: number of seconds to delay.
Output parameters: none

*************************************************************/

Void Delay1Sec (long lDelay)

{

auto unsigned long lStart;

lStart = SEC_TIMER;
    while ((SEC_TIMER - lStart) < lDelay)
          hitwd ();

}
```

6.6.4 Final Thought

The main section in Program 6.11 assumes a duty cycle of 50% for turning the LED on and off. In the main section, the programmer can experiment with unequal parameters for the calls to the "Delay1mS" subroutine in order to change the duty cycle.

Additionally, the programmer can use a bi-color LED and write a routine that will take an input parameter and perform the following operations on the LED:

- Turn the LED on in Solid Red, Green or Orange
- Turn the LED off
- Flash the LED in Solid Red, Green or Orange

6.7 Project 3: Debouncing a Switch

One may think that reading a switch is a simple matter. However, closing or opening a switch creates lots of "noise" because switch contacts do not close or open once, but rather they bounce open and close for a while before finally settling into their new state. The switch must be debounced so that the system sees only a single state change.

6.7.1 Theory: Debouncing a Switch in Software

Debouncing a pushbutton switch involves the following steps: recognizing the switch press, waiting a bit (around 50 milliseconds is good) for all the "bounce noise" to go away, and waiting for the switch to be released.

Chapter 6

6.7.2 Implementation: Debouncing a Switch

Program 6.14 (`debounce1.c`) contains code fragments to illustrate the debounce principle. This program runs on the Rabbit 3200 prototyping board and uses switches S2 and S3 to toggle status of LED1 and LED2, respectively. Switch S2 is debounced using the above method, while switch S3 is not debounced at all. When the program is executed, there is substantial difference in how the two LEDs behave when the switches are pressed. LED1 toggles on and off cleanly and smoothly each time switch S2 is pressed, while LED2 seems to have a mind of its own.

Program 6.14: Code fragment for debouncing a switch.

```
/****************************************************************
    Debounce1.c

    Description
    ===========
    This assembly program uses a non-debounced switch and a
    debounced switch to toggle LED states on the prototyping board.
```

<code deleted for brevity>

```
//////////////////////////////////////////////////////////////
// void ledControl(int LED, int state)
// DS1 led on protoboard is controlled by port G bit 6
// DS2 led on protoboard is controlled by port G bit 7
// LED turns on if state = 0
// LED turns off if state = 1
// Input: LED, state
//////////////////////////////////////////////////////////////
#asm root
ledControl::
    // (sp+2) should contain LED ID - DS1 or DS2
    // (sp+4) should contain LED state - ON or OFF

        ld  hl,(sp+4)           ; get LED state
        ex  de,hl               ; store in DE

        ld  hl,(sp+2)           ; get LED ID

        ld     a, 1
        cp     DS1
        jr     NZ, control_DS2  ; are we working on DS1?
                                ; no, jump to control DS2

control_DS1:
// if (state == ON) then turn LED on
        ld     a,d              ; DE contains 0 for ON
        or     a,e
        jr     nz,turn_DS1_off  ; turn LED off if DE not 0
```
(Program 6.14 continued on next page)

Introduction to Rabbit Assembly Language

Program 6.14: Code fragment for debouncing a switch (continued).

```
//use shadow register to keep other bit values
        ld      a,(PGDRShadow)
        res     DS1,a              ; clear bit 6 only

        jp      ledControl_done    ; ready to write to port

turn_DS1_off:
//use shadow register to keep other bit values
        ld      a,(PGDRShadow)
        set     DS1,a              ; set bit 6 only
        jp      ledControl_done    ; ready to write to port


control_DS2:
// if (state == ON) then turn LED on
        ld      a,d                ; DE contains 0 for ON
        or      a,e
        jr      nz,turn_DS2_off    ; turn LED off if DE not 0

//use shadow register to keep other bit values
        ld      a,(PGDRShadow)
        res     DS2,a              ; clear bit 7 only

        jp      ledControl_done    ; ready to write to port

turn_DS2_off:
//use shadow register to keep other bit values
        ld      a,(PGDRShadow)
        set     DS2,a              ; set bit 7 only

ledControl_done:
        ld      (PGDRShadow),a     ; update shadow register
   ioi  ld      (PGDR), a          ; write data to port g

        ret
#endasm


////////////////////////////////////////////////////////
////////////////////////////////////////////////////////
void main()
{
   int LED1state, LED2state;

   initPort();                     // initialize port G only
   LED1state = LED2state = OFF;    // start with both LEDs off

   while (1)
```
(Program 6.14 continued on next page)

Chapter 6

Program 6.14: Code fragment for debouncing a switch (continued).

```
    {
            if (!BitRdPortI(PGDR, S2)) //wait for switch S2 press
            {
                // toggle LED state
                if (LED1state == ON)
                    LED1state = OFF;
                else
                    LED1state = ON;

                // wait for 50 milliseconds and then
                // wait for switch to be released
                delay1mS(50);
        while (!BitRdPortI(PGDR, S2));

                ledControl(DS1, LED1state);

            } //S2 pressed


            if (!BitRdPortI(PGDR, S3)) //wait for switch S3 press
            {
                // toggle LED state
                if (LED2state == ON)
                    LED2state = OFF;
                else
                    LED2state = ON;

                ledControl(DS2, LED2state);

            } //S3 pressed


    } //while

} //main
```

A few points to note with the above code:

- When the ledControl function did not work at first, the "`#asm`" line above it was replaced with "`#asm debug root`" so that the code could be debugged with breakpoints and single stepping.
- The C program passes parameters to the `ledControl` function through the stack. Note the order in which the called function retrieves the parameters from the stack.
- When called, function `ledControl`, will find the first parameter, DS2, in the HL register pair.

6.7.3 Starting to Build a Custom Library

Since we are putting together some useful sections of code, wouldn't it be nice to have a library available that can be used from any program? While ANSI C programmers can use the `#include` directive to use code from other modules, Dynamic C does not support the `#include` directive. Instead, Dynamic C allows the programmer to put together custom libraries and use them through the `#use` directive.

Here are some guidelines to follow for creating custom libraries:

- Most libraries needed by Dynamic C programs have a `#use` statement in the file `lib\default.h`.
- Any library that is to be used in a Dynamic C program must be listed in the file `LIB.DIR`, or another `*.DIR` file specified by the programmer.
- The programmer can specify a different `*.DIR` file in the `Compiler Options` dialog box to facilitate working on multiple projects.

Dynamic C also has the concept of "modules"—a Dynamic C library typically contains several modules, which provide Dynamic C with the names of functions and variables within a library that may be referenced by program files with the `#use` directive.

Modules organize the library contents in such a way as to allow for smaller code size in the compiled application that uses the library. Here are some guidelines for defining modules:

- A module has three parts: the key, the header, and the body.
- A module begins with its `BeginHeader` comment and continues until either the next `Begin-Header` comment or the end of the file is encountered. The `Begin-Header` comment must be preceded by a forward slash, 3 astericks and one space (/***). The `Begin-Header` comment must end with an asterick and a forward slash (*/).
- The "module key" is a list of function and data names separated by commas, and is contained within the first line of the module header.
- An shown in the Dynamic C user manual, the programmer can define "Function Description Headers" to provide on-line help messaged through Dynamic C.

Chapter 6

We will define a library called MYLIB.LIB and put two function from Program 6.14 in that library: `delay1mS` and `ledControl`. For example, the module header and function descriptor for the `delay1mS` function would look like:

Program 6.15: Module header and function descriptor for delay1mS in MYLIB.LIB.

```
/*** BeginHeader Delay1mS*/
void Delay1mS(int mSdelay);
/*** EndHeader */

/* START FUNCTION DESCRIPTION ****************************************
Delay1mS          <MyLib.LIB>

SYNTAX:           void Delay1mS ( int mSdelay );

DESCRIPTION:      This routine creates a 1 millisecond delay.

PARAMETER1:       number of milliseconds to delay

RETURN VALUE:     none

*****************************************************************/

void Delay1mS(int mSdelay)
{

<Body of function>

}
```

Since function `initPort` will likely change from program to program, depending on usage of the I/O ports, there is no point in putting that function in a library. We will follow the following steps in creating the library:

1. Cut the functions `delay1mS` and `ledControl` from Program 6.14 and place them in a file MYLIB.LIB. As described above, create the module headers for the two functions.

2. Create a folder called `customLibs` in the Dynamic C installation folder and place the custom library file MYLIB.LIB in this folder.

3. Look for the Dynamic C directory file called "LIB.DIR" in the Dynamic C installation folder. Create a copy of this directory file in the Dynamic C installation folder and rename it to MYLIBS.DIR. Edit this file and put a relative path to the above library file as "customLibs\MYLIB.LIB."

4. From Dynamic C's "Options" and then "Project Options" selections, select the "Compiler" tab and click on "Advanced." Click on "use" and point to the new directory file MYLIBS.DIR.

Introduction to Rabbit Assembly Language

5. At the top of the source file that needs to use this library, place the statement `#use mylib.lib`.

In this manner, any number of user libraries can be attached to the default library that is distributed with Dynamic C. We have provided a custom library file on the CD-ROM.

The source file should now be able to use the custom library. We'll soon use it in the next section.

6.7.4 A Hidden Trap in the Above Program

Once a switch press is detected, the code waits for the switch to be released. What if the switch is damaged and gets stuck in the "pressed" position? The code will wait forever for the switch to come out of that state. Programmers sometimes assume that devices will work as expected and write code that can get into a loop and not get itself out of trouble.

There are ways to work around this issue. The programmer can use timers to time out if the switch is not released within a period of time. Alternately, the program can use a large counter to determine how many times the loop has executed and exit after a predetermined number of loop iterations. Either of these approaches would solve the problem.

The program can use a more drastic approach to get out of trouble—a watchdog timer that will reset the program if the watch dog does not get "hit" within a predetermined period of time. We will cover the watchdog timer in detail in Chapter 7.

Such an oversight by a programmer can sometimes have disastrous consequences. One of the authors had to fix an elusive bug in a major mainframe—the computer system had multiple CPU boards and a cluster of these systems was running a high profile stock exchange. A switch inside one of the CPU boards got stuck in a certain position when the board was being inserted. The result? A disastrous event during which the CPU boards crashed, one after another, and the stock exchange was shut down for 11 minutes. Millions of dollars were lost because one part of code was stuck in a loop waiting for the switch to be released.

6.7.5 Final Thought

Instead of debouncing just one switch, the programmer can write a routine to scan a keypad matrix. The same principle applies—debounce a key after it is pressed, and pass along values to a higher-level routine.

6.8 Project 4: Driving a Multiplexed LED Display

First came the LED, then display technologies improved and seven segment displays became common. Soon, people started combining seven segment displays to be able to display more digits, but each digit required eight port pins to operate, including the decimal, of course. Then the embedded world realized that keeping all those displays lit all the time required too much power from portable devices. Thus, the multiplexed display was born.

Chapter 6

6.8.1 Theory: What is a Multiplexed LED Display?

While the common seven segment display requires eight port pins to operate, the multiplexed display required one port pin per segment, plus one pin per digit. So, to display three digits, one could either use three seven segment displays, requiring 24 port pins, or a multiplexed display that needs just 11 pins. This is shown below in Figure 6.5:

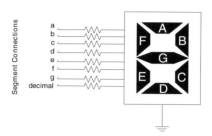

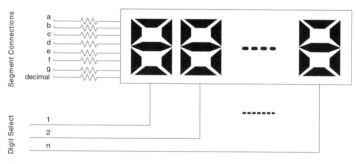

Figure 6.5: Seven segment single digit and multiplexed displays.

In order to use a multiplexed display, the microprocessor latches the relevant data on the segment lines, and enables the proper digit that has to display that data. The microprocessor has to keep switching between digits, faster than the human eye can tell. A characteristic of the human eye, called "retinal retention" (or visual persistence), tricks the human eye into thinking that all the digits are lit all the time, and the eye does not see the digits flickering as they are turned on and off in sequence.

Because not all the digits are lit at any one time, the multiplexed display has the added advantage of conserving power, as compared to the same number of digits being displayed with conventional seven segment displays. The advent of LCDs (Liquid Crystal Displays) has further reduced power consumption and has given programmers the additional ability to display graphics and download custom fonts into the displays.

Introduction to Rabbit Assembly Language

6.8.2 Implementation: How We Did It

We used the Fairchild MSQC4111C display, which has the connections shown in Figure 6.6:

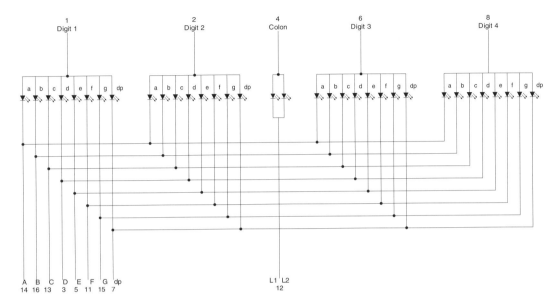

Figure 6.6: Connections for the Fairchild MSQC4111C multiplexed display.

As stated earlier, to make such a display work, the microprocessor has to set up the correct data on the segment lines, and light up the relevant display digit. The microprocessor can then turn that digit off and repeat the process for the next digit.

The display board receives its power from the J2 header on the RCM3200 prototyping board. Since parallel port "A" is straightforward to use, we decided to use it to provide data to the display's segment lines. Moreover, all bits of port A are available on the prototyping board connector J4, as opposed to those of ports G, D, and F. Using consecutive bits for port A to drive the display segments makes it easy to just write the right segment bits on port A; it would get more complicated if some of the segments had to be driven by a few bits of port G and a few from port D.

We used parallel port "C" to select the four display digits and to power the display colon that is commonly found blinking in clocks. Port C provides four input bits and four output bits, so the designer has to make sure the correct bits are chosen. Since we need to drive four display digits, the full output functionality of port C can be utilized. Using port C for parallel I/O eliminates the possibility of using serial ports A through D, so this design will have to be modified if those serial ports need to be used.

Chapter 6

We used commonly available PNP transistors to provide the current for each display digit. As shown in Figure 6.7, the display board uses two 34-pin ribbon cables to talk to the J2 and J4 headers on the RCM3200 prototyping board.

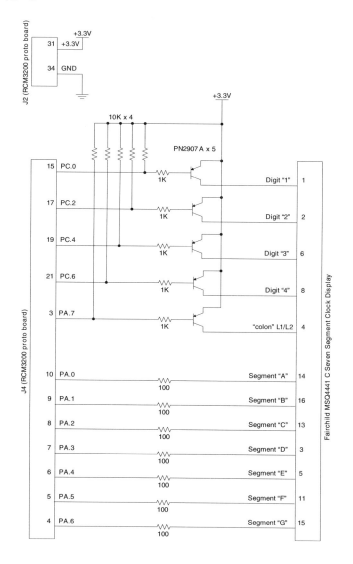

Figure 6.7: Seven segment multiplexed display wired to the RCM3200 prototyping board.

We used 100-ohm resistors to limit current to 10mA/segment, and a 1K resistor to limit base current on the PN2907As. We used 10K resistors to keep the I/O pin pulled up when the rabbit I/O pin is floating. Moreover, we removed the resistors that were used on the prototyping board to pull up the associated port pins.

All signals are active low. What this implies is when the rabbit powers up, we need to force the associated port pins high. Otherwise the display may come up in an unpredictable manner.

Introduction to Rabbit Assembly Language

Before using the display, we decided to write some test code to make sure that:
- Each digit and segment was wired correctly, per the schematic,
- Each digit and segment was addressable through the associated port pin, and
- We could measure the current drain per segment and per digit.

Program 6.16A (`testLED.c`) initializes ports and then turns on each segment and cycles through the digits. The following functions are used in this program:
- `Delay1mS`: uses Dynamic C's timer variables to induce a 1mS delay
- `display_off`: turns off the entire display by de-asserting all digits and segments
- `initialize_ports`: sets up ports A and C for outputs
- `turn_on_digit`: turns on a selected digit
- `turn_off_digit`: turns off a selected digit
- `cycle_digits`: turns on one digit at a time, with a two-second delay between digits
- `test_display`: calls the above functions to turn on one segment at a time, and then display that segment on each digit

The importance of writing a test program at first is to make sure the hardware is wired correctly. If the programmer jumps into more complex code without first writing such a test program, and things do not turn out as expected, it may take the programmer a long time to go back and check if everything was wired correctly. Also, functions developed and tested in writing such a test program can be used later. For instance, we will leverage most of the functions in other programs.

Program 6.16A: Testing the multiplexed display.

```
/*******************************************************************
    testLED.c

    This program tests the Multiplexed LED display.
    The program runs on the Rabbit 3200 prototyping board.

<code deleted for brevity>

/***************************************************************/

void main()
{
    initialize_ports();

    //printf(«Ports Initialized\n»);

    while (1)
    {
    display_off();
    test_display();
    Delay1mS(WAIT_TIME);
    }

}
```

Chapter 6

Now that we know the display has been wired correctly, and that we can access each segment and each digit, we can do something more useful. Simply turning on a segment is not enough... we need to map each hex digit that can be displayed to its related segment mapping, so that we know which segments to turn on for each digit. Using Table 6.22A, we can derive a digit-to-segment map as follows:

Table 6.22A: Digit-to-segment mapping.

Digit to Display	Segment "a"	Segment "b"	Segment "c"	Segment "d"	Segment "e"	Segment "f"	Segment "g"
0	1	1	1	1	1	1	0
1	0	1	1	0	0	0	0
2	1	1	0	1	1	0	1
3	1	1	1	1	0	0	1
4	0	1	1	0	0	1	1
5	1	0	1	1	0	1	1
6	1	0	1	1	1	1	1
7	1	1	1	0	0	0	0
8	1	1	1	1	1	1	1
9	1	1	1	0	0	1	1
A	0	0	0	0	0	0	0
B	0	0	0	0	0	0	0
C	0	0	0	0	0	0	0
D	0	0	0	0	0	0	0
E	0	0	0	0	0	0	0
F	0	0	0	0	0	0	0

For example, we need to turn on all segments except "e" and "f" in order to display digit "3." Note that for these projects, we do not need to display the hex digits "A" through "F," and for that reason, the display will be turned off. Moreover, although the table shows a "1" in the position where a segment has to be turned on, because the LED inputs are active low, we'll end up inverting the resultant values.

Since the LED inputs are active low, a logic "1" is realized with 0 volts.

The above table needs two adjustments:

- Examining Figure 6.7, we see that display segments "a" through "g" are connected to Port A bits 0 through 6 respectively. We have to redo the above table to accommodate the correct position of bits.

- Because the display is active low, we have to adjust the table accordingly.

The resultant table is shown in Table 6.22.B:

Table 6.22B: Two adjustments made to Table 6.22A.

Digit	Bit 6 "g"	Bit 5 "f"	Bit 4 "e"	Bit 3 "d"	Bit 2 "c"	Bit 1 "b"	Bit 0 "a"	Result	Inverted Result
0	0	1	1	1	1	1	1	0x3F	0x40
1	0	0	0	0	1	1	0	0x06	0x79
2	1	0	1	1	0	1	1	0x5B	0x24
3	1	0	0	1	1	1	1	0x4F	0x30
4	1	1	0	0	1	1	0	0x66	0x19
5	1	1	0	1	1	0	1	0x6D	0x12
6	1	1	1	1	1	0	1	0x7D	0x02
7	0	0	0	0	1	1	1	0x07	0x78
8	1	1	1	1	1	1	1	0x7F	0x00
9	1	1	0	0	1	1	1	0x67	0x18
A	0	0	0	0	0	0	0	0x00	0x7F
B	0	0	0	0	0	0	0	0x00	0x7F
C	0	0	0	0	0	0	0	0x00	0x7F
D	0	0	0	0	0	0	0	0x00	0x7F
E	0	0	0	0	0	0	0	0x00	0x7F
F	0	0	0	0	0	0	0	0x00	0x7F

We will use the table lookup code from Program 6.8B to map digits to their associated segments.

Program 6.16B (`displayLED1.c`) puts together code from the test program (6.16A), and adds segment mapping to cycle various digits through the LED display:

Program 6.16B: Digit-to-segment mapping and controlling each digit.

```
/**************************************************************
    displayLED1.c

    This program implements display segment mapping and digit control
    for the Multiplexed LED display.  The program runs on the Rabbit
    3200 prototyping board.

<code deleted for brevity>

void main()
{

    int i, display, counter;

    initialize_ports();
```
(Program 6.16B continued on next page)

Chapter 6

Program 6.16B: Digit-to-segment mapping and controlling each digit (continued).

```
        while (1)
        {
        display_off();
        Delay1mS(WAIT_TIME);

        for (i=0; i<10; i++)
        {

                display_digit (i, DISP_DIGIT_1);
                display_digit (i, DISP_DIGIT_2);
                display_digit (i, DISP_DIGIT_3);
                display_digit (i, DISP_DIGIT_4);

                Delay1mS(WAIT_TIME);
        } // for


        } // while

} // main
```

Since we are comfortable using the `t_lookup` function above, we will put into into the custom library `MYLIB.LIB`.

The next program (6.16C; `displayLED2.c`) puts together delay, switch debouncing, and display segment mapping to implement a dual-mode counter:

- The counter counts up or down each time switch S2 is pressed
- The counter reverses its behavior each time switch S3 is pressed. If the counter is counting up, it starts counting down, and vice versa
- The counter's range is 0 through 99. It wraps around to 0 from 99.

Program 6.16C runs in a loop and scans for keypresses. It refreshes the display every 20 mS to make sure the eye perceives all digits to be on all the time (visual persistence). If a button is pressed, an extra 10 mS is taken up to debounce the button. The program employs the following notable functions:

- `initialize_ports`: unlike the other initialization routines shown above, this one incorporates assembly instructions to illustrate how a programmer can access internal registers through assembly.
- `display_results`: displays two numbers; one on digits 1 and 2 and the other on digits 3 and 4. This routine will come in handy when we build a clock; the hours and minutes information will be passed into this routine as the two parameters.

Introduction to Rabbit Assembly Language

Program 6.16C: Putting it together: delay, switch debouncing, display segment mapping, and visual persistence with a multiplexed display.

```
/*******************************************************************
    displayLED2.c

    This program implements a counter for the LED display board:

    * The counter counts up or down each time switch S2 is pressed

    * The counter reverses its behavior time switch S3 is pressed.
      If the counter is counting up, it starts counting down, and
      vice versa

    * The counter's range is 0 through 99 and it is wrapped around
```

<code deleted for brevity>

```
********************************************************************/

void main()
{
    int counter, mode, colon_state, digit3, digit4;

    counter = 50;
    mode = COUNT_UP;
    initialize_ports();

    while (1)
    {
    // first save the current state of display colon
    colon_state = BitRdPortI(PADR, DISP_COLON);

    digit3 = counter/10;
    digit4 = counter - digit3*10;

    display_digit(digit3, DISP_DIGIT_3);
    delay1mS(10);
    turn_off_digit(DISP_DIGIT_3);

    display_digit(digit4, DISP_DIGIT_4);
    delay1mS(10);
    turn_off_digit(DISP_DIGIT_4);

    if (!BitRdPortI(PGDR, S2))        //wait for switch S2 press
        {
            // S2 changes the counter, depending on the mode

            if (mode == COUNT_UP)
                counter++;
```

(Program 6.16C continued on next page)

Chapter 6

Program 6.16C: Putting it together: delay, switch debouncing, display segment mapping, and visual persistence with a multiplexed display (continued).

```
                    else
                            counter--;

                    //if overflow happened, wrap counter
                    if (counter>99) counter=0;
                    if (counter<0) counter=99;

                    // wait for switch to be released
                    delay1mS(10);
            while (!BitRdPortI(PGDR, S2));

      } //S2 pressed


      if (!BitRdPortI(PGDR, S3))      //wait for switch S3 press
            {
                    // S3 flips the counter mode

                    if (mode == COUNT_UP)
                            mode = COUNT_DN;
                    else
                            mode = COUNT_UP;

                    // wait for switch to be released
                    delay1mS(10);
                    while (!BitRdPortI(PGDR, S3));

      } //S3 pressed

      // just for fun, invert the status of the colon
      if (colon_state)
            {
            colon_state = FALSE;
            BitWrPortI(PADR, &PADRShadow, BITRESET, DISP_COLON);
            }
      else
            {
            colon_state = TRUE;
            BitWrPortI(PADR, &PADRShadow, BITSET, DISP_COLON);
            }

      } //while

} // main
```

Introduction to Rabbit Assembly Language

6.8.3 Final Thought

While the above program achieves the desired goal, a good deal of the microprocessor's time is taken up just in delay loops, updating each display digit. Wouldn't it be nice if the microprocessor was busy with more important and time critical things such as fetching data, and then it came back once in a while to light up the display, as needed. In Chapter 7, when we start working with Interrupts, we will explore these ideas. Moreover, Dynamic C allows us to "costate" separate sections of code that will perform different things, and Dynamic C will keep track of distributing the microprocessor's time among separate functions. In Chapter 8, we will re-visit this project and use `costates` and `cofunctions` to do multiple things simultaneously: keep the display updated and scan for keypresses, while the processor is free to do other things.

6.9 Project 5: Setting Up a Real-time Clock

As the name states, a real-time clock keeps track of time. Depending on the application, a real-time clock can keep track of hours, minutes, seconds, and/or milliseconds, and perform am/pm and calendar functions as well. A number of semiconductor vendors have had real-time clocks available in chip form for decades, and some of these chips are sophisticated enough to provide leap year adjustment and alarm functions as well.

Since cost is usually a concern to systems designers, some applications have chosen to use a software-based clock instead of external chips. This project will implement a real-time clock using assembly language. In Chapter 7 we will implement a real-time clock using the Rabbit 3000's internal real-time clock.

6.9.1 Theory: What Needs to Happen for a Software-based Real-time Clock to Work?

It is quite easy to implement a real-time clock. First, we need to know the granularity and range of the real-time clock—for instance, does it have to measure hours, minutes and seconds or days (am/pm) and milliseconds as well? For this example, we will implement a clock that keeps track of hours, minutes and seconds.

6.9.2 Implementation: How We Did It

The listing is shown in Program 6.17 (`rabbitClock.c`). After initializing variables that keep track of hours, minutes and seconds, the processor simply updates the seconds counter once a second, and increments the minutes counter when the seconds counter overflows from 59 to 60. Similarly, the program increments the hours counter when the minutes counter overflows from 59 to 60. The overall loop is timed such that all of these operations happen just about once every second.

The clock also has a mode switch ("S2" on the RCM3200 prototyping board) that allows the user to select an hours / minutes or a minutes / seconds display. The switch can be pressed any time to change the display mode.

Program 6.17: Implementing a real-time clock.

```
/********************************************************************
    rabbitClock.c

    This program implements a software-based realtime clock that works
    with the LED display board:

    * The clock keeps track of hours and minutes

    * The clock can display hours and minutes or minutes and seconds

    * The clock is not precise because it uses software delay loops

<code deleted for brevity>

/*******************************************************************/

void main()
{
    int colon_state, counter, display_mode;

    static int hours, minutes, seconds;

    initialize_ports();

    hours = 12;
    minutes = 59;
    seconds = 40;

    display_mode = SHOW_HOURS;

    // first save the current state of display colon
    if (BitRdPortI(PADR, DISP_COLON))
        colon_state = COLON_OFF;
    else
```
(Program 6.17 continued on next page)

Introduction to Rabbit Assembly Language

Program 6.17: Implementing a real-time clock (continued).

```c
            colon_state = COLON_ON;


    while (1)
    {
    seconds += 1;

    if (seconds >59)
    {
            seconds = 0;
            minutes += 1;
    }

    if (minutes >59)
    {
            minutes = 0;
            hours += 1;
    }

    if (hours >12)
    {
            hours = 1;
    }


    // this loop is timed so that it takes about a second
    for (counter=0; counter<25; counter++)
    {

    if (display_mode == SHOW_HOURS)
       display_results (hours, minutes);   // about 40 mS taken here
    else
       display_results (minutes, seconds); // about 40 mS taken here


    if (!BitRdPortI(PGDR, S2))        //wait for switch S2 press
    {
            // toggle display state
            if (display_mode == SHOW_HOURS)
                display_mode = SHOW_SECS;
            else

                display_mode = SHOW_HOURS;

            // wait for switch to be released
            delay1mS(10);

            while (!BitRdPortI(PGDR, S2));
```

(Program 6.17 continued on next page)

Chapter 6

Program 6.17: Implementing a real-time clock (continued).

```
      } //S2 pressed


   } // for

   // flip colon state
   if (colon_state == COLON_ON)
   {
   BitWrPortI(PADR, &PADRShadow, BITSET, DISP_COLON);
      colon_state = COLON_OFF;
   }
   else
   {
   BitWrPortI(PADR, &PADRShadow, BITRESET, DISP_COLON);
      colon_state = COLON_ON;
   }


   } //while

} // main
```

6.9.3 Final Thought

Unless the delay loop is created with great precision, the clock will not be very accurate—each iteration of the delay loop will add an element of error to the clock. It is far better to implement a clock that uses the processor's timers to generate interrupts. Another way to yield a higher degree of precision is to use the Rabbit's internal real-time clock. Both of these designs are presented in the Interrupts chapter.

Additionally, it is easy to see from the code that every time switch S2 is pressed, the clock loses at least 10 milliseconds while it waits for the switch to be released.

The clock program can be modified to contain additional capabilities, such as:

- allow it to be set over the serial port,
- implement an alarm function that will flash an LED or flash the entire LED display, and more!

The last few projects utilized a number of common functions, for instance, for figuring out display segment mappings, for refreshing display digits, etc. Wouldn't it be nice to not have to paste every one of these functions in each program that needs them? This is where Dynamic C's library functions come in… we'll see in Chapter 8 how we can turn some of these functions into library functions so that they can be stored in just one place and can be accessed from any program.

CHAPTER 7

Interrupts Overview

The power of an embedded system is its ability to respond to real-world events; a system that is disconnected from the rest of the world is often not doing anything useful. We can consider everyday applications such as a microwave oven or a DVD player where an embedded system is deployed: in the former case, the smell of burning popcorn will prompt the user to hit the "stop" button, which should be able to turn off the microwave element. In the latter case, hearing the phone ring may prompt the user to use a remote control to pause the DVD player. Regardless of how busy the DVD player is in decoding images and audio from the disc, it has to be able to respond to user demands and take the appropriate action.

If we consider the above two examples, the devices are not much use if they cannot respond to real-time stimuli and take appropriate action, regardless of what they were doing at the time they received the input.

In industrial applications, an embedded system may monitor a machine and make appropriate decisions. However, in case of user input from a serial port or a critical condition identified by a sensor, the system may pause what it is currently doing, respond to the external event and make appropriate decisions, and then continue doing what it was doing earlier.

Another reason microprocessor-controlled systems use interrupts is because that is often the only way in a design to achieve high performance I/O. Such a design requires low-latency responses to external stimuli and the microprocessor simply cannot afford the time to finish what it is currently doing.

Most programs, whether written in assembly or "C," are highly linear[1]. Interrupts break the linearity of execution by temporarily halting normal code execution, jumping to other parts of code to handle critical events, and then resuming normal program flow from where the jump was made.

Embedded systems are often referred to as "event-driven" or "real-time" systems, because of their ability to respond to real-world events. It is also important to note that the real-time events are completely asynchronous to CPU timing and the CPU has to respond to these events regardless of where it is in program flow.

Take the example of a personal computer (PC), a common feature at home and work, which, over time, has learned to interact with humans. Today's PCs use mice, keyboards, touch

---

[1] Dynamic C provides "Costates" and "Cofunctions," which bring a flavor of "multitasking" in giving the programmer the ability to write nonlinear programs. We will work with Costates and Cofunctions in Chapter 8.

Chapter 7

screens, hard disk drives, and so forth. Each of these devices uses interrupts to signal real-time events to the processor. Without using interrupts, most PCs would simply not be able to respond to user inputs.

This chapter will discuss some of the important aspects of interrupts: what generates interrupts, what types of interrupts the CPU has to contend with, what happens when the CPU has to service an interrupt and what it does to handle an interrupt. We will look at a number of programming projects that highlight good practices for dealing with various types of interrupts on the Rabbit 3000 CPU.

Before proceeding, it is important to know the distinction between interrupts and polling. They are both useful techniques to keep track of real-world events and the programmer should be able to decide which of the two approaches to take and their associated pros and cons.

7.0.1 Polling vs. Interrupts

The behavior of an embedded system generally boils down to watching for some event, making a decision, and then performing some action. The events may originate in the external world or they may be generated within the embedded system itself. For example, a keypad may be used to accept human input. In another system, an internal timer may generate ticks every second. Both of these can be considered events that need watching.

In embedded systems, an input device (one that generates an event) can either be polled, or the system can be constructed so the event generates an interrupt.

A polled event is one in which the application code explicitly samples the input device and then executes specific code based on the state of that input. An interrupt event is one which suspends an application and forces the CPU to execute a special piece of code, called an interrupt service routine (ISR), after which the application is allowed to resume. Another key distinction between polling and interrupts is that polling happens under program control, while interrupts are completely asynchronous and can happen any time.

In a polled system, the application has to constantly check on the status of input devices. This is somewhat akin to a parent that is constantly running from room to room checking on the well being of children. While this might seem like a good idea, it more or less precludes the parent from getting much else done.

In an interrupt-driven system, the application can do whatever computational work it needs to, and if an input device requires service, an exception is generated and the CPU vectors off to service the ISR. This is akin to a parent performing some task—say, balancing the family checking account—and if a child needs attention, it yells. The parent stops the accounting, goes (vectors off) to see what event has occurred and, after settling the child's situation (servicing the interrupt), returns to balance the family's accounts.

Thus, an interrupt is a signal that indicates an event has occurred. The event could be an external one, or an internal one, such as one triggered by an onboard peripheral. Depending on how the code is written and what the processor is doing at the time, it can choose to ignore the interrupt. On the other hand, if it chooses to accept the interrupt, it suspends what it is doing, processes the interrupt, and returns to what it was doing when the interrupt occurred.

Much like screaming children, car alarms, telephones and smoke detectors, interrupts in embedded systems must be prioritized. At first it may seem that people run as exception driven devices. Embedded systems can be constructed to be the same. However, people have a lot more computational horsepower than an 8-bit controller. Care must be taken when designing exception-driven embedded systems. ISRs are notoriously difficult to debug, especially if interrupts are nested (meaning that an ISR has to be serviced while the code is already servicing another interrupt). Stacks can get out of hand. Variables can become corrupted. This is not an admonishment against such designs, only a caution.

While a parent running from room to room would quickly tire, an 8-bit CPU will not. Consider the design of an electronic thermostat. Let's say there is a keyboard that allows a human to change the temperature, and an LCD upon which the current temperature and the desired set-point are displayed. There will also need to be some temperature sensor and relay for turning on and off a heating element and fan.

This system could be designed to allow the keyboard to generate an interrupt when a key is pressed. The ISR would have to read the key and update either a "key buffer" or perhaps await a complete keyboard transaction and update the LCD and set-point.

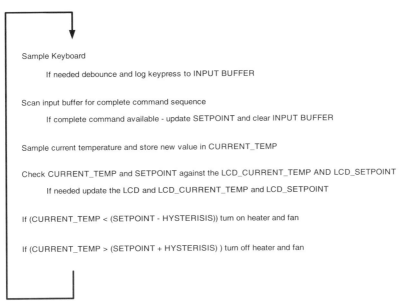

Figure 7.1: Flow chart of a polled heater thermostat.

Similarly, the system could be designed so the processor polls the keyboard. Figure 7.1 shows how this might be implemented. Both implementations might look straightforward, but consider what happens when a key is stuck closed.

Chapter 7

If the interrupt-driven system is designed such that an interrupt is always present at the CPU, then the ISR will be repeatedly executed. Depending on the system design, this may preclude the rest of the application from executing. If the heating element happens to be on, this means that the possibility exists that the embedded controller may never turn off the heating element. This may be a safety concern.

Likewise, in the polled version, the "debounce and log keypress" section should be designed with an intrinsic time out (discussed in Chapter 6). If this is done then the application will continue through the main loop and the worst a stuck key can do is to prevent a human from changing the set–point to a desirable temperature. Since the main loop is still executing, the heater will be turned on and off to maintain the last valid set-point contained in the SET-POINT variable.

Interrupts are powerful tools but, as can be seen from this example, they can be a little tricky to set up and use. There are both hardware concerns and software concerns. Systems should be designed so that "stuck" interrupts can be managed gracefully.

For example, if this interrupt-driven system were designed so that a piece of hardware only generated a momentary interrupt on a stuck key then the problem of a stuck interrupt consuming all the CPU cycles would be eliminated. Likewise, if the ISR were written to detect a stuck key, some defensive fail-safe strategy might be contrived. For example the offending interrupt might be masked—although this leads to issues of how, when and why might the interrupt be re-enabled.

There are many seasoned embedded engineers that advocate designing embedded systems as polled systems using a round-robin main loop to sample the various input devices. Sometimes interrupts are clearly a better choice than polling. For example, in power sensitive applications, running the processor just to sample a device like a keypad is a huge waste of power, if nothing else requires the CPU's attention. It would be far better to place the CPU in a low power mode and allow an interrupt to wake up the CPU to service a key press. Other widely accepted uses for interrupts include sending and receiving serial data and to work with timers that generate periodic events; this chapter contains various examples in these areas.

7.1 Interrupt Details

The embedded systems designer has to know the ability of the microprocessor to handle interrupts. What number of external interrupts are supported that the designer can tie to external logic? Which of these interrupts will always be handled, so that critical events can be connected to these inputs. The programmer has to know which internal and external interrupt sources exist, which ones have priority over others, which can be enabled by program control, and which ones are always enabled. This section will describe these ideas in general, and in particular how these are implemented in the Rabbit 3000.

7.1.1 Interrupt Sources

Interrupts can be generated from many sources, internal and external. In some cases, a microprocessor needs to respond to external events such as a warning from a sensor. In other cases, the microprocessor responds to internally generated interrupts such as periodic ticks from

Interrupts Overview

a timer, or notification from a serial port that a byte has been transmitted ("transmit buffer empty") or received ("receive buffer full").

While most of the Rabbit interrupts are internal, the processor supports two external interrupts, 0 and 1. Hardware can be connected to these two pins to trigger the processor to respond to external stimuli. These two interrupts are edge sensitive and not level sensitive; they can respond to rising, falling, or either edge. By this design, once the interrupt is triggered, then the interrupt line cleared and then reasserted, then the CPU would consider another interrupt to have occurred.

Both the Input Capture and the Quadrature Decoder are onboard peripherals like the UARTs and are considered internal interrupt sources. The commonly-used interrupt sources for the Rabbit 3000 and their associated ISR addresses are shown in Table 7.1.

Table 7.1: Commonly used Rabbit interrupt sources[2].

Interrupt Source	Interrupt Number	Associated Interrupt Register	ISR Starting Address
Input Capture	0x1A	ICCR (0x57)	{IIR[7:1], 1, 0xA0}
Quadrature Decoder	0x19	QDCR (0x91)	{IIR[7:1], 1, 0x90}
External Interrupt0	0x00	I0CR (0x98)	{EIR[7:0], 0x00}
External Interrupt1	0x01	I1CR (0x99)	{EIR[7:0], 0x10}
Timer A	0x0A	TACSR (0xA0)	{IIR[7:1], 0, 0xA0}
Timer B	0x0B	TBCSR (0xB0)	{IIR[7:1], 0, 0xB0}
Serial Port A	0x0C	SASR (0xC3); SACR (0xC4)	{IIR[7:1], 0, 0xC0}
Serial Port B	0x0D	SBSR (0xD3); SBCR (0xD4)	{IIR[7:1], 0, 0xD0}
Serial Port C	0x0E	SCSR (0xE3); SCCR (0xE4)	{IIR[7:1], 0, 0xE0}
Serial Port D	0x0F	SDSR (0xF3); SDCR (0xF4)	{IIR[7:1], 0, 0xF0}
Serial Port E	0x1C	SESR (0xCB); SECR (0xCC)	{IIR[7:1], 1, 0xC0}
Serial Port F	0x1D	SFSR (0xDB); SFCR (0xDC)	{IIR[7:1], 1, 0xD0}

The last column of the table shows whether the EIR or the IIR register needs to be set up for the particular ISR. The last few bits are the offset into the interrupt vector table that stores pointers to ISRs. More information on this subject is available on the enclosed CD.

7.1.2 Multiple Interrupt Sources

Depending on the hardware design, multiple sources may trigger the same interrupt. While the Rabbit 3000 has two external interrupts, these lines can be shared among multiple devices. The programmer has to configure the appropriate Interrupt Control Register (I0CR or I1CR) to trigger on the appropriate edge of the input.

---

[2] A couple of interrupts are not shown in the table, because they are not of much use to the programmer. These include the "system management" and "slave port" interrupts.

Such an example is shown in Figure 7.2. Once an interrupt happens, the state of the interrupt lines is determined by reading Parallel Port E.

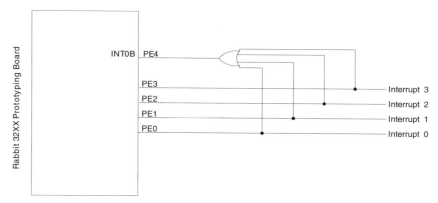

Figure 7.2: Working with multiple interrupt sources.

In order for the ISR to determine which device caused the interrupt, the level following the edge has to be long enough for the ISR to get started and sample the inputs, but shorter than the ISR execution time. There must also be an external arbitrator, otherwise it becomes tricky to handle them.

Another trick to use with the Rabbit is to use the two input capture channels as external interrupts. They have sixteen inputs and the designer has to choose one of sixteen. Round Robin sampling can be utilized to capture the edges.

There are three considerations to such a design:

- Since this scheme will work only with low-to-high transitions, the designer has to select interrupt sources that support such outputs. Rabbit 3000 supports rising, falling, or both edges for external interrupts, as well as for the input capture start/stop conditions[3].
- If one of the inputs rises and then falls, the software has to be able to catch the state of the attention request line before it changes. This scheme may not work very well for very narrow pulses that are beyond the latency of the ISR.
- If one of the Interrupt lines goes high and stays high, none of the other interrupt inputs will be able to trigger an interrupt.

Level sensitive interrupts can pose a challenge—consider a FIFO with a low-active request that is read in the ISR. Reading the FIFO will cause the line to go back high. If another byte is loaded on the other side at the same time, the line will immediately go low again. The Rabbit may miss the fast low-high-low transition because there is a filter at the input pin.

This can be solved by having the ISR read the status of the interrupt pin. It reads the FIFO until it sees the pin stay high. The time it takes to read the pin is long enough for the input filter to clear. An example is shown below in Program 7.1:

---

[3] For Input Capture, this is selected through the ICTxR register.

Interrupts Overview

Program 7.1: Code fragment for using level sensitive interrupts.

```
void interrupt FIFO_ISR()
{
    while ((RdPortI(PEDR)&0x2)==0) // repeat read of fifo
    {
          RdPortE(FIFO_DATA);     // will clear interrupt

// repeat loop to handle case where request reassert
// within 3 clocks of the above read.

    }
}
```

Although some of the ISRs presented here are in C, we recommend that ISRs be written in assembly language and not contain infinite loops.

7.1.3 Software Interrupts

Software interrupts are also called traps; they are a unique type of instruction and a well-accepted misnomer. While interrupts happen asynchronously, software interrupts are a programmed event.

Software interrupts on the Rabbit do not operate the same as interrupts, and the following differences should be observed:

- On the Rabbit processor, a software interrupt does not push the Interrupt Priority Level (IPL—see Section 7.1.7) on the priority stack. Some other processors do push the processor priority onto the stack for this type of instruction; that would be equivalent to a "real" software interrupt. The IPL is pushed on the priority stack (not the processor stack) by an internal or external interrupt. The priority stack is an 8-bit shift register and the IPL gets shifted into the IP register. The IPSET and IPRES instructions work on that register, while software interrupts do not work on that register.

- The code must return from a software interrupt with a "RET" instruction. The main reason for using an RST (one byte) is that it is more efficient than a CALL (three bytes). However, the RST instructions do use the internal interrupt vector table as the source for their associated functions. This may be why they are often referred to as "software interrupts." While software interrupts can call the same ISRs used by other interrupts, other ISRs may return control through the "RETI" instruction, which would create an issue since an RETI may pop the wrong number of items off the stack. In such a case, the programmer has to be careful in pushing the right number of elements on the stack.

Software interrupts can be generated by the program and can be thought of as high-priority function calls that can be invoked anytime by the program. In Rabbit terminology, software interrupts are referred to as "RSTxxx" or "resets."

Chapter 7

By issuing a "reset" instruction, the code can directly transfer control to an associated reset handler. A "RET" instruction at the end of the reset handler gives control back to the main program.

A software interrupt is simply an instruction that passes control to its associated handler. Dynamic C provides the reset instructions shown in Table 7.2:

Table 7.2: Rabbit reset instructions.

Interrupt	Purpose	ISR Starting Address
RST 10	Not used by Dynamic C	{IIR[7:0], 0x20}
RST 18	Used to enable Dynamic C debugging support	{IIR[7:0], 0x30}
RST 20	Used to enable Dynamic C debugging support	{IIR[7:0], 0x40}
RST 28	Used to enable Dynamic C debugging support	{IIR[7:0], 0x50}
RST 38	Not used by Dynamic C	{IIR[7:0], 0x70}

Dynamic C uses the reset instructions for debugging and programmers are advised to use them carefully. The ISRs for software resets can be set up just as ISRs for internal and external interrupts. The `SetVectIntern` calls can be used for this purpose, as shown in Section 7.1.4.

Just like the other interrupts, reset interrupts cannot be inhibited.

7.1.4 Interrupt Vectors

The interrupt vector defines the address of the ISR for each interrupt that happens. As shown in Table 7.1, the upper 7 or 8 bits of the address comes from a special register (IIR for internal and EIR for external interrupts) while the remaining bits are fixed for each interrupt.

The following Dynamic C functions exist to get information about interrupt vectors. These functions read the address of external and internal interrupt table entries, respectively:

- `GetVectExtern3000 (int interruptNum);`
- `GetVectIntern (int interruptNum);`

The "`interruptNum`" value can be found in Table 7.1.

The following Dynamic C functions are used to set up interrupt vectors for external and internal Interrupt Service Routines (ISRs), respectively:

- `SetVectExtern3000 (int interruptNum, void *isr);`
- `SetVectIntern (int interruptNum, void *isr);`

The second parameter, "`*isr`," is the ISR handler address, which must be in root memory.

Program 7.2A shows how the interrupt vector is set up in C program, taking into consideration whether we are operating with a separate instruction and data space.

Interrupts Overview

Program 7.2A: Setting up an interrupt vector.

```
// bind ISR to external interrupt0,
// taking into account the possibility of separate instruction and data
space
#if __SEPARATE_INST_DATA__ && FAST_INTERRUPT
        interrupt_vector ext0_intvec my_isr0;
#else
        SetVectExtern3000(0, my_isr0);
#endif
```

The ISR can be bound to the interrupt table with an alternate method that deals directly with the interrupt table, as shown in Program 7.2B:

Program 7.2B: Setting up interrupt vectors by directly modifying the interrupt table.

```
void setupTimerBisr()
{
#asm
    ld      a, iir              ; get internal interrupt register
    ld      h, a                ; copy to high nibble of HL
    ld      l, 0B0h             ; Timer B interrupt IIR address offset
    ld      iy, hl              ; IY now points to interrupt table for Timer B
    ld      hl, timerb_isr      ; Load address of ISR
    ld      (iy), 0C3h          ; Put JUMP command into interrupt table
    ld      (iy+1), hl          ; Follow JUMP command with address of the ISR
#endasm
}
```

7.1.5 Enabling and Disabling Interrupts

There are two types of interrupts: maskable and nonmaskable:

- Maskable interrupts can be enabled or disabled by the program. If an interrupt is triggered while it is masked, it becomes a pending interrupt and is serviced when the corresponding interrupt is un-inhibited. Programmers can ignore pending interrupts by clearing interrupts as soon as they are un-inhibited. Lower-priority interrupts are effectively inhibited until the IP is restored.
- Nonmaskable interrupts cannot be inhibited by the CPU and are always serviced.

The current Rabbit processors do not have a nonmaskable interrupt. Once the IPL is set to three then no interrupts can occur. In this case, the processor can only be interrupted by a RST instruction. Moreover, ISRs do not automatically disable interrupts upon entry; this must be done explicitly by executing an IPSET 3 instruction at the start of an ISR.

7.1.6 Interrupt Latency

Interrupt latency refers to the time required for an interrupt to take place after it has been requested. The latency is nondeterministic and depends on several factors:

- What instruction the processor is executing at the time the interrupt is requested. Clearing an internal register takes a different amount of time than a 16-bit indexed load or store.
- Latency is of course much higher if interrupts are pending because the interrupts are inhibited or the processor is already servicing an interrupt at the same or higher priority.

Programmers can use several approaches to reduce latency:

- Instead of following the common practice of inhibiting interrupts at the beginning of each ISR, leave the interrupts uninhibited. Interrupts with priorities equal to or less than the current interrupt priority are automatically inhibited. This will allow other ISRs to be launched in case more time critical interrupts need to be serviced. This complicates the code and the code has to be written carefully to account for the additional stack space taken by multiple ISRs and the shared resources the ISRs may have to use.
- Assign multiple priorities to various interrupts, and do not inhibit any interrupts. In this case, lower-priority ISRs will automatically give way to higher-priority interrupts that need to be serviced.

It is good programming practice to know the execution time for each ISR so that the programmer can guarantee the maximum interrupt latency. The Rabbit generally executes at the same speed from RAM or flash. One exception is the RCM3200 which uses relatively slow flash and fast RAM.

As a rule of thumb, designers should allow 100 µs for interrupt latency. This can allow for five active interrupt routines, and each disabling interrupts for at most 20 µs. The maximum interrupt latency in the Rabbit, assuming that the interrupt is not being blocked, is 29 clock cycles.

7.1.7 Interrupt Priorities and the IP Register

The interrupt levels are kept in the 8-bit IP register, which works as a 4-level stack for 2-bit values (0 through 3). The current interrupt level is located in the lowest two bits and the CPU keeps track of the three previous interrupt levels.

The IP register is modified each time an interrupt happens, or an IPSET or IPRES instruction is executed. The IPSET instruction sets up the interrupt level (priority) while the IPRES instruction restores the previous interrupt level.

When an interrupt occurs (or IPSET n is executed), the IP register is shifted left two bits, and the new interrupt priority is placed in the lowest two bits. When an IPRES is executed, the IP register is rotated right two bits (the lowest two bits become the highest two bits).

Since the priority stack holds only the current and 3 previous priorities, instructions are provided to push and pop the IP register on the regular stack. PUSH IP and POP IP simply

Interrupts Overview

push/pop the IP register onto the main stack. Note that this is a byte-wide push/pop, unlike all the other PUSH/POP instructions which are 2 bytes wide.

The RETI instruction combines POP IP and RET. It pops the IP register off the main stack, pops the return address off the main stack, then goes there. This instruction is not generally useful, since it does not restore the previous interrupt level.

> Two reasons why programmers should not use the RETI instruction at the end of an ISR:
> - The instruction does not restore the previous interrupt level.
> - The instruction pops a byte from the stack, which can create an inconsistent stack.

If the code needs to do a "POP IP" and "RET," the programmer should use "RETI" instead. The programmer should still restore the original priority because RETI is used in the debug kernel. For this reason, programmers are advised not to use the RETI instruction.

If an interrupt is requested, and the priority of the interrupt is higher than what the processor is currently working on, the interrupt will take place after the execution of the current assembly instruction is complete. The following privileged instructions do not allow an interrupt to occur after them: IPSET, IPRES, RETI, POP IP.

Interrupt requests of equal or lower priority are masked when an interrupt occurs. A programmer can manually change the masking by modifying the IPL, but it needs to be done with careful thought, since it can affect Dynamic C's timekeeping and other internal functions.

One reason to modify the IPL is when an ISR is expected to take such a long time that the system can miss arriving interrupts. In such a case, once the ISR has cleared the cause of the interrupt and the critical section of the ISR is complete, the programmer can re-enable interrupts by performing an IPRES.

For example, let us assume we entered an ISR at interrupt level 1. When the processor starts executing the ISR, interrupts of the same and lower priority will be disabled. However, early on in the ISR, an IPRES instruction would put processor priority down to 0, which means that other interrupts will be able to occur. This is typically not done but is shown here as an example.

Rabbit 3000 uses the interrupt priorities shown in Table 7.3:

Table 7.3: Rabbit interrupt priorities.

Priority	Effect on Interrupts
0	(LOWEST). All interrupts (priority 1, 2 and 3) take place after execution of current nonprivileged instruction
1	Only interrupts of priority 2 and 3 take place
2	Only interrupts of priority 3 take place
3	(HIGHEST). All interrupt are suppressed (except the RST instructions)

Here is another example:

1. We start with a cleared IP register: IP = 0x00 = (binary) 00:00:00:00. The least significant two bits {0,0} form the current priority

2. A level-1 interrupt occurs. IP is shifted left two bits and the new priority is appended. The IP register looks like IP = 0x01 = 00:00:00:01. The currently priority is thus level 1

3. That ISR is interrupted by a level-2 interrupt. IP is again shifted left two bits and the new priority appended. The IP register now looks like IP = 0x06 = 00:00:01:10. We are currently at priority level 2

4. The level-2 ISR finishes with an IPRES and RET. The IP is rotated two bits to the right. The IP register is now IP = 0x81 = 10:00:00:01, so we are back to priority level 1

5. The level-1 ISR finishes with an IPRES and RET. The IP is again rotated two bits to the right. The IP register now looks like IP = 0x60 = 01:10:00:00; we are back to priority level 0.

If there is concern for overflowing the IP stack (i.e., if an ISR reduces the priority level after handling some particularly critical portion of code), then PUSH IP should occur at the start of the ISR and POP IP should occur right before the IPRES.

> Because of Dynamic C's use of interrupts, programmers should consider the following aspects of using interrupts in their designs:
>
> - Most interrupting devices should use priority 1 interrupts. Devices that need extremely fast response to interrupts should use priority levels 2 or 3.
>
> - The processor priority should not be raised above 1, except after considering its impact on Dynamic C's internals. For example, Dynamic C's debug communications uses priority level 1 interrupts.

By default, an ISR can only be interrupted by an interrupt of higher priority, unless the programmer explicitly lowers the interrupt priority within the ISR.

7.2 Writing an Interrupt Service Routine

This section will first highlight the key things a programmer must do in an ISR, and will then discuss the implications of ISRs written in C or assembly. While the programmer can write an ISR in either C or assembly, there are trade-offs—this section will compare and contrast the two approaches.

7.2.1 What Needs to Happen in the Interrupt Service Routine

The rule of thumb about ISRs is "do no harm," meaning that the ISR should not adversely affect the normal operation of the CPU and the program it is executing. The programmer should follow these general guidelines in writing ISRs:

1. Upon entry, the ISR must save all the registers that it will use—usually the registers are pushed on the stack. Interrupt routines written in C save all registers automatically and assembly routines must explicitly save the registers.

2. When entering an ISR, the processor automatically masks interrupts of the same or lower priority. Only a higher priority interrupt will interrupt the ISR being executed.

3. The ISR may need to determine what caused the interrupt. In the case of internal interrupts, the ISR will interrogate the relevant on-chip registers to find out why the interrupt happened. In the case of external interrupts, the ISR may look at on-chip or external registers to determine what caused the interrupt.

4. So that the same interrupt does not keep triggering the ISR, the ISR must remove the cause of the interrupt. This is often done easily by clearing an interrupt flag in the appropriate register. If the interrupt has more than one probable cause, the ISR must check for all possible causes and remove all the causes at the same time.

5. On exiting the ISR, the code must restore the interrupt priority level so that other interrupts can get the attention of the CPU. ISRs written in C restore the interrupt priority level automatically when the function returns. However, assembly ISRs must restore the interrupt priority level explicitly with the IPRES instruction.

6. If interrupts were disabled earlier, the ISR must re-enable them; otherwise there will be no coming back to the ISRs. Several, but not all, of the Rabbit peripherals[4] must be re-enabled within the ISR.

7. Once the body of the ISR code has been executed, the ISR must restore the registers it saved earlier. C-based ISRs perform this automatically upon exiting the ISR, while assembly routines must explicitly restore registers.

8. Assembly-based ISRs must exit with a "RET" instruction instead of the familiar "RETI" instruction that programmers generally use to return from an ISR. For reasons discussed earlier, the RETI instruction is hardly ever used to return from an ISR. Dynamic C always uses the sequence IPRES followed by RET. RETI is virtually useless in the Rabbit architecture because it attempts to pop the interrupt priority from the normal stack—not the IP stack. RETI does save a couple of bytes of code and the curious programmer will notice that it is used in a few Dynamic C libraries.

---

[4] Timer B is one of these peripherals.

Chapter 7

7.2.2 Declaring an Interrupt Service Routine in C

Dynamic C requires an ISR written in C to be declared with the keyword "interrupt." A shell of a C-based ISR looks like the following:

Program 7.3: Code fragment for a C-based ISR.

```
// Set equal to 1 to use fast interrupt in I & D space
#define FAST_INTERRUPT 0

// define ISR prototype
void my_isr0();

/****** main ******/
void main()
{

// bind ISR to external interrupt0,
// taking into account the possibility of separate instruction and data
// space
#if __SEPARATE_INST_DATA__ && FAST_INTERRUPT
        interrupt_vector ext0_intvec my_isr0;
#else
        SetVectExtern3000(0, my_isr0);
#endif

// insert code here to enable external interrupt0

} // main


/****** Interrupt Service Routine  ******/

nodebug root interrupt void my_isr0()
{
   // ISR code goes here
} // my_isr0
```

It is perfectly permissible to have ISRs in xmem and do long jumps to them from the vector table. It is even possible to place the entire body of the ISR in the vector table if it is 16 byte long or less, but the `SetVectIntern` and `SetVectExtern3000` functions only set up jumps to 16 bit addresses in logical memory.

A complete C-based ISR is shown in Program 7.15.

There are a few recommendations for writing ISRs:

- For best success, ISRs should be written in assembly language. One of the issues with writing ISRs in C is that if the ISRs take too long to execute, Dynamic C can lose communication with the processor (which is also interrupt-based), and this will generate errors with the IDE. Time critical ISRs should always be written in

Interrupts Overview

assembly. A "rule of thumb" is that Dynamic C ISRs take about 10 times the execution time as those written in assembly language.

- Since printfs take a relatively long time to execute, programmers should not use printfs in ISRs.
- Because loops can take a relatively long time to execute, programmers should avoid using loops in ISRs.

7.2.3 Handling Interrupts in C

The advantage of using C-based ISRs is that upon entering and exiting the ISR, C does some of the grunt work for the programmer: saving registers upon entry and restoring registers upon exiting the ISR. The cost is added latency.

ISRs written in C restore the interrupt priority level automatically when the function returns. However, stand-alone assembly ISRs must restore the interrupt priority level explicitly with the IPRES instruction.

Traditionally, programmers have written ISRs in assembly, based on the execution speed and efficiency that assembly provides. The added latency due to a C-based ISR may not be an issue for that ISR, but it may become significant if many other ISRs have to be serviced as well.

A code fragment for a C-based ISR is shown below:

Program 7.4: Code fragment showing a C-based ISR.

```
nodebug root interrupt void timerb_isr()
{
    RdPortI(TBCSR);            //clear the interrupt


    /*************************************************************
    Body of ISR goes here
    *************************************************************/

    // set match registers to re-enable Timer B interrupts
    WrPortI(TBM1R, NULL, timerb_match_msb);
    WrPortI(TBL1R, NULL, timerb_match_lsb);

} // Timer B ISR
```

Note the use of the "`nodebug root interrupt`" declaration. It tells the compiler to treat this segment of code as an ISR in root space, and to not insert special debug code between C statements. If the programmer wishes to use breakpoints and single stepping, the "`nodebug`" should be changed to "`debug`."

Using C calls in an ISR requires saving all registers first. It is impossible to determine the register usage of the C statement.

Chapter 7

7.2.4 Handling Interrupts in Assembly

An ISR written in assembly doesn't have the luxury (or overhead) of Dynamic C doing all the housekeeping upon entering and exiting that is available with a C-based ISR. The programmer has to explicitly implement the following operations in assembly-based ISRs:

- Save and restore CPU registers. This is usually done by using the stack. Needless to say, the programmer has to be careful to match the order in which registers are pushed and popped, and that the same number of registers is popped as were pushed. While this may seem like a lot of work, only the registers that are used in the ISR need to be saved and restored, which reduces the interrupt handler's execution time.
- Disable interrupts (if desired).
- Clear the interrupting condition if it is not automatically cleared.
- Re-enable interrupts before exiting the ISR (if disabled).
- Restore the interrupt priority level.
- Along with saving registers on the stack, programmers can use exchange instructions to invoke alternate registers in ISRs, but the alternate registers need to be saved first as well.

The following code shows how an assembly-based ISR can be written.

Program 7.5A: Code fragment for an assembly-based ISR.

```
; External interrupt Routine #0
#asm nodebug
my_isr::

    PUSH AF            ; save registers that will be used in the ISR
                       ; disable the interrupt, if necessary

    /*************************************************************
    Body of ISR goes here
    *************************************************************/

                       ; re-enable the interrupt, if necessary
    POP AF             ; restore registers
    IPRES              ; restore interrupt priority
    RET                ; return from interrupt - see statements above about
                       ; RETI
#endasm
```

The typical method is to use C to set up the interrupt vector, and use standalone assembly to develop the entire ISR. An example is shown in program 7.5B.

Program 7.5B: Code fragment showing stand-alone assembly-based ISRs.

```c
void timera_isr();
void timerb_isr();



void main()
{

   #if __SEPARATE_INST_DATA__
   interrupt_vector timera_intvec timera_isr;
   interrupt_vector timerb_intvec timerb_isr;
   #else
   SetVectIntern(0x0A, timera_isr);     // Timer A int vector
   SetVectIntern(0x0B, timerb_isr);     // Timer B int vector
   #endif

   // initialize Timer(s)

   // Enable interrupt(s)

   // Stay busy with other useful stuff

} // main



#asm root nodebug
timera_isr::

   // save registers to be used

   // clear the interrupt

   /****************************************************************
   Body of Timer A ISR goes here
   ****************************************************************/

   // re-enable interrupts

   // restore registers

ret                        ; end of Timer A ISR


timerb_isr::

   // save registers to be used

   // clear the interrupt
```
(Program 7.5B continued on next page)

Program 7.5B: Code fragment showing stand-alone assembly-based ISRs (continued).

```
    /***********************************************************
    Body of Timer B ISR goes here
    ***********************************************************/

    // set up match registers

    // re-enable interrupts

    // restore registers

ret                     ; end of Timer B ISR
#endasm
```

The program first binds the ISRs to their respective interrupts after determining whether separate instruction and data spaces are being used. It then sets up the appropriate interrupt vectors. The program shows, at a high level, what else needs to be done to set up the timers, enable interrupts, and service the interrupts in the appropriate ISRs.

One of the advantages of using assembly-based ISRs is performance—there is no latency involved that is caused by the additional housekeeping done by C-based ISRs. The assembly-based ISRs do not get away from their own housekeeping, as shown before and after the body of each ISR.

> Note the use of double colons in declaring the two ISRs ("`timerb_isr::`" and "`timera_isr::`"). This is Dynamic C's standard method for calling assembly routines from C. The double colon construction declares that the label is of global scope.

A complete assembly-based ISR is shown in Program 7.14.

7.2.5 Handling Multiple Interrupt Sources

Sometimes multiple interrupt sources are shared by a peripheral and, after an interrupt happens, the ISR needs to find out which of the various sources caused the interrupt. As an example, a number of 8-bit microprocessors provide for a serial interrupt, shared by both the receive and transmit interrupts. After the program gets to the serial ISR, the program queries a status register to determine whether a "received data" or "transmitted data" operation caused the interrupt. The former happens when the receive buffer is full while the latter happens when the transmit buffer is empty.

Interrupts Overview

The following code fragment shows how the Rabbit serial ISR differentiates between transmit and receive interrupts:

Program 7.6: Code fragment for handling multiple interrupt sources.

```
#asm root

serial_ISR::
    push    af                  ; save any registers used by the ISR
    push    bc
    push    de
    push    hl

ioi ld      a,(SASR)            ; determine interrupt source (RX or TX)
    rla                         ; bit 7 is set if data has been received
    jr      c, isr_RX           ; bit 7 was rolled into carry flag

isr_TX:                         ; the transmit interrupt brought us here
ioi ld      (SASR), a           ; clear interrupt

    /***************************************************************
    Body of Transmit ISR goes here
    ***************************************************************/

    pop     hl                  ; restore saved registers
    pop     de
    pop     bc
    pop     af
    ipres                       ; restore IPL
    ret

isr_RX:                         ; the receive interrupt brought us here
ioi ld      a, (SADR)           ; read byte, clear interrupt


    /***************************************************************
    Body of Receive ISR goes here
    ***************************************************************/

    pop     hl                  ; restore saved registers
    pop     de
    pop     bc
    pop     af
    ipres                       ; restore IPL
    ret

#endasm
```

Upon entry, the ISR first does housekeeping, saving any CPU registers that will be used by the ISR. Next, since a serial ISR can be caused by either a transmit event (transmit buffer empty) or a receive event (receive buffer full), the ISR determines which of the events caused the interrupt. Examining the most significant bit of Serial Port A Status Register (SASR: 0xC3) determines the source.

If a transmit event caused the interrupt, simply writing to SASR would clear the interrupt flag. On the other hand, if a receive event cause the interrupt, simply reading the receive buffer (SADR: 0xC0) would clear the interrupt flag.

After the program is done executing the body of the transmit or receive interrupt, the ISR does the rest of the housekeeping, restores interrupts and saved CPU registers before relinquishing control.

Timer A is another example of where the ISR needs to determine why the interrupt took place. Since TMRA1 through TMRA7 can trigger a Timer A interrupt, the ISR needs to check their associated bits in the Timer A Control / Status Register (TACSR: 0xA0) to find out which timer(s) generated the interrupt. This is shown in Program 7.12A.

7.3 Project 1: Polled vs. Interrupt-Driven Serial Communication

Various types of serial interfaces exist on 8-bit microprocessors, including RS-232, RS-485, HDLC (High-level Data Link Control), SDLC (Synchronous Data Link Control), SPI (Serial Peripheral Interface), I²C (Inter-IC bus), USB (Universal System Bus), IrDA® (Infrared Data Association), Microwire™ and Dallas 1-Wire™. Even though the physical interface is different for most of these protocols, they can use the same hardware on the microprocessor—a device called a USART, Universal Synchronous/Asynchronous Receiver/Transmitter.

The essence of serial port processing is to service the receive buffer as soon as a complete byte has been received and, if any data needs to be sent out, to send it as soon as the microprocessor has finished sending out the previous byte.

There are two common ways for the microprocessor to communicate with a serial port:

- Polled I/O: the program keeps checking the appropriate registers of the serial port to see whether a complete byte has been received (meaning that the receive buffer is full) or whether the previous byte has been transmitted (meaning that the transmit buffer is empty).
- Interrupt-Driven I/O: Rather than continuously checking the I/O device, the microprocessor keeps going about its business and does other things that keep it busy. When the I/O device needs attention (i.e., when the receive buffer is full or the transmit buffer is empty), it generates an interrupt and the serial ISR knows what to do in either case.

Serial I/O is an important aspect of the Rabbit 3000's communication ability, since the microcontroller is equipped with six serial ports. In this section, we will discuss the Rabbit's serial I/O capability in some detail, illustrating both polled and interrupt-driven I/O with some software examples.

Interrupts Overview

7.3.1 Rabbit Serial Ports From the Programmer's Perspective

The Rabbit 3000 has six on-chip serial ports designated A, B, C, D, E, and F. All the ports can perform asynchronous serial communications at high bit rates. Given below are special aspects of some of these serial ports:

- Ports A-D can operate as synchronous clocked ports (SPI). The transmit and receive signals of Ports A and B can be switched to alternate I/O pins.
- Ports E and F support SDLC/HDLC synchronous communications.
- Port A has the special capability of being used to remote boot the microprocessor via asynchronous, synchronous, or IrDA (asynchronous serial).

These serial ports are clocked by the following timers, as shown in Table 7.4.

Table 7.4: Serial port driven by Rabbit timers.

Serial Port	Driven by Timer	Timer Constant Register
Serial Port A	A4	TAT4R (0xA9)
Serial Port B	A5	TAT5R (0xAB)
Serial Port C	A6	TAT6R (0xAD)
Serial Port D	A7	TAT7R (0xAF)
Serial Port E	A2	TAT2R (0xA5)
Serial Port F	A3	TAT3R (0xA7)

> Dynamic C libraries exist for managing serial ports. These libraries are written in assembly language and use circular buffers. Section 7.3.7 lists the appropriate Dynamic C library functions that support serial I/O.

There are several options for clocking the UART timers:

- PCLK/2—this is the default
- PCLK—activated via TAPR and is normally enabled by editing the bios
- Timer A1—used when "low" bit rates are required

PCLK stands for peripheral clock, whose rate is controlled by the Global Control / Status Register (GCSR: 0x00). Section 7.4.1 provides an example using PCLK as a clock source for Timer A.

Using a serial port requires the following elements to be in place:

- Setting up the appropriate clock source for the associated timer (A2 through A7); whether PCLK, PCLK/2, or prescaled with timer A1. Timer A Control Register (TACR: 0xA4) is modified for this operation.
- Setting up the appropriate bit rate for the serial port. This is done by loading the associated Time Constant Register (TAT2R through TAT7R) with the correct divisor value for the appropriate bit rate.

Chapter 7

A Dynamic C system variable, `freq_divider`, can be used to calculate the appropriate divisor value:

divisor = (long)(freq_divider * 19200.0 / (float) required_bit_rate+ 0.5) − 1L;

- Enabling the appropriate function on the port pin. Since parallel ports and serial ports share their pins with various other I/O functions, the appropriate port pins have to be selected for their alternate function (in this case, as serial port transmit and receive pins).
- The Serial Port Control Register (SxCR) has to be set up to select the serial mode (asynchronous, internally or externally clocked), the number of data bits (7 or 8), and the interrupt priority for the associated serial port.
- If the serial I/O is interrupt-driven, the appropriate ISRs have to be declared.

As an example, let us configure Serial Port E with the following specifications:

- Asynchronous Mode.
- 19,200 baud, 8 data bits, no parity, 1 stop bit.
- Use PCLK/2 for clock source.

Program 7.7 below shows the code fragment for configuring Serial Port E.

Program 7.7: Code fragment for setting up a serial port.

```
// set up serial port E

// set bit rate divisors
// (divisor+1) for 19200 baud is stored in BIOS variable "freq_divider"

    divisor = (long)(freq_divider * 19200.0/(float)baud + 0.5) - 1L;

    WrPortI (TACR, &TACRShadow, TACRShadow & ~0x04); // TA2 to use Pclk/2
    WrPortI (TAT2R, &TAT2RShadow, (char)divisor);    // set baud generator
    WrPortI (PGFR, &PGFRShadow, PGFRShadow | 0x40);  // enable TxE on PG6
    WrPortI (SECR, &SECRShadow, 0x01 );              // async mode, 8N1, IPL=1
```

Once the serial port is configured in the above manner, it can be used for either polled or interrupt-driven communication. As an example, the last line of the code establishes a priority 1 interrupt for Serial Port E. The interrupt must not be enabled until the ISR has been "installed."

7.3.2 Theory: Polling Serial Ports

Polling serial ports is less efficient than using interrupts, because polling requires the CPU to keep checking for whether data has been received or whether data can be transmitted, even though there may be no serial activity or no change in the serial port's state since the last poll. Therefore, polling can be CPU intensive and in extreme cases, the serial port can start losing data if the system cannot read and process incoming data faster than it is arriving.

Interrupts Overview

Almost all of today's microprocessors use *buffered* serial ports. This means that there is more than one register on the receive channel. An incoming byte is first shifted into the first register. When that is complete, the received byte is transferred to the another register and if enabled, an interrupt is generated. This gives the system the ability to receive multiple bytes at the same time the ISR is pulling the first byte out of the buffer register. Serial ports A-D on the Rabbit 3000 have a double-buffer but ports E and F have a four byte FIFO.

It is important to clarify a couple of things here:

- Since SxAR, SxLR and SxDR use the same register for transmitting and receiving, the processor employs separate transmit and receive buffers internally.
- When using a 9-bit protocol there is some distinction between transmitting an address byte (SxAR) and transmitting a data byte (SxLR)—writing a byte to the SxAR appends a low ninth bit (the byte value is interpreted as an address by the receiving device), while writing a byte to the SxLR appends a high ninth bit. Writing a byte to SxDR doesn't append *any* ninth bit.
- There is a huge difference between an operation being "slow" and being "CPU intensive." The vast majority of applications use a main loop with a lot more latency than that of an ISR. Interrupts are going to be the way to go when outputting a serial stream.

As an example, let us determine what it will take to use Serial Port B in polled mode:

- The program checks bit 7 of the Serial Port Status Register (SBSR:0xD3) to see if a byte has been received
- If so, the received byte is read out of the Serial Port Data Register (SBDR:0xD0)
- Now the program checks bit 2 of SBSR to see if the transmit register is empty and a byte can be transmitted
- If so, the program places the byte to be transmitted in SBDR

7.3.3 Implementation: Polling Serial Ports

Program 7.8A runs on the RCM3200 prototyping board and demonstrates polling of receive and transmit status bits.

The following steps should be performed to run this program:

1. Connect the serial cable to jumper block J5 on the RCM3200 proto board.
2. Plug the other end of the serial cable into an available serial port.
3. Use a terminal emulation program such as HyperTerm.
4. Type some characters in the terminal emulator and see output echoed back on its window and the stdio window.

> The RCM3200 prototyping board uses jumper block J5 for serial I/O. A suitable cable is available from Rabbit Semiconductor (Part number 540-0009) that will facilitate hooking up a connection to a standard 9 or 25-pin serial connector.

Chapter 7

The program waits till it receives a character from Serial Port A, and then echoes back received characters, while printing them on the stdio window.

Program 7.8A: Code fragment showing polled serial port.

```
/*************************************************************
    serialP.c

<code deleted for brevity>

while (1)
{

  // bit 7 set means receive buffer full
  if (BitRdPortI(SBSR, SER_RCV))
  {
  received_data = RdPortI(SBDR);
  printf("%c", received_data);

  // bit 2 reset means transmit buffer empty
  if (!BitRdPortI(SBSR, SER_XMIT))
  {
  WrPortI(SBDR, NULL, received_data);
  }
  }

} // while
```

A couple of points to observe here:

- The program polls a bit from register (SBSR: 0xD3) to know if a character has arrived on the serial port. Only then would it go further and take appropriate action
- Either parallel port C or D can be used for serial port B I/O. The following code is needed to set up parallel port C appropriately:

Program 7.8B: Setting parallel port C for use with serial port B.

```
        ld      a, (PCDRShadow)
        set     BDRIVE_TXD, a       ;Tx should initially be idle high
        ld      (PCDRShadow), a
ioi ld          (PCDR), a
        ld      a,(PCFRShadow)
        set     BDRIVE_TXD, a       ;choose port pin functionality
        ld      (PCFRShadow),a
ioi ld          (PCFR),a
        xor     a
        ld      (SBCRShadow),a
ioi ld          (SBCR),a            ; disable interrupts, use parallel port C
                                    ; for serial port B
```

Interrupts Overview

7.3.4 Theory: Interrupt-driven Serial Ports

Polled I/O often affects performance and, in extreme cases, can cause data over runs, meaning that the system is slow to the point that it does not receive all the data being sent to it. One way to deal with this issue for serial communication is to use interrupt-driven serial I/O, especially on the receive side.

In the next section, we will look at two cases of interrupt-driven serial I/O: a straightforward, no-frills ISR and one which uses some additional handshaking for better control over transmit and receive buffers.

7.3.5 Implementation: Interrupt-driven Serial Ports

The following program runs on the RCM3200 prototyping board and sets up interrupts to work with Serial Port A. Similar to Program 7.8A, it waits to receive a character, and echoes it back. Just to confirm that a serial ISR happens every time a character is received, the ISR flips the state of LED DS1.

Program 7.9: ISR for interrupt-driven serial port B.

```
/***************************************************************
    serialI.c

<code removed for brevity>



/***************************************************************
Serial B ISR
***************************************************************/
#asm nodebug root

serB_ISR::
    push    af                  ; save used registers
    push    hl

    ioi ld  a, (SBSR)           ; get status register value
    rla                         ; byte received? bit 7 to carry
    jr      nc, SerB_TX         ; jump if no - check for Tx


    /*************************
    Serial ISR Receive Routine
    *************************/
SerB_RX::
    ioi ld  a, (SBDR)           ; read the byte and clear the interrupt
    jr      z, serial_done      ; Nothing received? we're done


    //echo the byte
```
(Program 7.9 continued on next page)

Chapter 7

Program 7.9: ISR for interrupt-driven serial port B (continued).

```
    ioild    (SBDR), a           ; send the byte back to echo it
    jr       serial_done



/************************
Serial ISR Transmit Routine
************************/
SerB_TX::
    ioild    (SBSR), a           ; clear the TX interrupt


serial_done:

    // just to show we are in the ISR,
    // flip the LED state
    ld       a,(PGDRShadow)
    xor      ~MASK_DS1           ; invert the bit
    jr       nz, led_on

led_off:
    set      DS1,a               ; set bit to turn LED off
    jp       led_done

led_on:
    res      DS1,a               ; reset bit to turn LED on

led_done:
    set      DS2,a               ; make sure LED DS2 remains off
    ld       (PGDRShadow),a      ; update shadow register
    ioild    (PGDR),a            ; write data to port g


ISR_exit:
    pop      hl                  ; restore used registers
    pop      af

    ipres                        ; restore interrupt priority
    ret
#endasm
```

The "setupSerial" routine sets up Serial Port B to user PCLK/2 and then sets up the TAT5R Time Constant Register with the right value to obtain a bit rate of 19,200 baud. It then sets up Port C to use the alternate port function to enable the Serial Port B transmit pin. Next, the ISR vector is set up using a method we have seen earlier in this chapter.

The serial ISR is fairly straightforward and performs the following functions in the given order:

- Saves the registers it is going to use.
- Examines the "Received Data" bit of the Serial Port B Status Register to determine what caused the interrupt; was it due to a "receive buffer full" event? If so, the code reads the Serial Port B Data Register to obtain the received byte; the act of reading this data register clears the interrupt as well. The code simply places the data back in the Serial Port B Data Register, which sends the data back out the transmit pin.
- If the interrupt was not caused by a "receive buffer full" event then the interrupt must be due to a transmitter buffer empty. The Tx interrupt is cleared by writing to the status register or any of the registers that send data (SxDR, SxAR, SxLR).
- Finally, the ISR performs some simple housekeeping before exit: it restores the interrupt priority and the saved registers before exiting.

There are two cases when we get a transmit interrupt:

- The transmit register writes to the shift register.
- When the last bit leaves the shift register. The system is telling us that the complete message has been transmitted and the transmitter is idle. We can always write to the status register and that will clear out the pending transmit interrupt.

Programmers are sometimes tricked by the way receive and transmit interrupts occur. In case of a "Received Data" interrupt, the code simply reads the data register, and this event clears the serial interrupt. However, when a "Transmitted Data" interrupt occurs, the interrupt remains asserted until the program transmits another byte. In order to avoid this situation from happening, the program needs to read the interrupt to clear it until it is ready to transmit another byte. Reading SASR again does not clear the transmit interrupt –writing to the status clears the interrupt.

7.3.6 Using Better Serial Handshaking

While Program 7.9 works for slow data, production quality code should use more robust handshaking to avoid data under-run and over-run issues with serial communication. The following measures should be implemented for higher quality data communication:

- Before transmitting another byte the program should insure that the Transmit Data Register is empty.
- The code should take into account whether nine bits are being transmitted or clear-to-send (CTS) flow control are being employed on the transmit side.
- On the receive side, the code should test for receive over-run conditions and whether request-to-send (RTS) flow control is being utilized.
- For both transmit and receive, the code should check to see whether or not it needs to employ parity control.
- One technique that can be employed is to use a circular buffer operating as a FIFO so that the program can operate on data at a different rate than it is being received.

Chapter 7

Fortunately, Rabbit Semiconductor has done the hard work of writing these low-level routines. Instead of reinventing the wheel and spending a lot of time testing low-level code, the programmer can use Dynamic C's built-in serial routines that incorporate the above handshaking mechanisms. These higher level routines are described in the next section.

7.3.7 Using Dynamic C's Serial Routines

Instead of writing their own low-level interrupt code, programmers can use a number of high-level serial routines provided by Dynamic C. Although counterparts for these routines exist for serial ports A through F, only the routines for Serial Port B are shown below:

- `serBopen (bit rate)`: Opens the serial port B for communication.
- `serBputs ("string\r")`: calls serBwrite.
- `serBwrite(string, strlen(string))`: Transmits bytes through serial port B.
- `serBwrUsed()`: Returns the number of characters in the Serial Port B output FIFO.
- `serBread(&data, length, timeout)`: Reads "length" number of bytes from Serial Port B or until timeout (in milliseconds) transpires between bytes.
- `serBwrFlush()`: Flushes the Serial Port B transmit (output) FIFO.
- `serBrdFlush()`: Flushes the Serial Port B receive (input) FIFO.
- `serBgetc()`: Returns the next available character from the Serial Port B read FIFO.
- `serBwrFree()`: Returns the number of characters of unused data space in the Serial Port B transmit FIFO.
- `serBpeek()`: Returns the first character of the data in the Serial Port B receive (input) FIFO without removing it.
- `serBclose()`: Disables Serial Port B.

Program 7.10 is a simple implementation of the above functions, using Serial Port B on the RCM3200 prototyping board. The program echoes received characters back through the serial port:

Program 7.10: Serial communication using Dynamic C functions.

```
/*****************************************************************
   DCserial.c

<code removed for brevity>


void main()
{
   int c;

   c = 0;
   serBopen(BITRATE);
```
(Program 7.10 continued on next page)

Program 7.10: Serial communication using Dynamic C functions (continued).

```
      while (c != CH_ESCAPE)          // Exit on Escape
      {
         if (((c = serBgetc()) != -1) && (c != CH_ESCAPE))
            {
            serBputc(toupper(c));
            if( c == '\r' )             // add "new line" to CR
                  {
                  serBputc('\n');
            } //if

         } //if

      } //while

      serBputs("\nDone\n");

      // allow transmission to complete before closing
      while (serBwrFree() != BOUTBUFSIZE);

      // close serial port
      serBclose();
} //main
```

7.3.8 Final Thought

Polled I/O should not be completely disregarded, and there are reasons why it makes sense to use it:

- Polled I/O is inherently faster than interrupt I/O, so if VERY high speed is required then polled I/O is preferable.

- If the programmer is unsure about whether the ISRs are working, some diagnostic code, using polled I/O, can be written to verify if the code will respond to I/O requests as planned. Once the code works, it can put in an ISR and tested further.

- Polled I/O is used a lot in embedded systems simply because it is simpler to write. Programmers may use interrupts for tick timers and serial ports, but polling for everything else. This is usually driven by the fact that the polled code is quicker to write and develop. For example, polled I/O is usually used to acquire data from analog-to-digital converters (ADCs). Either ADC data is slow enough to be polled or the ADC data is screaming fast and a CPLD collects the code in a dedicated RAM or an FPGA munches the data into some sort of stream (such as video). Embedded systems are often not all that "real time" (we'll read about this more in Chapter 8) and the use of polled I/O, being faster to develop, often reduces time-to-market.

Chapter 7

7.4 Project 2: Using Timer Interrupts

Unlike the software timing loops we worked with in Chapter 6, hardware-based timers can be used to generate precise periods. The Rabbit 3000 provides Timer A and Timer B, which can be set up to measure specific durations, and generate interrupts. Let's examine the two timers from the perspective of interrupting devices.

Timer A provides seven sub-timers, TMRA1 through TMRA7, that can each generate an interrupt upon counter underflow. Interrupts for each of these timers are controlled by Timer A Control / Status Register (TACSR: 0xA0), and the priority is controlled by Timer A Control Register (TACR: 0xA4). The Time Constant Registers, TAT1R through TAT7R, contain the 8-bit reload register for each of the sub-timers, TMRA1 through TMRA7.

Timer B is a 10-bit up counter that runs continuously and is driven by PCLK/2, PCLK/16, or by the output of Timer A1. Two 10-bit match registers (TBM1R: 0xB2 / TBL1R: 0xB3 and TBM2R: 0xB4 / TBL2R: 0xB5) are used to monitor the counter—when the counter reaches the value of one of the match registers, an interrupt can be generated. Each time an interrupt is generated due to a match, the interrupt is disabled and TBLxR must be reloaded to re-enable the interrupt. TBMxR does not need to be reloaded every time. If both match registers need to be changed, the most significant byte needs to be changed first.

Interrupts for Timer B are controlled by Timer B Control / Status Register (TBCSR: 0xB0), and the priority is controlled by Timer B Control Register (TBCR: 0xB1). TBCR also determines whether Timer B is driven by PCLK/2, PCLK/16, or Timer A1.

7.4.1 Theory: Determining Timer A Periods

Timer A is actually 10 separate 8-bit down counters that are primarily used to provide clocks to peripherals. The separate timers can be driven by PCLK or PCLK/2. As shown later in this section, the rate of the peripheral clock can be set via the GCSR register. Moreover, timers A2 through A7 can individually be driven by the output of timer A1 for even lower rates.

These timers all have an 8-bit divider that determines how many incoming clock pulses go by before an output pulse occurs; i.e., if the divider is set to 37 then 38 pulses will enter the timer before a single pulse comes out (if set to 0, every input produces an output).

The following equation can be used to determine Timer A frequencies:

$$F_{TimerA} = Clock_{TimerA} / (N + 1) \text{ where N is the divider}$$

In order to achieve a certain frequency, one can solve for N in the above equation.

Let's do some simple math for sub-timer TMRA7, and let's make the following assumptions for our calculations:

- Timer A Prescale Register (TAPR: 0xA1) is set up to clock all Timer A timers by PCLK. Note that this is NOT the Dynamic C default[5].
- The Global Control / Status Register (GCSR: 0x00) is set up so that CPU=OSC; PCLK=OSC. This means that PCLK is the same as the processor clock, which is operating at the oscillator frequency.

---

[5] By default, Dynamic C clocks all Timer A timers by PCLK/2.

- The processor is operating at a base frequency of 29.4912 MHz.

Let's load Timer A1's reload register (TAT1R: 0xA3) with 249. What period do we get?

$$F_{TimerA1} = 29.4912 \text{ MHz} / (249 + 1) = 117.9648 \text{ KHz}$$

which equates to a period of 8.477 μsec.

Let's make a couple of changes to the above assumptions and see what we get:

- Set up TACR to clock TMRA7 by Timer A1.
- In addition to loading Timer A's reload register (TAT1R: 0xA3) with 249, load Timer TMRA7's reload register (TAT7R: 0xAF) with 117. What period do we get?

As shown earlier, Timer A1 will divide sub-timer TMRA7 at 117.9648 KHz.

TMRA7 will further divide 117.9648 KHz by 118[6] to arrive at an interrupt frequency of 999.702 Hz, which is very close to a period of 1 millisecond.

Here is an example of using timer A for a serial port:

The programmer wants serial port B to work at 2400 baud, and the processor clock is 22.1184 MHz. To determine the divider for timer A5, which controls the serial port B baud clock, the possible dividers can be calculated using the following formulas:

$$\text{Divider}_{TimerA} = \text{Clock}_{TimerA} / (32 * \text{baud}) - 1, \text{ and}$$

$$\text{baud} = \text{Clock}_{TimerA} / (32 * (\text{Divider}_{TimerA} + 1))$$

So the two possible dividers are (when using PCLK and PCLK/2):

$$22.1184 \text{ MHz} / (32 * 2400) - 1 = 287 \quad \text{(when using PCLK)}$$

$$(22.1184 \text{ MHz}/2) / (32 * 2400) - 1 = 143 \quad \text{(when using PCLK / 2)}$$

Since the second value is a valid divider (within the 0–255 range), we could use that value directly by loading that value into the timer A5 divider when driving timer A by PCLK/2.

For the sake of doing a complete calculation, let's say that the programmer wants to drive timer A by PCLK instead (perhaps some other peripheral requires a high clock rate). In that case, we'll have to use timer A1 as a prescaler for timer A5 since the divider in that case is outside our valid range. Let us pick some good values for that—we need to produce two dividers whose product is 288 (since the full cycle is divider + 1). A bit of math shows that 288 = 12 * 24 (other combinations are obviously possible), so we can set the timer A1 divider to 11, and the timer A5 divider to 23. This should produce:

$$\text{Timer A1 output} = 22.1184 \text{ MHz} / (11 + 1) = 921,600 \text{ Hz}$$

which can then be fed into timer A5, producing a bit rate on serial port B of

$$\text{bit rate} = 921600 \text{ Hz} / (32 * (23 + 1)) = 2400 \text{ baud}$$

---

[6] These counters also divide by N + 1, not N.

7.4.2 Theory: Determining Timer B Periods

Timer B is a 10-bit free running counter that can be driven from one of three sources: PCLK/2, PCLK/8, or the output of timer A1. It also has two match registers, B1 and B2. When the counter matches either of those values, it outputs a pulse that can be set up to trigger an interrupt and/or clock the output values on parallel ports D-G.

The period of Timer B is fixed at 1024. Using the match register, Timer B can be made to divide by any number from 1 to 1023. The match registers generate an interrupt when Timer B reaches that value.

> Programmers should be careful about two aspects of Timer B:
> - Each time the Timer B ISR ends, if the same value is loaded each time in the match registers, Timer B will interrupt every 1024 clocks.
> - Timer B is a free running counter and will continue to count unless the ISR disables it.
> - Timer B cannot be reset to zero.

For example, if the ISR continuously updates Timer B with the same value, the interrupts will always occur at the same frequency: Clock/1024. Thus, a processor clock of 29.4912 MHz gives us:

$$F_{TimerB} = Clock_{CPU} / (2*1024) = 14,400 \text{ Hz.}$$ (assuming that the input clock is PCLK/2)

As we will see in this section, generating any other period requires more work.

The programmer can select the clock source for Timer B—there are three options available:
- Timer A1
- PCLK/2
- PCLK/16

Let us look at a real-world example: we want to generate a tick value of 28.8KHz with Timer B. How do we go about doing it?

Let us consider the key assumptions for our calculations:
- The processor is operating at a base frequency of 29.4912 MHz
- Use PCLK = $Clock_{Oscillator}/4$. This is set up with GCSR
- Use PCLK/16 to be the clock source for Timer B. This is done with TBCR

The above scenario gives us 460.8 KHz to be the clock source for Timer B.

Dividing 460.8 KHz by 16 gives us a tick value of 28.8 KHz. This means that the Timer B interval should happen every 16 counts. We need to load Timer B's Match register with this count value and update it each time in the Timer B ISR.

The pseudocode to make this happen is shown in Program 7.11:

Program 7.11: Pseudocode for setting up Timer B.

```
// set timer B to interrupt every 16 counts
#define TIMER_B_INTERVAL 16

main
{ MATCH_VALUE = TIMER_B_INTERVAL
  set up interrupt vector for Timer B ISR
  set the match register to MATCH_VALUE  ; can set it to zero first time
  enable Timer B interrupt
  run an infinite loop
}

TimerBISR::
{
  clear the interrupt
  MATCH_VALUE = MATCH_VALUE + TIMER_B_INTERVAL (ignore carry beyond 10 bits)
  set the new MATCH_VALUE into the match registers
  restore priority levels
}
```

Working with this math, we can derive the following interval values for Timer B's match registers:

Table 7.5: Interval values for Timer B match register for the above example.

Required Timer B Tick Value	Interval Value for Timer B Match Register
500 Hz	921
1 KHz	460
5 KHz	92
10 KHz	46

Another technique to get longer periods is to use Timer A1 to clock Timer B. Let us look at an example here, and see if the delay is long enough to flash an LED.

- The processor is operating at a base frequency of 29.4912 MHz.
- Set up TAPR to drive Timer A1 by PCLK/2
- Set up TBCR to clock Timer B by Timer A1
- Load Timer A1 with the maximum value of 0xFF
- Load Timer B's match registers with the interval value of 0x00

What flash rate do we get?

Timer A1 is driven by PCLK/2, and dividing that by 256 to give us 57.6 KHz ticks. Therefore, if Timer B is being driven at 57.6 KHz, using the maximum interval value of 1023 in the match register would give us a Timer B tick rate of 56.3 Hz.

Chapter 7

This is still too fast for a flashing LED and we need to use another divider to slow it down further. We can implement another counter that increments each time the ISR is called, and then examine that counter within the ISR; if the counter reaches a certain value, we can decide that the LED needs to be flashed.

For instance, if we wait for a rollover of bit 5 of the counter, it has the same affect as multiplying the period by 64. Let us work backwards and see what some of the above settings have to be in order to get a flash rate of ½ Hz. To flash an LED at ½ Hz, the base frequency needs to be 1 Hz. A simple calculation shows us that we need to use a reload value of 900 for the Timer B match register in the ISR to arrive at the required base frequency.

The resultant base frequency is 0.999983 Hz, giving us a flash rate that is pretty close to ½ Hz. We'll use these calculations later in Program 7.14.

Here's an example of using timer B to produce a PWM output on the parallel port E pins:

Assumptions:

- The processor clock is 14.8456 MHz
- We'll use PCLK/2 = 7.3728 MHz
- We want a pulse of width 30 μsec.

This means that our pulse should be 7.3728 MHz * 30 μsec = 221 clocks long.

Let's pick appropriate values for the match registers; for instance, we can turn on at count = 512 (0x200) and off at count = 512 + 221 = 733 (0x2DD).

We will set up the parallel port E pins to be clocked by timer B1, and write the first value (high) into the data register—it will not appear on the pins until the timer B1 match occurs. We then load the B1 match registers with the value 0x200 by first writing 0x80 to the MSB and 0x00 to the LSB, in that order. We then enable timer B and watch for the first match to occur, either by monitoring the timer B status register or writing an interrupt handler.

When the first match occurs, we will load the next value for parallel port E (low) and the next match value, 0x2DD, by writing 0x80 to the MSB and 0xDD to the LSB. We wait for that match to occur, then go back to loading the original match value for the rising edge of the pulse.

Note that this example will produce a 30 μsec pulse every (PCLK/2) / 2^{10} = 7200 Hz, which equates to a period of 138.8 μsec.

If we wanted a slower pulse rate, we could run timer B from PCLK/16 (906 Hz, meaning a tick every 1.1 msec), or even run it off of the output of timer A1, producing rates as low as

Timer A1 output = (PCLK/2) / 256 = 28.995 KHz

Timer B rollover = 28.995 KHz / 2^{10} = 28.3157 Hz –> 35.31 milliseconds

when the maximum timer A1 divider is used.

Interrupts Overview

To summarize, there are a few tricks to using Timer B properly:
- Once enabled, Timer B runs continuously unless it is explicitly stopped.
- New match values are marked valid by writing the LSB, so the programmer needs to be certain to write the MSB first.
- When the match registers are written, the values are stored in preload registers. When the next match occurs, the preload register values are transferred to the actual match registers.
- When a match occurs, the current match registers are marked invalid until the interrupt flag is cleared by reading TBCSR. To re-enable the same match values, only the LSB needs to be rewritten. TBLxR must be reloaded to re-enable the interrupt, and TBMxR does not need to be reloaded every time.
- If Timer B match registers are always updated with the same value, the interrupts will always occur at the same frequency: Timer B Clock / 1024. In order to achieve the required period, each time we are in a Timer B ISR, we need to add the count value for the desired interval to the current match value and then update the match registers with this value.

7.4.3 Theory: Setting Up a Timer A Interrupt

Sub-timers in Timer A require the appropriate Timer Constant Register to be set up, interrupt priority set up through TACR, and then interrupts enabled through TACSR. The programmer should make sure that the main clock for Timer A is enabled through TACSR. Upon power up, since the reload registers may contain an unknown value, it is advisable to follow this sequence. Program 7.12A sets up timer A to work with PCLK/2 and interrupt priority 1. It then sets up timers A6 and A7 to trigger the interrupt. The code then sets up the interrupt vector for Timer A, enables timer A6 and A7 interrupts, and the Timer A clock. Finally, it enables Timer A interrupts at priority 1.

Program 7.12A: Code fragment for setting up Timer A interrupts.

```
/**************************************************************
    timerA.c

<code removed for brevity>

    WrPortI ( TAPR,  NULL, TAPR_PCLK2);             // set clock source

    WrPortI ( TAT6R, NULL, TAT6R_interval); // set Timer A6 interval
    WrPortI ( TAT7R, NULL, TAT7R_interval); // set Timer A7 interval

    SetVectIntern (INT_TIMERA, TimerAISR ); // set interrupt vector

    // enable interrupts on Timers A6 and A7, and enable main clock
    WrPortI ( TACSR, &TACSRShadow, TACSR_ENA_TMRA7 | TACSR_ENA_TMRA6 |
    TACSR_MAIN_CLK_ENA);

    // enable Timer A interrupts at priority 1
    WrPortI ( TACR, &TACRShadow, TACR_INT_PRI_1 );
```

Chapter 7

Each time an interrupt happens, the ISR needs to determine what caused the interrupt. In this case, since TMRA6 or TMRA7 can trigger the Timer A interrupt, the ISR should check their associated bits in the Timer A Control Register (TACR: 0xA4) to find out which timer generated the interrupt.

Program 7.12B: Checking interrupt source in the Timer A ISR.

```
/***************************************************************
// code fragment from Timer A ISR in timerA.c

    ioild   a, (TACSR)              ; find out who interrupted
                                    ; and clear the interrupt
        ld  b, a                    ; save for processing
        ld  a, (PBDRShadow)         ; get shadow value

        bit TACR_INT_TMRA6, b       ; was it Timer 6?
        jr  z, Try7                 ; jump if no

        xor TACSR_ENA_TMRA6         ; invert bit 6

Try7:
        bit TACR_INT_TMRA7, b       ; was it Timer 7?
        jr  z, ExitISR              ; jump if no

        xor TACSR_ENA_TMRA7         ; invert bit 7

ExitISR:
        ld  (PBDRShadow), a         ; update shadow register
    ioild   (PBDR), a               ; output new value on port
```

Let us also determine the period for the above square waves:

- We have set up TMRA6 with a reload value of 100.
- We have set up TMRA7 with a reload value of 210.
- We have set up Timer A Prescale Register to clock all Timer A timers by PCLK/2.
- The Global Control / Status Register (GCSR: 0x00) is set up so that CPU=OSC; PCLK=OSC. This means that PCLK is the same as the processor clock, which is operating at the oscillator frequency.
- The processor is operating at a base frequency of 29.4912 MHz.

The interrupt frequency for Timer A6 is calculated as follows: $F_{TimerA6}$ = PCLK / 2 / (Reload value + 1) = 29.4912 MHz / 2 / 101 = 145,996 Hz. This equates to a period of 6.8495 μsec.

The period of the square wave will be twice that value since the bit is inverted at each interrupt. The values for Timer A7 can be calculated in the same way.

7.4.4 Theory: Setting Up a Timer B Interrupt

Timer B requires the appropriate Timer Constant Register to be set up, interrupt priority set up through TBCR, and then interrupts enabled through TBCSR. The programmer should make sure that the main clock for Timer B is enabled through TBCSR. Upon power up, since the reload registers may contain an unknown value, it is advisable to follow this sequence.

Unlike Timers A1-7, Timer B does not reset to 0 when the match occurs. So, the match registers will have to be changed on every interrupt. Keeping in mind that TimerB is 10 bits and wraps to 0, we will have to calculate the next match register. Suppose that we wanted an interrupt on every 900 counts. The match values would be as follows:

900
776   (900 + 900 = 1800; 1800 − 1024 = 776)
652   (again, 776 + 900 = 1676, so it wraps around to 1676 − 1024 = 652)
528
404
… and so on

The match registers are only 10-bit. The registers are TBL1R and TBM1R (xxx2R for the second match registers.) The MSB register is nonintuitive to use, since bits 8 and 9 of the 10-bit count are stored in bits 6 and 7 of the TBM1R respectively. This makes using a 16-bit value more difficult as carries out of the lower byte need to be moved to the upper bits. This can be done by filling the intermediate bits with ones. The sequence to add a count to the match values is as follows:

Program 7.13A: Code fragment for reloading Timer B match registers.

```
#define TIMERB_INC 900          // use 900 as an example increment
unsigned timerb_match;          // global with last match

// this is constant;
// shuffle the bits around to match the timerb match registers

const timerb_increment = TIMERB_INC&0xFF | ((TIMERB_INC<<6) & 0xC000)

// calulate next timerb value
void next_timerb()
{
// this will cause low 8-bits to carry to bits 14 and 15
    timerb_match |= 0x3f00;
    timerb_match += timerb_increment; // calculate next match value
}
```

The 'timerb_match' value can be loaded in the ISR to the TBM1R and TBL1R registers. This is best done in assembly language.

If a delay of 1024 clocks is needed, then just use the same values to reload the counter.

Chapter 7

An alternate method, shown in Program 7.13B, sets up shadow registers for the Timer B match register:

Program 7.13B: Alternate method that sets up shadow registers.

```
#define TIMERB_INC 900
unsigned TimerBvalue;
char TBL1Rshadow, TBM1Rshadow;          // create shadow registers

TimerBvalue += TIMERB_INC;              // update count
TBL1RShadow = TimerBvalue&0xFF;// the mask operation is not really necessary
TBM1RShadow = (TimerBvalue>>2) & 0xC0;  // put bits 8 & 9 into 6 & 7
```

The code fragment shown in program 7.13C sets up timer B to work with PCLK/2 and interrupt priority 1. It then sets up the match counters TBM1 to zero. Finally, the code enables the main interrupt clock and match counter 1.

Program 7.13C: Code fragment for setting up Timer B interrupts.

```
// define match condition for TBM1
// #define MATCH_TBM1_LOW      0x00
// #define MATCH_TBM1_HIGH     0x00

// masks for Timer B Interrupt Priority
#define TBCR_INT_PRI_1 0x01

// masks for Timer B Interrupt Enable
// #define TBCSR_INT_MATCH1    0x02
// #define TBCSR_INT_MATCH2    0x04

#define TIMERB_INCREMENT 300    // this is the delay in clocks
// masks for Timer B Main Clock
#define TBCSR_MAIN_CLK_ENA    0x01

<code removed for brevity>

// clock timer B with (perclk/2) and set interrupt level to 1
WrPortI(TBCR, &TBCRShadow, TBCR_INT_PRI_1);

// set initial match condition
WrPortI(TBL1R, NULL, MATCH_TBM1_LOW); // 0 is simplest to start with
WrPortI(TBM1R, NULL, MATCH_TBM1_HIGH);

// enable matches and main clock
WrPortI(TBCSR, &TBCSRShadow, TBCSR_INT_MATCH1 | TBCSR_MAIN_CLK_ENA);
```

Interrupts Overview

When an interrupt happens due to a match from either of the match registers, the ISR needs to detect which match caused the interrupt. The ISR must set up the next match value, which will trigger the next interrupt.

The following code fragment (Program 7.13D) illustrates what needs to happen in a timer B ISR. As early in the ISR as possible, we should read TBCSR to determine whether a match from TBM1or TBM2 caused the interrupt; this read operation also clears the interrupt flag so that the ISR doesn't keep getting triggered indefinitely. Since we had set up just TBM1 to generate interrupts, it doesn't matter what the read from TBCSR tells us; we know the flag will point to TBM1 as having triggered the interrupt.

Before exiting the ISR, it is important to set up the match condition for the next trigger of the Timer B interrupt. In addition, the ISR restores interrupts before exiting.

Program 7.13D: Code fragment for finishing up the Timer B ISR.

```
; should be done as soon as possible in the ISR
    push    af
ioi ld a, (TBCSR)           ; load interrupt flags (clears flag)


/*******************************
Body of ISR goes here
*******************************/


// before exiting the ISR, it's important to set up the next match
// condition
/// this will always use a constant of 1024,
/// just use 0 if the values do not change
    xor     a
ioi ld (TBM1R), a

ioi ld (TBL1R), a
    pop     af
ipres                       ; restore interrupts
ret                         ; return
```

Chapter 7

Another approach to using timer B interrupts is to clock them from timer A1. This requires setting up TBCR accordingly, as shown in the code fragment program 7.13E:

Program 7.13E: Clocking Timer B from Timer A1.

```
// masks for Timer B Clock Source
#define TBCR_CLOCK_PCLK2      0x00
#define TBCR_CLOCK_PCLK16     0x08
#define TBCR_CLOCK_A1         0x04

// masks for Timer B Interrupt Priority
#define TBCR_INT_PRI_1        0x01
#define TBCR_INT_PRI_2        0x02
#define TBCR_INT_PRI_3        0x03

Main()
{
// miscellaneous initialization goes here

// set up timer B interrupt vector

// initialize Timer B
// clock timer B with (Timer A1) and set interrupt level to 1
   WrPortI(TBCR, &TBCRShadow, TBCR_CLOCK_A1 | TBCR_INT_PRI_1);

// Load Timer A1 with max time constant
   WrPortI(TAT1R, &TAT1RShadow, 0xFF);

// set up timer B match registers

// enable timer B and match interrupts using TBCSR

} // main
```

Note that for this approach to work as planned, it is important to load timer A1 with the right time constant.

The advantage of this approach is that relatively long time periods can be obtained, which are not attainable by clocking timer B directly through a PCLK/2 or PCLK/16.

We will use this approach in deriving relatively long periods that can flash an LED.

7.4.5 Implementation Using an Assembly-based ISR

We flash the LED at a duty cycle of 50% by changing its state each time the LED needs to be flashed. We read the associated shadow register to find out the previous state of the LED, reverse it, and then write it back to the port register, after writing to the associated shadow register.

We have used the math from Section 7.4.2 to flash the LED at 1Hz. Here are the required settings to get that rate:

Interrupts Overview

- Timer A1 clocks Timer B
- Set up TAPR to drive Timer A1 by PCLK/2
- Load Timer A1 with the maximum value of 0xFF
- Load Timer B's match registers with 900
- Use internal counter to divide by 64 i.e., wait for a rollover of bit 5
- The processor is operating at a base frequency of 29.4912 MHz

We will divide the CPU clock (29.4912 MHz) by 2, then by 1024, and then by 256 because we are using Timer A as a prescaler. Using a Timer B interval value of 900 and a software divider of 64 will get us a base rate very close to 1 Hz and the LED blink rate will be twice this value: 0.5 Hz.

Program 7.14 works on the RCM 3200 prototyping board and implements the above scheme. Only the initialization section of the code is shown here:

Program 7.14: Initializing Timer B for flashing an LED.

```
/*****************************************************************
    intled1.c

<code removed for brevity>


void main()
{

    initPort();                         // initialize ports

#if __SEPARATE_INST_DATA__
    interrupt_vector timerb_intvec timerb_isr;
#else
    // initialize Timer B interrupt vector
        SetVectIntern(timer_B_int_num, timerb_isr);
#endif

#asm
    // initialize Timer B
    // clock timer B with (Timer A1) and set interrupt level to 1
            ld      a, TBCR_CLOCK_A1 | TBCR_INT_PRI_1
    ioi     ld      (TBCR), a
            ld      (TBCRShadow), a

            // Load Timer A1 with max time constant
            ld      a, 0xFF
    ioi     ld      (TAT1R), a
            ld      (TAT1RShadow), a

            // set initial match
```

(Program 7.14 continued on next page)

Chapter 7

Program 7.14: Initializing Timer B for flashing an LED (continued).

```
            // start at 0, first interrupt will be random time, but
            // it is easier than trying to determine the first value

            xor  a
    ioi     ld    (TBM1R), a
    ioi     ld    (TBL1R), a

            // enable timer B and Match1 interrupts
            ld    a, TBCSR_INT_MATCH1 | TBCSR_MAIN_CLK_ENA
    ioi     ld    (TBCSR), a
            ld    (TBCSRShadow), a
#endasm

    // nothing left to do now; just wait for interrupts to do all the work
    while (1);

} // main
```

Some observations about this program:

1. The programmer can define a constant that will be added to the timer reload value. The following code will do the trick, and is used in program 7.15:

```
    // definition section of code

    #define TIMERB_INCREMENT 900     // delay in clocks
    const int timerb_inc= TIMERB_INCREMENT&0xff |
    ((TIMERB_INCREMENT<<6) & 0xc000);

    <code removed for brevity>

    // end of Timer B ISR

    //setup to load the match register

    ld    hl,(timerb_match)
    ld    a,h

    // will cause carry to bit 8 to propagate to bit 14
    or    0x3f
    ld    h,a
    ld    de,(timerb_inc)
    add   hl,de
    ld    (timerb_match),hl
    ld    a,h
ioi ld (TBM1R), a  ; time_ to reload the match
    ld    a,l      ; high first, then low
ioi ld (TBL1R), a  ; set up new B1 match register
                   ; in every interrupt!
```

Interrupts Overview

2. Since the ISR is structured as a standalone piece of assembly and not encapsulated as a C shell, two colons are used to define the ISR as "`timerb_isr::`"
3. The ISR has to run from root space, hence the need for "`#asm root`" in the definition of the ISR.
4. The initialization section in main() has the following line of code:

   ```
   ld a, TBCR_CLOCK_A1 | TBCR_INT_PRI_1
   ```

   While a C programmer will likely be tempted to code it as

   ```
   ld a, (TBCR_CLOCK_A1 | TBCR_INT_PRI_1),
   ```

   it would have a very different affect on the program, since the parenthesis would change a "load immediate data" to "load data from address."
5. The ISR for Program 7.14 (intled1.c) routine takes 234 clock cycles, which is about 7.93 microseconds, when the CPU clock is set to be the same as Oscillator clock. Since the timer calls it once a second that means the processor is in this interrupt routine 7.93 microseconds per second, or 0.08% of the time.
6. The first version of the code did not work—the ISR section of the code did not flash the LED as expected. To debug the code, we used the quickest and simplest debugging method and stepped through the code to determine whether the logic of "flipping the LED state" worked as planned. While Dynamic C allows the programmer to set breakpoints and then "trace into" code by pressing the <F7> key, the ISR has to be told explicitly to allow debugging. The "`#asm root nodebug`" has to be changed to "`#asm root debug`" so that Dynamic C can enable breakpoints and single stepping in the ISR.

7.4.6 Implementation Using a C-based ISR

Not only does this program implement the ISR in Dynamic C, it illustrates how most of the initialization can be done in C as well.

Since the ISR is called as a C function, there is no need for the ISR to save and restore CPU registers upon entry and exit, or to restore interrupts upon exit. A complete program that includes initialization and the ISR is shown in Program 7.15:

Program 7.15: Dynamic C-based ISR for flashing two LEDs at different rates.

```
/***************************************************************
    intled2.c

<code removed for brevity>

// values for Timer B Match Registers
#define TIMERB_INCREMENT 900   // this is the delay in clocks
const int timerb_inc = TIMERB_INCREMENT&0xff |
```
(Program 7.15 continued on next page)

Chapter 7

Program 7.15: C-based ISR for flashing two LEDs at different rates (continued).

```
        ((TIMERB_INCREMENT<<6) & 0xc000);

<code removed for brevity>

nodebug root interrupt void timerb_isr()
{

    RdPortI(TBCSR);      //clear the interrupt

    count++;

    //decide if we should flip state of LED DS1
    if (!(count % FLASH_SLOW))
          if (BitRdPortI(PGDR, DS1))
                BitWrPortI(PGDR, &PGDRShadow, ON, DS1);
          else
                BitWrPortI(PGDR, &PGDRShadow, OFF, DS1);

    //decide if we should flip state of LED DS2
    if (!(count % FLASH_FAST))
          if (BitRdPortI(PGDR, DS2))
                BitWrPortI(PGDR, &PGDRShadow, ON, DS2);
          else
                BitWrPortI(PGDR, &PGDRShadow, OFF, DS2);

    // set match registers to re-enable Timer B interrupts
    timerb_match|=0x3f00;
    timerb_match+=timerb_inc;
    WrPortI(TBM1R, NULL,timerb_match>>8);
    WrPortI(TBL1R, NULL, timerb_match);

} // Timer B ISR
```

Some observations from this program:

1. Since the ISR is declared as a C function, it needs to be given a return type, void in this case.
2. The ISR has to run from root space, hence the need for "`#asm root`" in the definition of the ISR.
3. Similar to the assembly case above, when the ISR code did not work at first, changing the ISR definition from "`nodebug root interrupt void timerb_isr()`" to "`debug root interrupt void timerb_isr()`" allowed us to use breakpoints and do single stepping through the ISR code.
4. Since Timer B Match Registers do not have their corresponding shadow registers, the "WrPortI" function is called with a NULL as the parameter for the associated shadow registers.

Interrupts Overview

7.4.7 Final Thought

Timer B configuration can get complicated. There are numerous clock combinations; the programmer will need to determine the count to use based on the CPU crystal frequency, the prescale value and clock mode. Using Timer A1 to clock Timer B will likely conflict with a serial port that will use Timer A1 (the serial ports use Timer A1 if their bit rate divisor can not be represented in just a single 8-bit timer). Moreover, changing to a new board with a different clock frequency will require the times to be recalculated.

Provided on the CD-ROM is SHDesigns Timer B library. This will call a user C or ASM function periodically. It can scale the interrupts as needed to almost any value. For example, on the RCM2200 (22 MHz), the user routine can be called from once every 20 microseconds, down to only once every 1500 seconds[7]. This library is given a single parameter, the delay in microseconds between calls to the user routine. The library will automatically determine the Timer B increment value. It will use Timer A and prescale values if needed to try to generate the most accurate time. All configuration is determined automatically regardless of the board type. This makes using Timer B much easier to use.

We are still not done with flashing LEDs... in Chapter 8, we will illustrate other ways to flash LEDs "in four lines of code" using Dynamic C's multitasking features.

While flashing LEDs may seem trivial at first, it is oftentimes an important part of the user interface for an embedded system. For example, one of the authors developed a Fibre Channel-based hard drive controller for a major storage vendor. This controller kept track of status and activity for each hard drive and flavors of the controller were built to support drive arrays of 2, 8 and 16 hard drives. The controller talked with the rest of the storage subsystem via I²C for communication and IPMI™ (Intelligent Platform Management Interface) for the higher-level protocol. The system provided for two bi-color status and activity indicators for each drive. These indicators were each implemented with bi-color LEDs and had settings similar to those shown in Table 7.6:

Table 7.6: A practical implementation using bi-color LEDs.

Drive Condition	LED Signal	Visual Indication
Access	Activity	Flashing Green
Critical Condition	Activity	Solid Red
Fault	Status	Fast Flashing Red
Predicative Fault	Status	Flashing Amber
Identify Disk Drive	Status	Slow Flashing Red
Slot Normal (Drive Present)	Status	Solid Green
Slot Normal (Drive Not Present)	Status	OFF
Power Fault	Status	Solid Amber

The storage vendor sold a billion dollars worth of these systems each year in the mid to late 1990s, and these LEDs served as the only user interface on the drive bay.

---

[7] Fast Timer B interrupts will consume much of the CPU and MUST be coded in assembly language. The limit for a simple Timer B handler written in C is about 50 microseconds between interrupts. An ISR written in C will likely cause missed serial port interrupts and make the debugger unstable.

7.5 Project 3: Using the Watchdog Timer

Embedded systems often have the obligation to perform properly under adverse conditions without being able to fall back on humans to reset them when things go awry. The best embedded systems are the ones that people never have to think about. For example, pilots don't care that their aircraft may have more computational horsepower than the Apollo space capsules. The pilots just care that the plane turns when so steered. Under no conditions does the pilot want to have to press CTRL-ALT-DEL on a keyboard to reset an errant system.

When engineers discuss building reliable systems, one of the first subjects raised is watchdogs. Just the presence of a watchdog timer often allows a designer better sleep at night. Unfortunately the protection offered by a watchdog timer can be destroyed by poor implementation of either the hardware or software.

Designing reliable embedded systems requires a relentless attention to detail. Watchdog timers are no different, but the programmer must be careful about being lulled into a false sense of security by the mere presence of a watchdog timer.

7.5.1 Theory: What is a Watchdog Timer Good For?

Embedded systems are subject to all manner of electronic, physical and user abuse. Electromagnetic interference (EMI), especially electrostatic discharge (ESD), can cause a CPU to lose track of reality in a microsecond. For high reliability, an embedded system should have a guardian, a watchdog, looking out for the CPU. If the CPU fails to operate correctly, then the watchdog should reset the CPU.

A watchdog timer is simply a free-running counter that can be reset to zero by a CPU and upon rollover the counter resets the CPU. If the CPU fails to reset the counter before the counter rolls over, then the CPU is deemed errant and is reset.

This simple scheme has been implemented in many ways, and there are two characteristics worth mentioning. First, the counter's reset must not be level sensitive. If level triggered reset existed, an errant CPU could simply hold the watchdog in reset, thus making the watchdog ineffective. Second, the watchdog's rollover period (also called timeout) must be long enough for the CPU to reset the watchdog under all nominal conditions.

Well-behaved embedded systems should be able to recover from a watchdog trip in less time than it takes for a human to notice a major problem. For example, many automobiles have several embedded controllers constantly fine-tuning engine performance. If one of these CPUs goes berserk and is reset by its watchdog, it would be nice if the driver experienced only a momentary loss of horsepower or a check-engine light. It would be bad form for the controller to take 10 seconds to reboot, causing the driver to completely lose engine function and to have to restart the engine while the car is moving at 60 miles an hour.

Watchdog timers have become so ubiquitous that many modern microcontrollers and CPUs offer an onboard watchdog timer. This saves money, board space and development time over an external device. The Rabbit microprocessor has an on-board watchdog timer.

Interrupts Overview

For the vast majority of systems an internal watchdog timer is adequate. For some hardened applications, the watchdog may have to be galvanically isolated from the CPU and in control of the system's power rails, as well as the CPU's reset line. For example, high-energy transients can cause CMOS latch-up in silicon devices. There are many ways to manage high-energy transients with transorbs, MOVs (metal oxide varistors), zener diodes, current limiting resistors and spark-gap suppressors (also called gas discharge tubes—GDTs). However, some systems need that extra safety factor of having the power completely cut to the embedded system and brought back up. For these kinds of systems, a simple internal watchdog will not be suitable and an external supervisor will need to be implemented.

Consider a weather station located on Antarctica. This system is devoid of human contact and is subject to high-energy events in the normal course of operation. Nearby lightning strikes can induce high voltages in sensors for reasonably long periods of time. At low levels, the lightning suppression on the sensor lines will shunt or dissipate the energy.

Higher energy events may just cause a CPU to lose track of where it is supposed to be executing code. In this case an internal watchdog that hits the CPU's reset line will cause the weather station to come back online.

At even higher energy levels, the CPU, or some peripheral device like an ADC (analog-to-digital converter) may go into CMOS latch-up. If an external watchdog can cycle power to the entire system, then there is an excellent chance the weather station will come back to life.

Of course if lightning directly strikes a communication line or sensor, all that will be left is carbonized fiberglass and splattered silicon. The chance of this happening can be minimized by deployment of lightning rods and other devices that will cause "near" lightning strikes as opposed to direct lightning strikes.

The selection of a watchdog depends on many system level factors, not just the digital electronics in the system. Managing these risks and making trade-offs is an engineer's stock in trade.

7.5.2 Watchdogs—A Deeper Appreciation

We know that a watchdog timer is a simple hardware counter. That's only half the story. The software on the CPU makes up the other half.

For our purposes of discussing watchdog software, we will denote the simplest function used to reset the watchdog timer as `hitwd()`. When this function is called, the appropriate action is taken to reset the watchdog's hardware counter.

Generally there are two methods for software to talk to hardware. The first is the "direct approach," in which the application software wiggles bits directly in the hardware. The second method inserts a layer of software (a driver) between the hardware and the application.

This additional layer of abstraction appears to be more complex than the direct approach. It might appear wise to keep the system of hardware and software that implements the watchdog as simple as possible. This seems to imply that the application should call `hitwd()` directly as often as need to keep the watchdog hardware from timing out and resetting the system.

The difficulty with this approach is that it often leads to code with calls to `hitwd()` sprinkled in every loop. Code written this way usually runs just fine. The watchdog is reset frequently enough that it never times out while the code is running correctly. Unfortunately this type of code compromises the watchdog's ability to function when a fault occurs.

The situation to consider when working with watchdogs is how the watchdog timer is treated when the CPU goes errant. To consider this situation, we first have to decide how an errant CPU might behave.

Many forms of EMI can cause a CPU to deviate from normal behavior. The CPU's resulting spasmodic behavior can be roughly categorized into four classes.

- Special Function Registers (SFRs) or RAM contents corrupted
- Program counter (PC) corrupted
- Internal state machine(s) (FSMs) corrupted
- CMOS latchup

When the contents of RAM are corrupted, a CPU may continue running a program from flash memory without deviating from the application's algorithms although with bogus data, the behavior of the code from the user's point of view maybe unacceptable. If the application has calls to `hitwd()` sprinkled about every twenty lines or so, the watchdog is unlikely to reboot the system.

If the CPU's SFRs are corrupted, odd behavior can result. For example, UART SFRs being corrupted might upset communication bit rates, while the MMU's SFRs being upset might cause the CPU to behave as if RAM were corrupted or the PC were corrupted. Depending on which SFR bits were corrupted, the behavior of the CPU might become errant enough that the application code either hangs or the CPU begins executing instructions out of sequence. Corruption of CPU SFRs has a slightly higher probability of allowing our application with haphazard calls to `hitwd()` to fail to reset the watchdog.

Corruption of the PC is a special case of the "corruption of CPU SFRs" case. In this instance, the CPU may begin fetching instructions from a memory space that doesn't contain application code. If this occurs, the watchdog is almost certainly going to reset the CPU. However, there is no guarantee the PC will become corrupt enough to cause instructions to be fetched outside of the application framework. Therefore there are no guarantees that the watchdog will reset the system if the application framework is littered with calls to `hitwd()`.

If the CPU's internal FSM(s) become corrupt, then the CPU will be unable to fetch and process instructions. Depending on how the FSM(s) are implemented, it is possible that they may convulse back to a valid state, but the instruction pipeline and data path will likely be in a confused state. If the CPU's FSM(s) become corrupt there is an excellent chance that the watchdog will reset the system even if the application code has multiple calls to `hitwd()` sprinked throughout.

In the case of the CMOS latch-up, the CPU is incapable of operating properly. In fact, if the situation is allowed to persist, the CPU will heat up enough to become permanently damaged.

Interrupts Overview

The watchdog will certainly rollover, assuming the watchdog is still running independently of the CPU. Of course, a simple "reset" will be inadequate to rescue a CPU from latch-up. A complete power down and proper power up sequence will be required to salvage this situation.

The system can suffer multiple failure modes simultaneously. For example, RAM and the PC can become corrupted at the same time.

Some situations can lead to others. For example, a corrupted stack in RAM may cause the PC to later become corrupt.

The art to designing a "reliable" system by using a watchdog is to only call `hitwd()` after careful consideration of the system's overall health. This requires only having one place where `hitwd()` is actually called, and then only after certain sanity checks have been done. This implies a driver. While this abstraction does increase complexity of the software, it is also a more sophisticated and robust approach.

Consider a system that contains a `SanityCheckHitDog()` driver that is the only place that calls `hitwd()` and only after doing a sanity check on the application framework. The application code calls `SanityCheckHitDog()` as often as it needs. Now consider the cases of corruption described above.

CMOS latch-up had no change in behavior. The watchdog will rollover.

Corrupt CPU FSMs will almost certainly prevent the CPU from successfully executing code, and even if the system does return to some semblance of normal operation, the sanity check may detect abnormal conditions within the system.

A corrupt PC will be very unlikely to stumble across a call to `hitwd()`. A call to `SanityCheckHitDog()` may be somewhat more likely to be tripped into and again, the sanity check will be executed before the call to `hitwd()`.

A corrupt set of SFRs may also be detected if the sanity check in the `SanityCheckHitDog()` function is well designed. Errors in even minor SFRs may be caught and the `hitwd()` not executed, thus allowing the watchdog to reset the CPU.

In the event that RAM is corrupt, it is still possible that the sanity check can identify a problem, but this is the least likely situation to cause the watchdog to reset the CPU, unless a major corruption occurs or the sanity check is particularly rigorous.

We have seen that the mere presence of a watchdog does not ensure that an errant system will be reset. We can increase the possibility that a corrupt system will be detected and that the watchdog will not be "hit" by adding a sanity check before `hitwd()` is called.

7.5.3 Watchdogs and Interrupts

Many engineers have a horrified knee jerk reaction to resetting a watchdog from inside an ISR, particularly an ISR called as part of a periodic interrupt. This stems from that fact that regardless of how far off the beaten track a PC may be rampaging through memory when the periodic interrupt occurs, the ISR will be called and the watchdog hit. Thus, even the most

Chapter 7

errant system is unlikely to be reset by a watchdog unless somehow the interrupt vector is corrupted or the interrupt masked.

Turns out that a knee jerk reaction isn't the best opinion upon which to base a design decision. ISRs in fact can be used quite effectively for servicing watchdogs, but only if they are well designed and adequate sanity checks are performed.

There are three ways for a CPU to execute code inside of an ISR. The first method is for the CPU to have vectored to the ISR as matter of legitimately servicing the ISR. In this case, the stack will be properly set up with a return address. Depending on the processor, there may be readable SFRs that indicate an interrupt has occurred. Either of these can be incorporated into the sanity check.

The second way for a CPU to execute code inside an ISR is for the PC to approach the start of the ISR linearly, and probably errantly. In this case, the stack my have a return address (or data that appears as an address) that is out of bounds for an application function. Furthermore, SFRs may indicate that no interrupt is pending. The sanity check can look for either of these conditions.

A common trick to handle this case is to place a "jump to self" endless loop just ahead of the ISR. If a rampaging PC approaches the ISR, the endless loop will capture the attention of the CPU and eventually the watchdog will timeout.

The third way a CPU can execute ISR code is errantly branch the execution path into the middle of the ISR. In this case, again the stack is not likely to have a known return address, nor will SFRs indicate an interrupt is pending. So the sanity check can preclude the call to `hitwd()`.

If an ISR wants to be sure it was entered from the front, as opposed to being branched into by a corrupted PC, the beginning of the ISR can use a static variable (dedicated memory location) to store a sentinel value at the start of the ISR. The sanity check can test for the proper sentinel value. Upon exiting the ISR, the ISR must clear the sentinel value.

This scheme lends confidence to the sanity check that the ISR was at least entered from the beginning and not branched to in the middle of the ISR by a rampaging CPU.

Another technique for managing watchdogs in an ISR is the virtual watchdog. A virtual watchdog is simply a software counter that the application increments and the ISR decrements. If the counter reaches zero, the ISR determines that the main application is errant. The ISR then disables interrupts and enters an infinite loop to wait for the hardware watchdog to reset the system.

If the system has multiple concurrent tasks running, each task can be allocated a virtual watchdog. If any task fails to reset the assigned virtual watchdog, then that task can be declared failed. The ISR can either restart the failed task, or allow the hardware watchdog to reset the entire system.

Another technique is to use an array of virtual watchdogs within an application's framework. A corrupt application is unlikely to properly service all of the assigned virtual watchdogs. The

Interrupts Overview

ISR will detect the tripped virtual watchdog(s) and can thereby allow the hardware watchdog to reset the system.

Notice the phrase "lends confidence" used above. This is indicative of the terminology used when designing reliable software systems. Because there are so many faults that may occur, it is impossible to check for all possible system faults. This is true in part because the functionality of a CPU subjected to a power interruption or other EMI event may be impaired.

The technique of using sentinel values can be compromised by a RAM location becoming corrupted with a value that corresponds to the proper sentinel value indicating the ISR was entered at the beginning of the code and then an errant branch to the middle of the ISR. These two conditions are intuitively unlikely to occur at the same time. However, the possibility does exist.

The engineer's job is making the trade-offs so that the risk of failures is weighed appropriately against the requirements for system reliability, speed and cost. Sometimes these decisions are based on "intuition," and sometimes a more quantitative approach is necessary.

Dynamic C allows the application developer the freedom to treat the watchdog either as a direct access device by directly calling the function hitwd(), or by using virtual watchdogs. The application can even disable the watchdog.

The designer can always implement an external watchdog if the Rabbit's internal watchdog is deemed inadequate. This will add cost and complexity to the system. Additionally, the service functions for the external watchdog will have to be written.

Rabbit's watchdog timer is built into the CPU. A 32.768 kHz crystal provides the clock for the watchdog. Although it is generally considered unwise to defeat a watchdog by design, the code fragment shown in Program 7.16A can disable the watchdog timer. Doing so may be desirable for debugging.

Program 7.16A: Code fragment that disables the watchdog timer.

```
#asm
    ld a,0x51
ioi ld (WDTTR),a

    ld a, 0x54
ioi ld (WDTTR),a
#endasm
```

On power-up the Rabbit's internal watchdog is enabled. Dynamic C automatically enables an ISR to service the hardware watchdog. This is part of what Dynamic C refers to as their virtual drivers. This is done after the BIOS executes, but before the application is started.

Before the application's main() function is called, Dynamic C calls a function called premain() found in PROGRAM.LIB. Inside of premain() a call is made to VdInit(), which

Chapter 7

is found in `VDRIVER.LIB`. This starts a number of timer related functions. For example the global timer based variables `SEC_TIMER`, `MS_TIMER` and `TICK_TIMER` are initialized and the relevant timer resources are assigned to keep these variables updated. `VdInit()` also enables a periodic interrupt used to call the watchdog ISR. This ISR implements virtual watchdogs that are available to the application and calls `hitwd()` to reset the hardware watchdog.

If the developer wishes to address the hardware watchdog manually or otherwise forgo the services provided by and overhead of the virtual drivers and periodic interrupt, then the call to `VdInit()` must be removed from `premain()`. If this is done, the hardware watchdog must be hit manually.

The `hitwd()` function found in `VDRIVER.LIB` is shown in Program 7.16B:

Program 7.16B: The hitwd() function that hits the watchdog.

```
   void hitwd(void);
   /*** Endheader */
   #asm
   hitwd::
        ld a,0x5a
   ioi ld (WDTCR),a
        ret
   #endasm
```

The hex code 0x5a tells the hardware watchdog to reset with a 2 second timeout. If a shorter timeout is required then other codes from Table 1 can be substituted in the above function.

Table 7.7: Codes applicable to the WDTCR register and their effect.

Value	Watchdog's Behavior
0x5a	Restart watchdog timer 2-second timeout period
0x57	Restart watchdog timer 1-second timeout period
0x59	Restart watchdog timer 500 ms timeout period
0x53	Restart watchdog timer 250 ms timeout period
All others	No effect

For most applications, the developer will want to leave `PROGRAM.LIB` and `VDRIVER.LIB` unmodified and simply use the virtual watchdogs provided by Dynamic C.

The virtual driver provides ten virtual watchdogs. The number of virtual watchdogs can be changed by defining the value N_WATCHDOG in the program. Two functions are used to allocate and release a virtual watchdog. These are `VdGetFreeWd(timeout)` and `VdReleaseWd(ID)`. The function `VdHitWd(ID)` is used to reset a virtual watchdog.

The procedure to use a watchdog involves allocating an available virtual watchdog and configuring its timeout. This is done by calling `VdGetFreeWd(timeout)`. The function's return value will be the ID (an `int`) of the newly allocated virtual timer. The ID will be used to hit the timer and to de-allocate (free) the timer. The parameter `timeout (a char)` sets up the timeout period of the timer. The timeout period is computed using the following formula.

$$virtual\_watchdog\_period = \frac{TIMEOUT - 1}{16} \cdot seconds$$

The parameter `timeout` is supposed to have values from 2 to 255. If `timeout` is 0 or 1 the behavior of the watchdog is not clear. A little math shows us that `timeout=2` gives the virtual watchdog a 62.5 mS timeout period. Using `timeout=255` gives the virtual watchdog a 15.875 second timeout period.

If all ten virtual watchdogs have been allocated, calling `VdGetFreeWd()` will cause a runtime error.

Once a program has allocated and configured a virtual watchdog it must keep the virtual watchdog reset by calling `VdHitWd(ID)`, where ID was provided by `VdGetFreeWd()`.

The last thing that can be done to a virtual watchdog is to release it. By freeing up the virtual watchdog, another process or function can reuse the virtual watchdog. Many applications are simple enough that once one or more virtual watchdogs are allocated, there is no reason to free them. Dynamic C provides `VdReleaseWd(ID)` to de-allocate a virtual watchdog.

Program 7.16C is an example of using a virtual watchdog on an RCM3400 prototyping board:

Program 7.16C: Using a virtual watchdog.

```
// Program: ch7vwd.C -
//
// An example of using a virtual watchdog on an RCM3400
// development board

#class auto

//port D bit 6 has LED DS1 connected (RCM3400 dev board)
#define DS1 6

void delayMs(long);
void ledOn(int);
void ledOff(int);

void main()
{
   static int wdID;
   static unsigned long tm0;

   brdInit();                    // Initialize the RCM3400

   wdID = VdGetFreeWd(33);       // Allocate a virtual watchdog.
```
(Program 7.16C continued on next page)

Chapter 7

Program 7.16C: Using a virtual watchdog (continued).

```
                                //
                                // After grabbing this virtual watchdog
                                // there are only nine left
                                //
                                // Configure the watchdog for a 2 second
                                // timeout.  (33-1) / 16 = 2 seconds

    while(1)                    // main loop
      {
        VdHitWd(wdID);          // hit the virtual watchdog

        ledOn(DS1);             // Blink DS1
        delayMs(250);           //
        ledOff(DS1);            //
        delayMs(250);           //
      }
} // end main()

// This function uses the shared system variable MS_TIMER to
// count down milliseconds.  This function polls MS_TIMER until
// "waitTime" milliseconds have elapsed.  Notice this function
// does NOT hit the watchdog.
void delayMs(unsigned long waitTime)
{
 unsigned long timeToQuit;

 timeToQuit = MS_TIMER + waitTime;

    while ((long)(timeToQuit - MS_TIMER) >= 0);

}

// This function turns on an LED on port pin on portD.
// Note that LEDs are active low on the RCM3400.
void ledOn(int led)
{
    BitWrPortI(PDDR, &PDDRShadow, 0, led);
}

// This function turns off an LED on port pin on portD.
// Note that LEDs are active low on the RCM3400.
void ledOff(int led)
{
    BitWrPortI(PDDR, &PDDRShadow, 1, led);
}
```

This program blinks the DS1 LED on an RCM3400 development board. Although hidden from the external world, a virtual watchdog is running to enhance the system's ability to recover from CPU upsets.

Interrupts Overview

Behind the scenes, the virtual driver (started in `premain()` by a call to `VdInit()`) is servicing the hardware watchdog. If no virtual driver were allocated in `main()`, the watchdog would still be running and providing some level of protection. The allocation and use of the virtual driver adds an extra layer of protection to the code.

The `delayMs()` function does not hit the virtual watchdog. This means the maximum amount of time the program can spend in `delayMs()` is less than 2 seconds. Otherwise the virtual watchdog will expire and the virtual driver will in turn let the hardware watchdog expire and reset the system.

If this program were modified to eliminate the virtual watchdog then the virtual driver would just service the hardware watchdog in the background as a matter of course. This means that under no circumstances would the hardware watchdog reset if `delayMs()` were executing. The program could spend a long time in `delayMs()`.

If an errant CPU were to inadvertently jump into `delayMs()`, perhaps because the PC were corrupted, then the `waitTime` parameter would not be set up. If no virtual watchdog is used, the CPU could spend a long time just waiting in `delayMs()`. If a virtual watchdog is used, then the maximum time the corrupt system could spend in `delayMs()` would be less than 2 seconds before the system reset.

By employing a virtual driver, the program sets up a more defined framework for the virtual driver calling `hitwd()`. This technique improves the system's ability to recover from an errant CPU.

However, if this technique is incautiously applied the results can be a precariously balanced system prone to undesired resets. As an example, consider that our program grows to 100 times the current size. If a call to `delayMs()` is made the programmer has to keep in mind that the function purposely doesn't hit the virtual watchdog. In a large program it is easy to lose sight of these details. If the program were allowed to make calls to `delayMs()` with a user specified parameter or even a parameter computed based on some environmental input, if the passed parameter causes the CPU to spend enough time in `delayMs()` then the system will reset. Tracking these sorts of bugs down can be a nightmarish task.

Attention to detail and a considered approach to the design of the watchdog system is required to build robust systems.

7.5.4 Final Thought

The design of a suitable watchdog for electronic systems requires careful attention to detail. The mere presence of a watchdog guarantees nothing. Both the hardware and the software must be crafted such that the watchdog is as sensitive to perturbations in the application's framework.

Generally speaking, sensitivity comes at the price of speed and cost. The more machine cycles spent in evaluating sanity checks the slower the application runs. The more hardware added to a system, the higher the system cost.

Chapter 7

Rabbit Semiconductor has provided a reasonable watchdog mechanism for most systems in which a failure does not constitute a hazard to a human. The Rabbit's internal watchdog saves money and board space. Dynamic C's virtual watchdogs give the application developer great flexibility in how much sensitivity the system has to failures.

7.6 Project 4: Setting Up a Real-time Clock

The real-time clock (RTC) in an embedded system keeps track of time, hours, minutes, seconds, milliseconds, days, months, or years, depending on the application. For instance, while the clock in a videocassette recorder needs to keep tracks of hours, minutes, seconds and days of the week, the clock in an automobile does not usually care about days of the week. An RTC can be implemented using the following methods:

- Software-based timing loops: as we saw in Section 6.5, software loops that execute within a known time period can be used to implement an RTC. The extent to which the loops are timed depends on the precision expected from the RTC. The advantage of this approach is its low cost and ease of implementation, since no external hardware is required.

- External RTC chips: such chips have been in use since the early 80's. These chips are available in various flavors, from those that have calendars that keep track of leap years, to those that generate alarm interrupts, to those that boast on-chip RAM that the CPU can use. Some of these clock chips are so packaged that they fit below popular static RAM chips and coexist with memory accesses to the RAM. Such chips are usually clocked by their own low-cost 32 kHz crystal and work independently of the CPU's oscillator. Using these chips requires external port pins and increases cost and design complexity of the system.

- Internal peripherals: microprocessor manufacturers have started to implement on-chip RTCs. The advantage is that the systems designer does not have to buy external chips, and the programmer can access the internal RTC in a manner similar to communicating with other internal peripherals. Such clocks often use a 32 kHz crystal independent of the CPU's own clock source and power supply. In our case, the Rabbit RTC and its associated 32 kHz oscillator are powered from a separate power pin that can have its own power source while the CPU is powered down.

This section will describe the Rabbit 3000's on-chip RTC and will describe how to control various aspects of the clock through software. It will also introduce the programmer to Dynamic C's functions that access the RTC.

7.6.1 Theory: Description of the Rabbit Real-Time Clock

The Rabbit RTC is a 48-bit modified ripple counter. The RTCs individual bytes may be read from the six holding registers, RTC0R through RTC5R. Any of these registers or the entire 48-bit counter can be set to any value—although not directly.

Interrupts Overview

The CPU registers relevant to the RTC are shown in Table 7.8.

Table 7.8: Rabbit 3000 on-chip registers used to implement a real-time clock.

Register Name	Mnemonic	I/O Address	R/W	Reset
Global Control/Status Register	GCSR	0000h	R/W	11000000
Real Time Clock Control Register	RTCCR	0001h	W	00000000
Real Time Clock Byte 0 Register	RTC0R	0002h	R/W	xxxxxxxx
Real Time Clock Byte 1 Register	RTC1R	0003h	R	xxxxxxxx
Real Time Clock Byte 2 Register	RTC2R	0004h	R	xxxxxxxx
Real Time Clock Byte 3 Register	RTC3R	0005h	R	xxxxxxxx
Real Time Clock Byte 4 Register	RTC4R	0006h	R	xxxxxxxx
Real Time Clock Byte 5 Register	RTC5R	0007h	R	xxxxxxxx

The Global Control/Status Register (GCSR) selects the CPUs clock source, which could be the same 32 kHz oscillator that clocks the RTC. If this oscillator is chosen to clock the CPU, the clock has to be read differently in order to make up for timing errors caused by the slower-running processor reading the RTC running at the same clock rate. The recommended practice in this case is to read only the upper 5 bytes of the RTC.

The Real Time Clock Control Register (RTCCR) selects which registers are to be written to, or to issue a reset to the RTC.

7.6.2 Implementation 1: Using Dynamic C's Functions to Access the RTC

Dynamic C already provides a set of useful functions that can be used to access the real-time clock. The following functions are provided:

- void `write_rtc(unsigned long int time)`: writes seconds to RTC
- unsigned long int `read_rtc(void)`: reads seconds from RTC
- unsigned long int `read_rtc_32kHz(void)`: reads seconds from RTC when using the 32 kHz clock

Dynamic C uses the standard C structure called "tm" to communicate with the relevant functions. The tm structure looks like the following:

```
struct tm
{
   char tm_sec;            // seconds 0-59
   char tm_min;            // 0-59
   char tm_hour;           // 0-59
   char tm_mday;           // 1-31
   char tm_mon;            // 1-12
   char tm_year;           // 80-147      (1980-2047)
   char tm_wday;           // 0-6 0==sunday
};
```

Chapter 7

The following Dynamic C functions use the `tm` structure to perform useful conversions:

- unsigned long `mktime(struct tm *timeptr)`: converts time structure to seconds
- unsigned int `mktm(struct tm *timeptr, unsigned long time)`: converts seconds to structure

The key functions that read and write time are as follows:

- int `tm_wr(struct tm *t)`: writes tm structure to the RTC
- int `tm_rd(struct tm *t)`: reads tm structure from `SEC_TIMER`, not from the RTC

The two functions `read_rtc` and `tm_rd` may seem similar, but the former reads time directly from the RTC, while the latter reads time from the system variable `SEC_TIMER`, which contains the same result. Rabbit Semiconductor advises the programmer to use `tm_rd` and not `read_rtc` to read time. One exception to this rule is that `read_rtc` and not `tm_rd` should be used to read time for the first time after resetting the clock.

There is something important to consider at this point:

- While Unix routines use 1900 as the start of history, Dynamic C time functions use 12 AM on January 1, 1980, as the start of history, and all time is measured in seconds from that point on (this provides an additional 80 years of time before the 48-bit counter rolls over).
- However, the Dynamic C libraries convert to/from 1900, so the code should add 1900 to the year it reads from the RTC, and the code should subtract 1900 from the calendar year it writes to the RTC.
- Although Dynamic C uses 1980 as the reference time, the programmer can use whatever reference is desired—as long as it is consistent.

As an example, let us implement a real-time clock with the following specifications:

- Leverage the multiplexed LED driver from the displayLED1 and displayLED2 programs in Chapter 6
- Use Rabbit's internal real-time clock for timekeeping
- Establish a background task with interrupts to make sure the display remains updated with the current time
- Establish a foreground task that allows the user to print the time on the screen

The period of the ISR depends on the granularity of the display. Since we will be displaying seconds and not milliseconds, it is sufficient to use an ISR that runs at 1Hz. Using the same math that we saw before in Section 7.4.2, we arrive at the following steps for generating Timer B interrupts at 1Hz:

- The processor is operating at a base frequency of 29.4912 MHz.
- Set up TBCR to clock Timer B by Timer A1.
- Load Timer A1 with the maximum value of 0xFF.
- Use a reload value of 900 for Timer B. This will be added to the Timer B match registers each time the ISR runs.

Interrupts Overview

- Use a divider to further divide the counter by 128.
- Define the constant that will be added to the Timer B reload value in the ISR. Doing the necessary math tells us the constant has to be 900.

Program `RTC1` meets the above criteria. We will highlight a few sections of code here.

The `main()` program first calls the familiar `initialize_ports()` routine to set up Port A to drive the segment bits for the multiplexed display, and Port C to drive the digit select bits. Port G is set up in a familiar fashion as well, in order for the program to be able to drive the two LEDs on the prototyping board. Next, the program binds `timerb_isr` to the Timer B interrupt. Since Timer B is used to display the time, the program first sets up the RTC before enabling Timer B interrupts, so that incorrect data is not displayed.

Program 7.17A: Setting up the RTC in main().

```
/*****************************************************

    RTC1.c

<code removed for brevity>

    rtc.tm_sec = 58;                        // change the time
    rtc.tm_min = 59;
    rtc.tm_hour = 23;
    rtc.tm_mday = 31;
    rtc.tm_mon = 12;
    rtc.tm_year = 103;                      // 2003-1900=103

    // set clock
    tm_wr(&rtc);
    t1 = mktime(&rtc);
```

Just to keep things simple, the initial time for the RTC is hard coded in `main()`, and the program does not give the user an option to set the time. It is easy to write code that will get user input and set the RTC.

Timer B is driven in a method we have seen before, where Timer A1 is used as a prescaler for Timer B. Once the main program sets up Timer B's clock source and interrupt level, loads the Timer A1 reload value, and establishes the Timer B initial match register, we are ready to enable the Timer B match register and interrupts, and off we go to the ISR.

The ISR simply reads time from the RTC, establishes which digit needs to display what, and reloads the match register again so that we can visit the ISR again. The code that updates one of the display digits is shown in Program 7.17B:

Chapter 7

Program 7.17B: Updating one display digit in the ISR.

```
/*********************************************************

    RTC1.c

<code removed for brevity>

    // digit 1: figure out which segments to turn on
    segment_map = get_segments(dig1);

    // and transfer the segment map to the LED segments
    WrPortI(PADR, &PADRShadow, segment_map);

    // turn on digit, wait, and then turn it off
    BitWrPortI(PCDR, &PCDRShadow, BITRESET, disp_digit_1);
    delay1mS(DIGIT_DELAY);
    BitWrPortI(PCDR, &PCDRShadow, BITSET, disp_digit_1);
```

When bringing in data from a routine, converting it, and sending it to another routine, the programmer can inadvertently insert bugs in any step of the process, thus resulting in unexpected behavior. A good programming habit is to test the data at each step, to confirm known data is supplied to the next step. This approach will help point to the source of errors as soon as the programmer encounters unpredictable data. In case of Program 7.18, although the `print_time` function is neither required nor used in the final version of code, it serves an important purpose; it allows the programmer to print the contents of the "time" variable at any step of the process, to make sure the results are as expected.

Since the ISR manages displaying the data on the LED display, a couple of tricks have been utilized to make the ISR more efficient:

1. Unlike in the original design of the routine that displays data on the LED (Chapter 6), the ISR refreshes the display. To make the code more efficient to run, subroutine calls are minimized and replaced with inline code as much as possible.

2. As described in Section 7.4.5, defined values are loaded in Timer B match registers TBM1R and TBL1R rather than computing their reload value at the end of each iteration of the ISR. Recomputing the value is NECESSARY if Timer B needs to divide by the same value every time—unless that value is 1024.

3. The `display_results` function extracts the units and tens digits from a number, so that each digit can be individually displayed on an LED segment. The math to do so is quite simple:
   - digit_tens = number/10;
   - digit_unit = number%10;

   However, these calculations result in time-consuming code, if this math is performed each time in the ISR. Functions `get_digit_tens()` and `get_digit_unit()` speed up the conversion in Program 7.17B by using lookup tables instead of actually doing the

Interrupts Overview

math. Program 6.16B showed a method to implement table lookups in assembly.

4. Timer B can be used to generate a synchronized clock output. For example, it can be used to talk to a SPI-based analog-to-digital converter (ADC), where we can set up Timer B to schedule the next time we will get a sample from the ADC.

7.6.3 Implementation 2: Display Time on the Rabbit LCD

Once Program 7.17A is debugged and tested, it is fairly straightforward to modify it so that data from the RTC is displayed on the Rabbit LCD module. Program 7.18 runs on the RCM3200 prototype board, reads time from the RTC, and refreshes the Rabbit LCD module each time in the Timer B ISR, which is configured to trigger at 1Hz.

This program is similar to Program 7.17A, except that it contains some extra code in function `initialize_LCD()` to initialize the integrated LCD / keypad module. The "`#memmap xmem`" preprocessor is needed since the LCD fonts are stored in XMEM space.

Because the top line of the display needs to read "The time is:", the main program puts this on the LCD instead of printing this line with each iteration of the ISR.

Unlike `RTC1.C`, the ISR is written in assembly. The ISR needs to simply read the time from the RTC and update the bottom line of the LCD to reflect the time. Before refreshing the time on the LCD, it always clears the bottom line. This is important in case the new text to be displayed has fewer characters than the previous text, in which case the remaining characters from the previous text would still show up on the LCD screen. Program 7.18 shows a code fragment that reads time from the RTC and refreshes the LCD:

Program 7.18: ISR code fragment to read the RTC and display time on the LCD.

```
/***********************************************************

    RTC2.c

<code removed for brevity>

    //Start of ISR body

    // read time in seconds since 1980
    t2 = read_rtc();
    // convert to time structure
    mktm(&t3, t2);

    // the code displays minutes and seconds
    // it's very easy to change it to display
    // hours and minutes instead

    // first blank the line that displays the time
    glPrintf (0, 16, &fi6x8, "                    ");
    glPrintf (0, 16, &fi6x8, "%02d/%02d/%04d %02d:%02d:%02d",
              t3.tm_mon, t3.tm_mday, 1900+t3.tm_year,
              t3.tm_hour, t3.tm_min, t3.tm_sec);
```

Chapter 7

The Timer B ISR also flashes an LED on the prototyping board by flipping its state for each iteration of the ISR. This is yet another quick debugging aid to visually establish whether the ISR is executing once a second. Looking at the LED would tell the programmer if the ISR is triggering in the ballpark of the 1Hz rate.

This example illustrates just the basics of what one can do with a real-time clock. The program can be improved in many ways, including:

- Giving the user the ability to set the time, perhaps through the serial port.
- Not using the gl… Dynamic C libraries in the ISR. These can impede performance since they are building a string from bitmapped fonts for the LCD.
- Using NTP (Network Time Protocol) to automatically synchronize the clock with some external networked entity.

7.6.4 Brief Description of Dynamic C's LCD Library

Dynamic C's LCD library provides a number of useful functions; the following will be of interest to most programmers. To use these functions, the programmer should refer to the library GRAPHIC.LIB.

- `dispInit()`: initializes the display and integrated keypad. `dispInit()` is for LCD displays, and `glInit()` is for graphical displays.
- `dispLedOut(int led, int value)`: turns individual LEDs on and off.
- `glPlotDot(int x, int y)`: draws a single pixel in the LCD buffer, and on the LCD if the buffer is unlocked.
- `glPlotLine (int x0, int y0, int x1, int y1)`: draws a line in the LCD buffer, and on the LCD if the buffer is unlocked.
- `glBlock(int x, int y, int Width, int Height)`: draws a rectangular block in the page buffer, and on the LCD if the buffer is unlocked.
- `glPlotCircle(int xcenter, int ycenter, int radius)`: draws a circle in the LCD page buffer, and on the LCD if the buffer is unlocked.
- `glFillCircle(int xcenter, int ycenter, int radius)`: draws a filled circle in the LCD page buffer, and on the LCD if the buffer is unlocked.
- `glXFontInit(fontInfo *pInfo, char pixWidth, char pixHeight, unsigned startChar, unsigned endChar, unsigned long xmemBuffer)`: initializes the font descriptor structure; the font is stored in Xmem.
- `glPrintf(int x, int y, fontInfo *pInfo, char *fmt, ...)`: similar to printf; uses a given font set to prints a formatted string on the LCD screen.
- `glBlankScreen()`: blanks (to white) the LCD screen.
- `glFillScreen(int pattern)`: fills the LCD screen with a pattern.
- `glBackLight(int intensity)`: sets the intensity of the backlight.

> Programmers should be careful with calling gl... library functions (or any of the Dynamic C library functions, for that matter) from an ISR. This can be a performance impediment and can lead to trouble. The best thing to do is not to call ANY Dynamic C library functions from an ISR.

7.6.5 Final Thought

This chapter introduced us to several important concepts: the types of interrupts that exist on the Rabbit 3000 microprocessor and how to use them; how to write interrupt service routines in C and assembly language; how to establish serial communication and use some of Rabbit's other peripherals such as timers, the real time clock and the watchdog timer.

In the next chapter, we will start looking at some of Dynamic C's enhancements, including costates and cofunctions, which give the code the feeling of "multitasking."

CHAPTER 8

Multitasking Overview

8.1 Why Use Multitasking?

Before discussing multitasking, the question to ask is "why use multitasking?" As long as the embedded system is achieving the desired results, why would a programmer want to delve into multitasking? The question can be answered in a few ways:

- Sections of code (called "tasks") can run independently of each other and share the CPU's time. This can give programmers the impression that more than one program is running simultaneously.
- Many embedded projects have hardware interfaces that need attention on a short turn-around time. Multitasking facilitates serving interfaces in this manner.
- Multitasking can allow programs to be written in a more structured manner, where each task[1] that needs to be carried out can be separated from others. This practice can simplify the structure of the programs, and make them easier to understand, debug, and test.
- As long as the interfaces between tasks are defined, tasks can be developed by separate members of a team, which allows several tasks to be developed and tested independently and in parallel.
- In most embedded systems, a controller often communicates with more than one external device. Multitasking allows the application to talk to the devices through separate tasks, without worrying about one big program having to manage all the devices at once. This can make the code easier to understand, develop and debug.
- The choice of an appropriate RTOS (real-time operating system) and properly designed tasks can result in a system that is highly responsive to real-world events.
- Programs usually spend most of their time in idle loops, waiting for user input, I/O cycles to complete, or interrupts. The idle time that is otherwise wasted can be channeled into running sections of code other than the ones that are waiting, thereby increasing CPU utilization.

This chapter will describe various aspects of multitasking, the key elements of an RTOS, and how multitasking is achieved with Dynamic C. Before we start to explore multitasking, let us look at a real-world example—a program used to schedule events for a finite state machine.

---

[1] The terms *task*, *process* and *thread* are often used interchangeably. Depending on the operating system in use, there may be a distinction between them.

Multitasking Overview

The example will illustrate, among other things, a CPU that wastes most of its time just waiting for things to happen.

8.1.1 Project 1: Scheduling Timed Events for a State Machine

This project implements a simple state machine with the following three states and associated rules:

State 1: stay in this state for 5 seconds, then go to State 2

State 2: stay in this state for 2 to 10 seconds, then go to State 3

State 3: stay in this state for 2 to 10 seconds, then go back to State 1

- For each state, display a message indicating which state the program is in, wait, and then transition to the next state.
- Spend a fixed amount of time in State 1, but for a random period between 2 and 10 seconds in the other states.
- While the program is waiting in any state, a key press (S2 on the prototyping board) will force transition to the next state.

Figure 8.1 presents a state diagram for this project:

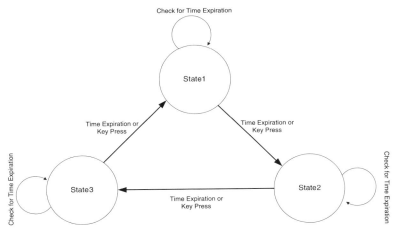

Figure 8.1: State diagram with three states.

Such a state machine has many applications in real-world control scenarios. For example, a control system can be in an initialization state, then turn on a device for a period of time, turn it off and move to the next state where it would turn another device on for some time. If the program received an external trigger during that time, it could move to yet another state.

Finite state machines, sometimes called finite state automata (FSA) by software engineers, are one of several tools available to engineers designing a system.

In the early days, unstructured or so called spaghetti coding was the order of the day. Wild flowcharts covering over-sized sheets of butcher paper passed for design documents. While this design technique is currently frowned upon, people still use this technique.

Chapter 8

Another technique in vogue in the late 1980's was a rigid top-down design methodology. Nassi Schneiderman flow-charting was a highly praised practice. Code segments only had one way in and one way out. Code was designed as a huge shell containing little shells, where each shell only had one entry and one exit point. There are purists who still design systems using this technique.

The practice of using FSA allows a fairly free form approach (much like the early and now despised spaghetti coding approach) while also providing a rigid framework. Designing code using an FSA is in some ways the best of both worlds.

An FSA is an excellent way to implement the system's main loop. Each device is polled sequentially and serviced if needed. Each state can have multiple entry conditions and multiple exit conditions. For states with multiple exit conditions, the rule is that **the logical sum of the exit conditions must equal one**. This condition is imposed so that the designer is forced to evaluate all possible exit conditions and paths. This condition is the little bit of formality required to keep an FSA from degenerating into an unfortunate bit of spaghetti code.

FSAs are often used to implement menu systems, round robin type polling loops, and even control loops. Frequently an FSA will be used to implement a round robin main loop while ISRs are used to handle counter updates, and serial ports.

The ability to use an RTC, such as shown in Program 8.1, is valuable. The technique illustrated of changing states based on the RTC can be used to timeout from an otherwise stuck state.

Program 8.1 runs on the RCM3200 prototyping board and implements the above state machine. In addition to implementing the states and associated rules, the program illustrates a number of useful concepts:

- It uses a random number generator
- Since the RTC internally represents time in seconds, the program shows how simple math can be done to calculate different time periods

Program 8.1: Code segment for a simple state machine implementation. Complete program available on the accompanying CD.

```
/*******************************************************

    RTC3a.c

    This program runs on the RCM3200 prototyping board and
    uses RTC, LCD and Timer B routines from RTC2.c
    to illustrate time-based state machine programming.

<code removed for brevity>

    /***************************************************/
    // start of state machine
    state = 1;
```
(Program 8.1 continued on next page)

Multitasking Overview

*Program 8.1: Code segment for a simple state machine implementation.
Complete program available on the accompanying CD (continued).*

```
while(1)
{
    switch (state)
    {
        case(1):
            trans_time = 5;
            glBlankScreen();
            glPrintf (0,  0, &fi6x8,  "In State1 for");
            glPrintf (0,  8, &fi6x8,  "%d seconds", trans_time);

            while (tm_rd ( &rtc ));     // Read Current Time
            lTime1 = mktime ( &rtc );  //Convert to Univ Time

            // calculate transition time to next state
            lTime2 = lTime1 + (trans_time);

            // keep checking to see when to
            // transition to the next state
            while (lTime1 < lTime2)
            {
                while (tm_rd ( &rtc ));
                lTime1 = mktime ( &rtc );

                // allow a keypress to transition
                // us to the next state
                if (!BitRdPortI(PGDR, SW2))
                {
                   // if key has been pressed,
                   // wait for key release
                   do {delay1mS(50);}
                      while (!BitRdPortI(PGDR, SW2));
                      break; //abort if switch pressed
                };
            } //while

            state = 2;             // signal transition to next
            state break;

        case(2):
            // calculate transition time to next state
            trans_time = random (8,2);
            glPrintf (0,  0, &fi6x8,  "In State2 for");
            glPrintf (0,  8, &fi6x8,  "%d seconds", trans_time);

            while (tm_rd ( &rtc ));     // Read Current Time
            lTime1 = mktime ( &rtc );  //Convert to Univ Time

            // calculate transition time to next state
            lTime2 = lTime1 + (trans_time);
```

(Program 8.1 continued on next page)

Chapter 8

Program 8.1: Code segment for a simple state machine implementation. Complete program available on the accompanying CD (continued).

```
                    // keep checking to see when to
                    // transition to the next state
                    while (lTime1 < lTime2)
                    {
                            while (tm_rd ( &rtc ));
                            lTime1 = mktime ( &rtc );

                            // allow a keypress to transition
                            // us to the next state
                            if (!BitRdPortI(PGDR, SW2))
                            {
                               // if key has been pressed,
                               // wait for key release
                               do {delay1mS(50);}
                                   while (!BitRdPortI(PGDR, SW2));
                                   break; //abort if switch pressed
                            };
                    } //while

                    state = 3;   // signal transition to next state
                    break;

             case(3):
                  // calculate transition time to next state
                  trans_time = random (8,2);
                  glPrintf (0,   0, &fi6x8,  "In State3 for");
                  glPrintf (0,   8, &fi6x8,  "%d seconds", trans_time);

                  while (tm_rd ( &rtc ));    // Read Current Time
                  lTime1 = mktime ( &rtc ); //Convert to Univ Time

                  // calculate transition time to next state
                  lTime2 = lTime1 + (trans_time);

                  // keep checking to see when to
                  // transition to the next state
                  while (lTime1 < lTime2)
                  {
                          while (tm_rd ( &rtc ));
                          lTime1 = mktime ( &rtc );

                          // allow a keypress to transition
                          // us to the next state
                          if (!BitRdPortI(PGDR, SW2))
                          {
                             // if key has been pressed,
                             // wait for key release
                             do {delay1mS(50);}
                                 while (!BitRdPortI(PGDR, SW2));
```

(Program 8.1 continued on next page)

Multitasking Overview

Program 8.1: Code segment for a simple state machine implementation. Complete program available on the accompanying CD (continued).

```
                                    break;  //abort if switch pressed
                        };
                  } //while

                  state = 1;       // back to first state
                  break;
      } //switch

} //while
```

Once the initialization has taken place, the main program executes the state machine forever. The behavior of states 1 through 3 is identical except that State 1 spends a fixed amount of time (five seconds) before moving on to State 2. Let us examine what happens in State 2:

- When the program enters the state, it establishes how much time it will spend in this state; using a random number generator, the program arrives at this value. Next, the program identifies the state on the LCD
- The program now calls the `tm_rd()` function to get the current time from the RTC in a time structure. Using the `mktime()` function, the program coverts the time to the number of seconds and then performs some simple math to determine when a transition will happen to State 3

The advantage of maintaining time in seconds is that simple math can easily be performed, provided everything is expressed in seconds. For example, the following statements perform various time calculations, where lTime represents current time:

- half an hour ago: `lTime - (30*60L)`
- half an hour later: `lTime + (30*60L)`
- 6 ½ minutes ago: `lTime - (6*60L+30)`
- a day later: `lTime + (24*(60*60L))`
- a week later: `lTime + (7*(24*(60*60L)))`
- `if (lTime1 < lTime2)`: compares two time values
- The program now repeatedly calls the RTC and checks to see whether it has spent the required time period in the state. Moreover, it scans the S2 key on the prototyping board to for a key press and, when a key has been pressed, it does some debouncing and waits for the key to be released before moving on.
- Once a state transition is determined through time expiration or a key press, the program simply updates the variable `state` to signal the next state; the state machine invokes that particular state during the next pass

Chapter 8

A couple of observations:

- Some debugging code, although not used in the final application, has been left in the program: the `LCD_print_time()` and `print_time()` functions allows us to see the time on the LCD and stdio window, respectively. The functions are slightly different since `print_time()` expects "the number of elapsed seconds since 1980" as its input parameter, while `LCD_print_time()` calls the RTC and fetches time itself. Having such debugging code handy will be an asset to the programmer, if one has to fix the program many years after deployment, and will be helpful to other people who may inherit the code in the future.
- A simple modification for this program could be to define a separate state for initialization, and put all the initialization code in there. Once the program is done with this state, it never enters it again.

It is clear from this program that most of the CPU time is spent waiting for time to pass so that a transition can be made to the next state. Since the CPU spends most of its time waiting, it has no time left to work on anything else. In this chapter, we will look at a "tasked" approach, where each task (or state) gives up control during its waiting time, freeing up the processor to work on other tasks. This approach provides for huge improvement in efficiency and higher utilization of the CPU.

In Section 8.12, we will revisit the same program from a multitasking perspective.

Embedded systems are generally classed vaguely as interrupt (or exception) driven or polled. A purely exception driven system has a main loop that does nothing—all the action takes place in the ISRs. On the other hand, a purely polled system has no ISRs—all the action happens in the main loop.

Real world systems are often weighted in one direction or the other. Either a system's architect favored exception driven design or polling, but almost always some mix is found.

8.1.2 A More Formal Definition of Multitasking

Most programs run in a linear fashion, with the CPU having only a single "context," which is a combination of all of the CPU's resources (registers, memory, ports, etc.) being in a certain state. Multitasking allows the CPU to change context so that the resources are available to multiple tasks and it appears that the CPU is servicing multiple tasks at the same time. How often the context changes and what causes the context to change depends on whether multitasking is taking place on a *cooperative* or *preemptive* basis. Other details aside, the key difference between the two types of multitasking is:

- In cooperative multitasking, a task gives up control of the CPU voluntarily, when it deems appropriate. Compared to preemptive multitasking, cooperative multitasking is easy to understand and implement, requires fewer CPU cycles, and is easier to debug. On the other hand, cooperative multitasking does not guarantee when a task gets executed, and usually requires more upfront design work than preemptive multitasking.
- In preemptive multitasking, the operating system (also referred to as the "multitasking kernel") switches tasks without the express permission of the running task. The

Multitasking Overview

RTOS time slices the CPU's time and allows each task to run for a given time slice. While preemptive multitasking usually requires less up-front design work than cooperative multitasking, it is moderately complex to implement and requires more CPU cycles for locks and semaphores. Preemptive multitasking does guarantee latency, but is significantly more difficult to debug than cooperative multitasking.

So, in its essence, multitasking in a single-processor embedded system allows one process to run for a slice of the CPU's time, and then give way to another process, either voluntarily or not, depending on whether cooperative or preemptive multitasking is being utilized. The processes (also called *tasks*) appear to run simultaneously.

In case of preemptive multitasking, the RTOS rapidly switches between tasks at predetermined intervals. The switching has to be fast enough that the user doesn't notice any degradation in the task's responsiveness, and each task has to be given enough CPU time to get its work done. We will later see how Dynamic C allows the programmer to vary each task's time slice. In fact, the time slice can even be altered dynamically in runtime.

With the application being divided into various tasks, the issue of inter-process communication needs to be addressed—how are foreground (time-critical) tasks going to communicate with the background tasks? Flags can be utilized to implement a mechanism where appropriate functions are called if their flags are set by the foreground task. Later in this chapter, we will examine the mechanisms Dynamic C utilizes for inter-task communication.

8.2 Some More Definitions

Before proceeding to program with multitasking, it will be valuable to get a handle over the key elements of multitasking: the RTOS that provides the housekeeping, the task that needs to run, and how context switching happens.

8.2.1 The RTOS

Loosely defined, the RTOS does for an embedded system what a full-blown operating system does for a personal computer—manage resources. In this context, since multiple "programs" (i.e., tasks) need to run on the same processor, the RTOS provides the following key functions:

- The RTOS decides which task needs to run.
- The RTOS assigns the available resources (such as CPU cycles, sections of memory, and I/O) to each task.
- Just as importantly, an RTOS provides facilities to insure tasks don't clobber each other by stepping on each others' resources. When a task is running, it has exclusive access to its resources and thinks it is the only program running on the system.

8.2.2 The Task

A task is a collection of three things:
- A list of code statements that need to execute sequentially.
- The resources being used by the task. The key task resources for an embedded system are memory (used for storing variables for the task), the stack, CPU registers, and I/O devices being used by the task.

- The bookkeeping information that the RTOS creates and maintains about each task. If the programmer wants to take a peek into this, most RTOSs allow this information to be obtained.

When several tasks are involved, the objective of multitasking is to make it appear that the tasks are running in parallel. Since the CPU can execute only one instruction at a time, the two multitasking approaches (cooperative and preemptive) utilize two different ways to make tasks "appear to run in parallel."

- Cooperative multitasking takes advantage of embedded delays in the running task to grant control to tasks waiting to run.
- Preemptive multitasking rules by maintaining authority over CPU time and grants time slices to tasks.

8.2.3 Context Switching

One of the key roles of the RTOS is to "switch tasks," which involves saving the context (i.e., all the resource and housekeeping information) for the running task and restoring the context for the next task that is ready to run. When this is done fast enough, the CPU gives the illusion that it is running several tasks at once. The process of saving the context for the running task and replacing it with the "ready to run" tasks's context is called the context switch.

The software architect has to consider several factors, such as how often the context has to be switched, which tasks are expected to run longer than others, and which tasks have to run more often than others, resulting in increased context switching. Since context switching is mostly overhead, doing it frequently can cause the CPU to spend more of its time managing housekeeping details instead of running the programmer's code.

As we will see in the following sections, context switching cannot be avoided entirely, depending on whether cooperative or preemptive multitasking is being utilized. In the former case, a context switch will occur only when the currently running task will willingly give up control, while in the latter case, context switching will likely occur at predetermined intervals, generally triggered by interrupts.

8.2.4 Resource Sharing

The RTOS and various tasks work hand in hand to pull a smoke and mirrors show over the user, giving the illusion of various tasks running at the same time. However, the RTOS has its own bag of tricks and it makes each task think that it is the only task running in the system and has complete control over the system's key resources, the CPU and memory.

Sharing I/O devices becomes a bit tricky… what if one task sets up a serial port to work at a certain bit rate and another task wants to use the same serial port but wants to configure it differently? The programmer has to be careful about sharing I/O resources. Independently running tasks can cause problems for these resources. It may be easier to use cooperative multitasking in such a situation, since a program using cooperative multitasking explicitly yields control to the other task. This can easily be synchronized with I/O.

8.3 Cooperative Multitasking

In cooperative multitasking, the running task yields control either when it completes execution or when it waits for an event or a resource. This type of multitasking is called *cooperative* because the tasks in the system must cooperate with each other to share the resources. It is expected that every task will voluntarily give up the CPU without keeping the other tasks waiting for too long. If a task chooses to not cooperate, it can maintain control over the CPU and not allow other tasks the run, in effect *starving* them. In fact, if a task unexpectedly gets into an infinite loop, it may starve others without realizing that it is.

Figure 8.2 shows three tasks vying for CPU control under cooperative multitasking. For example, while Task A has control, Tasks B and C can simply wait for their turn.

When Task A gives up control, the other tasks can proceed. It is not easy to determine when a certain task will run.

Figure 8.2: Task timing for cooperative multitasking.

The reader can think of cooperative multitasking as an unsupervised group of children eager to share a video game. While they can agree to share the game among each other, the child currently playing the game must give up control of the game so that the next child can play. If a child gets greedy and decides not to share the game, the other children can be deprived as they wait for the poorly behaved child to yield the game.

Cooperative multitasking is the simplest form of multitasking. It is fast and requires low overhead because the RTOS does not decide on the context switch—the running task decides when the switch happens (usually when it has nothing to do or has reached a good "stopping point"). Unlike preemptive multitasking, the RTOS does not require a supervisor task to run that doles out CPU time.

From the perspective of personal computers, early versions of MacOS and Windows used cooperative multitasking. Preemptive multitasking started to appear from MacOS 8 and Windows 95/NT, and Linux also uses preemptive multitasking.

8.3.1 Advantages of Cooperative Multitasking

Advantages of cooperative multitasking include:

- A natural approach to multitasking. Programmers often find cooperative multitasking to be very intuitive, since it models human behavior—"finish what you are doing and then work on the next chore."
- Low CPU overhead, since a supervisor isn't required to perform task switching. A small amount of system code is required for context switching, and cooperative systems usually switch between tasks much faster than preemptive ones.

Chapter 8

- Tasks that give up control when they are done, with clean entry and exit points. This makes the tasks synchronous and subject to fewer problems from asynchronous interruptions.
- Less interrupt latency, and no CPU time wasted doing preemptions—cooperative multitasking is very gentle on the task switching time.
- Pre-defined control over task interactions, and less formal communication between tasks, which leads to simplified programming of many problems.

8.3.2 Disadvantages of Cooperative Multitasking

Cooperative multitasking does have some disadvantages:

- Tasks are expected to be good citizens and give up CPU control so that other tasks can run. The system can be held hostage to greedy tasks that may hold on to the CPU longer than expected.
- The tasks are not as loosely interacting as they would be if they were strictly prioritized. This can makes certain system design problems tricky to debug and tune.
- The system is less likely to have strictly deterministic behavior since it is difficult to ensure that high-priority tasks will always get to run promptly. It often cannot be determined when the currently running task will give up control of the processor[2].

8.4 Preemptive Multitasking

A preemptive multitasking system needs a kernel (the "scheduler") to start and stop tasks. An interrupt-based interval timer is often utilized to provide a "tick timer;" the scheduler uses this timer to switch tasks at a periodic interval, known as the time slice. The periodic interval is generally provided by an interrupt; hence context switching is almost always triggered by hardware.

The scheduler does the context switch and replaces the running task with the next ready task, without the permission of the running task. Unlike cooperative multitasking, the task switch happens asynchronousy to the running task.

The reader can think of preemptive multitasking as a group of children eager to share a video game, but an adult is in charge of making sure each child gets to play for a fair share of time. The adult supervisor decides when the child playing the game must give up control of the game so that the next child can play. Depending on certain factors (such as the age of the child), the supervisor may allow a child to play for a bit longer than others.

The time slice is small, in the order of dozens or hundreds of milliseconds, so that the system gives the impression that a number of tasks are executing simultaneously.

---

[2] Strictly speaking, this is a problem for both cooperative and preemptive multitasking, for different reasons. Preemptive multitasking has nondeterministic interleaving, while cooperative multitasking has difficult-to-predict scheduling.

Figure 8.3 shows three tasks running under preemptive multitasking. The scheduler grants control to each task, based on its scheduling rules. For an RTOS it is based on the task priority. When a task's time is up, it has to relinquish control. In a situation where all tasks have the same priority, they are all guaranteed to have the same slice of time to do their thing. For example, task B takes longer to execute than the time slice, so the task is left incomplete and will have to wait to run again to complete execution.

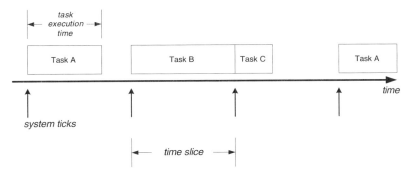

Figure 8.3: Task timing for preemptive multitasking.

From the perspective of the personal computer, operating systems based on preemptive multitasking have the ability to assign priority to the processes in real-time. Most mainframe OSs ran preemptive multitasking—time slicing (sharing) on mainframes was and remains a well-known concept.

8.4.1 Task Priorities

Unlike cooperative multitasking, preemptive multitasking associates a priority level with each task, and RTOSs generally allow the programmer to set a number of priority levels. An RTOS provides this service, but there are other types of scheduling systems available, the most common one being Round Robin scheduling where each task gets an equal shot at the CPU.

The CPU should always run the highest priority task—if a task was ready to run, then the RTOS performs a context switch and replaces the running task with one having the highest priority. If needed, RTOSs usually allow the high priority task to "suspend" itself so that lower priority tasks can execute. The suspension can happen for a specified length of time or until the high priority task receives a "wake up" signal from other tasks.

We need to be careful about priority inversion, described in Section 10.12.

8.4.2 Task Synchronization Using Semaphores

If two or more tasks depend on each other, it can lead to three known issues:

- While cooperative multitasking has clean entry and exit points, preemptive multitasking does not allow us to predict when a task is executed, since a task is switched out when its time slice is up.
- A "deadlock" condition can occur when there is a circular chain of tasks that lock on each other. For example, task C waiting on task B, which is waiting on task A, which

Chapter 8

is waiting on task C is a deadlock condition. In such a case, no tasks will continue because none has received a signal from the one it is waiting for.

- If two tasks are sharing data, a "race condition" can occur between them. This happens when one task (often called the "producer" task) starts to write data to memory and another task (often called the "consumer" task) starts to read the data before the producer task is done writing the data. The same can happen with a state. For example, if there is a state variable that needs to be updated, there can be a race to update it, leaving it in an unexpected state for one of the tasks.

Semaphores can be utilized as an interface between tasks so that they can synchronize their execution and know whether the other task has reached a certain point in its execution. The tasks utilize a messaging mechanism to keep each other informed. In section 8.8.3, we will look at how Dynamic C utilizes inter-task communication.

8.4.3 Advantages of Preemptive Multitasking

There following key advantages exist with preemptive multitasking:

- Tasks can be assigned priorities so that highest priority tasks can run for longer or more often. If all tasks are given the same priority, each task will get control in its assigned time slice, all time slices being equal.
- Preemptive multitasking provides for deterministic servicing time for tasks. Since the tick timer is utilized to generate the time slice, a programmer can ascertain the maximum response time for tasks.
- No single task can monopolize[3] the CPU indefinitely and starve the other tasks. When its time slice will be up, the running task will be switched out with another one. The RTOS retains control of the system, and gives or takes away control from tasks as it sees fit.

8.4.4 Disadvantages of Preemptive Multitasking

The added management cost of preemptive multitasking cannot be overlooked. The following are the key disadvantages of using preemptive multitasking:

- The added power of preemptive multitasking comes at a cost of increased complexity. The RTOS needs to have a task scheduler running that performs context switching, depending on allocated time slices and task priorities.
- Because context switching may happen more frequently than it does in cooperative multitasking, it can add additional CPU overhead. Moreover, the use of the scheduler increases the RTOS footprint.
- Unlike cooperative multitasking, tasks do not have clean entry and exit points. Task switching happens unpredictably and asynchronously. This can lead to complications if tasks depend on each other or share common resources. Coordinating between the tasks can cause added complexity to the system.

---

[3] To be precise, the highest priority task can always monopolize the system on an RTOS.

8.5 What to Be Careful About in Multitasking

Complex, real systems aren't always very deterministic. The systems designer has to be careful about what level of response is acceptable for a given application. Simply using multitasking may not always improve the response time.

The key to a successful multitasking implementation is to service interrupts in a timely manner, and to choose an RTOS with low software overhead. Since external interrupts can happen at any time, a large number of external interrupts can temporarily overload the processor and cause some interrupts to be missed. One way to deal with this situation is to divide the application between time-critical tasks (that run infrequently but require immediate attention) and less time-critical tasks (that run frequently); the former can run with interrupts disabled, while the latter would usually run with interrupts enabled. Several operating systems refer to these as the upper and lower handlers.

When using cooperative multitasking, it is important to know the loop time, i.e., how long it takes to do something. With Dynamic C, the loop time is the sum of the worst case of all the `costatements` (tasks) that are active within a given loop. Knowing the loop time will help us determine how long it will be before the other costatements will have control of the CPU.

With preemptive multitasking, programmers have to be careful about variables or shared resources that are partially updated by one task. Not knowing what gets updated by whom and when can lead to unexpected issues that can take time to trace.

With some RTOSes, the programmer should be careful about using preemptive multitasking with serial communication, since communication can be abruptly interrupted due to a context switch. Dynamic C has a useful solution to this issue—it uses the RTC (real time clock) for multitasking, which works at interrupt priority level 1, and serial port interrupts also work at priority 1. Thus, if we are in a serial ISR, an RTC interrupt will not affect serial communication. On the other hand, if a task switch happens, a serial ISR will not happen unless the task switch has reduced the priority level or has exited the ISR.

Serial interrupts should be handled fine within the RTOS if the user is not disabling interrupts during the execution of the task where serial data should be handled. Interrupts may add some jitter to the processing of the tasks.

Similarly, `printfs` have been implemented in Dynamic C to avoid communication issues with preemptive multitasking. The "`stdio.lib`" library is fully reentrant and the debug kernel takes care of serializing transfers of stdio output back to Dynamic C. In version 8.51, a new macro, `STDIO_ENABLE_LONG_STRINGS`, has been added so that strings longer than 127 bytes can be printed to the `stdio` window.

All library Dynamic C functions are marked as re-entrant or non-reentrant, and the programmer should check each function to determine whether it is suitable for use with multitasking.

Virtual watchdogs should be used in connection with cooperative multitasking. They insure that the cooperative system does not get hung up for too long a period.

8.5.1 When Not to Use Multitasking

Multitasking is not the panacea for all ills. While it offers certain advantages over traditional linear programming, there are certain cases when multitasking should be avoided:

- When the system is having performance problems, multitasking could add to the issues.
- When the system is already overcommitted and stretched to the point of having high interrupt latency and low response time for the application, it is not likely that multitasking will improve the situation. The solution may be to rethink the application design and related algorithms, not do as much, or pick a faster clock speed or different CPU.
- Many tasks can be solved without using multitasking. A lot of the choice has to do with personal preference. Some of it has to do with how independent the tasks are from other another.
- Programmers should not allow multitasking to complicate the system unnecessarily. A lot of people pick an RTOS because they can, not because they need it.

8.5.2 Using *Cooperative* or *Preemptive* Multitasking

Using cooperative or preemptive is often based on programmer preference. For small systems where the programmer has complete control of every task, cooperative multitasking can be superior to preemptive multitasking. In other situations, preemptive multitasking can be a better choice. There are a few general rules to consider when selecting cooperative or preemptive multitasking:

- Cooperative multitasking relies on applications making certain calls frequently in order to allow the other processes a chance to become active. Preemptive multitasking gives all processes an equal chance to get control on a regular basis. How the task are developed and how often they need CPU control can determine what type of multitasking should be used.
- Cooperative multitasking is typically simpler to program than preemptive multitasking. When using preemptive multitasking the programmer has to be careful about how data and resources interleave.
- When using cooperative multitasking, if the loop time is so high that it keeps other tasks from running that need to achieve something in a given time window, we should consider preemption. We can certainly use interrupts with cooperative multitasking to force preemption. This will help us handle code that requires low latency.
- Preemptive multitasking is more suited for several tasks that are independent to one another. In such cases, the tasks may not be sharing variables or resources, and that would simplify such a design.
- If we are doing a lot of work with a given piece of data and separate tasks are reading and updating the data frequently, there is a good change we will get into trouble if we use preemptive multitasking. In such a case, the programmer should give careful consideration to how often the data is updated, who updates it, and who reads it back.

- Tasks that need to be prioritized should use preemptive multitasking, since cooperative multitasking does not support guaranteed task prioritization.
- Tasks that need to run for a long time should use cooperative multitasking. The programmer should look at interrupts as preemptive code that is running, whether the programmer uses preemptive multitasking or not.

8.5.3 Using Both *Cooperative* and *Preemptive* Threads

Some systems can be designed to run cooperative and preemptive tasks simultaneously. For example, tasks that need high priority can be designed to run cooperatively while tasks with normal priorities can run preemptively.

Since cooperative tasks that need lots of processing can keep other tasks from running, the program should be designed so that processor intensive tasks would yield control at given periods. The system will find tasks with the same or higher priority as the running task that are ready to run. After finding such tasks it will block the current task and run the next task, waiting in the queue. In case there are no ready tasks with the same or higher priority, then control will be given to the task that was just blocked.

8.5.4 Initialization and Task Setup

Since preemptive multitasking requires a certain amount of CPU overhead for task switching, the task response time can get affected if the CPU is overburdened with the context switch overhead. A context switch rate should be chosen for the required response time to the application, without wasting CPU cycles in the context switching overhead. For example, the CPU overhead is very different if context switching happens at a rate of 64 kHz versus 1 kHz.

Tasks can have permanent and temporary data assigned to them. The kernel grants some temporary data (usually in stack space) when the task is started, and the task can explicitly ask the RTOS for data. The temporary data does not get saved if the task is killed, while permanent data remains in place if the task gets killed and then restarted. Unlike temporary data, permanent data is usually not assigned at task start up time but only when the task asks for it.

Upon initialization, each task should be granted enough stack space for its needs.

8.5.5 Avoiding Deadlock and Race Conditions

A "deadlock" situation occurs when tasks end up in a mutual wait condition—when tasks end up waiting for each other indefinitely.

As mentioned in Section 8.4.2, "race" conditions can be caused between producer and consumer tasks, especially if they are running at different priorities. If the producer task is running at a higher priority than the consumer task, it must know when the consumer task is busy processing data, so that the producer task can hold off and not send any more data until the consumer task is ready. On the other hand, if the consumer task is running at a higher priority than the producer task, the consumer task should know whether the producer task has data ready, so that it will not spend unnecessary CPU cycles looking for data that doesn't exist yet.

Atomic updates should also be considered. In this case, a variable or structured data could have a series of dependant updates that need to be atomic and two tasks could destroy the dependencies. This could happen with a long variable when it takes two machine instructions to do an update and the updating task gets interrupted between those instructions.

8.5.6 Resource Management

When shared resources are involved, such as interrupts, serial communication, or other peripherals, two tasks accessing the same resource can produce unexpected results. Programmers should put in a locking mechanism so that a task waiting to use a resource must find out that the resource is being utilized by another task.

If global variables are used and shared among tasks, such variables should be saved and restored properly when a context switch happens.

8.5.7 Inter-Process Communication

Processes or tasks generally need a synchronized method of communications. The relationship between the tasks can include one-to-one, one-to-many, or many-to-many. An RTOS needs to provide mechanisms to permit all of these kinds of communications. Common mechanisms include mailboxes, queues, and events.

8.5.8 Knowing the RTOS

The programmer should become familiar with various details of the RTOS in use. For instance, the programmer must know about interrupt latency[4] (i.e., the delay from the time an interrupt happens to the time the associated ISR gets executed. Note that this delay could be nondeterministic, for instance when interrupts are disabled in sections of code), the maximum execution times for system calls, the maximum response times for events, as well as the maximum time for which interrupts are disabled by the RTOS and ISRs. Guaranteeing the maximum response time for an event is the essence of real-time programming

Ideally, the RTOS in use must support features such as cooperative and preemptive multitasking, task priority, and predictable task synchronization.

---

[4] Interrupt latency and throughput are two different things.

Multitasking Overview

8.6 Beginning to Multitask with Dynamic C

By providing several C language extensions, Dynamic C supports both cooperative and pre-emptive multitasking. Dynamic C provides:

- `costatements` and `cofunctions` to support cooperative multitasking.
- `slicing` to support preemptive multitasking.
- Support for µC/OS-II real-time kernel as an add-on module. There is ample information available for µC/OS-II, including some excellent books. This RTOS will not be covered any further in the book.

> When working with multitasking, Dynamic C makes a strong distinction between how variables are shared between tasks: when using cooperative multitasking, variables can be shared between different tasks, but when using preemptive multitasking, variables cannot be shared between tasks.

8.6.1 Important Dynamic C Functions and Variables for Multitasking

The following system variables are shared and atomic when being updated. These are updated by the system's periodic ISR and used by some special functions[5], and should not be modified by user code. The `waitfor` statements used by Dynamic C rely on these timer variables.

- `TICK_TIMER`: this counter is updated every 1/1024th seconds
- `MS_TIMER`: this counter is updated every millisecond
- `SEC_TIMER`: this counter is updated every second

For example, `MS_TIMER` is a long and has the range 1 to 4294967296.

4294967296 / (1000 milliseconds * 60 seconds * 60 minutes * 24 hours) =

4294967296 / 86400000 = 49.71 days.

Thus, `MS_TIMER` and `TICK_TIMER` have less than two months before they roll over. The `costate` Delay functions handle this problem.

The following functions are used by the costatement "`waitfor`" constructs. The initial call to these functions starts the timing. The functions return zero when first called and continue to return zero until the number of specified ticks has passed, each tick being 1/1024 second. Once the number of specified ticks has elapsed, the functions return 1, and then the program can continue beyond the `waitfor` statement.

- `int DelayTicks(unsigned ticks)`: counts until the number of ticks specified has passed
- `int DelayMs(long delayms)`: counts until the number of specified milliseconds has passed
- `int DelaySec(long delaysec)`: counts until the number of specified seconds has passed

---

[5] These are `firsttime` functions that run just once in a given context, and can be made to run again.

Chapter 8

There are `firsttime` functions that can only be called from inside of a `costatement`, `cofunction`, or `slice` statement.

The functions below are the analogues to the above functions, and provide a periodic delay based on the time from the previous call:

- `int IntervalTick(long tick)`: Similar to DelayTicks but provides a periodic delay based on the time from the previous call
- `int IntervalMs(long ms)`: Similar to DelayMs but provides a periodic delay based on the time from the previous call
- `int IntervalSec(long sec)`: Similar to DelaySec but provides a periodic delay based on the time from the previous call

8.7 Dynamic C's Implementation of Cooperative Multitasking

Dynamic C uses *costatements* and *cofunctions* to implement cooperative multitasking. The essence of this approach is to think of the entire application as a large state machine run as an endless loop. Inside the loop, separate tasks are created; each with a *costate* statement, which run in sequence, one after the other. A high-level view is shown in Program 8.2:

Program 8.2: High level view of cooperative multitasking.

```
main()
{
    while (1) // endless loop
    {
            costate // first task
            {
                  ... // code statements
            }

            Costate // second task
            {
                  ... // code statements
            }
    } // while
} // main
```

A `costate` is therefore a list of code statements, and it uses an internal statement pointer to keep track of the running statement. When a task is created, the statement pointer points to the first statement in the `costate`. As the task gives up and recovers control of the CPU, the statement pointer directs it to execute the next statement that needs to run.

Dynamic C uses the following syntax for defining a costate:

```
costate [ name [state] ] { [ statement | yield; | abort; |
waitfor( expression ); ] . . .}
```

Multitasking Overview

The keyword `costate` identifies the statements enclosed in the curly braces that follow as a costatement. name can be one of the following:

- A valid C name not previously used. This results in the creation of a structure of type `CoData` of the same name.
- The name of a local or global `CoData` structure that has already been defined.
- A pointer to an existing structure of type `CoData`.

Costatements can be named or unnamed. If name is absent the compiler creates an "unnamed" structure of type `CoData` for the `costatement`.

state can be one of the following:

- `always_on`

The `costatement` is always active. This means the `costatement` will execute every time it is encountered in the execution thread, unless it is made inactive by `CoPause()`. It may be made active again by `CoResume()`.

- `init_on`

The `costatement` is initially active and will automatically execute the first time it is encountered in the execution thread. The `costatement` becomes inactive after it completes (or aborts). The `costatement` can be made inactive by `CoPause()`.

If state is absent, a named `costatement` is initialized in a paused `init_on` condition. This means that the `costatement` will not execute until `CoBegin()` or `CoResume()` is executed. It will then execute once and become inactive again. If the costatement is `init_on` it will not require a `CoBegin`.

Unnamed `costatements` are `always_on`. A `costatement` with `init_on` cannot be specified without specifying a name.

8.7.1 Dynamic C's Task States for Cooperative Multitasking

The state diagram in Figure 8.4 shows the possible states a task can be in and the routines used to make transitions to other states. Only one task can be in the running state at a time, and, of course, it makes the function calls to change its own state or the state of another task.

The dormant state applies to tasks that might be in program memory but have no assigned stack frame. When a task initially starts, it is associated with a frame and given control. As the diagram shows, only a running task can stop its own execution altogether and become dormant again. It is not possible to terminate other tasks directly.

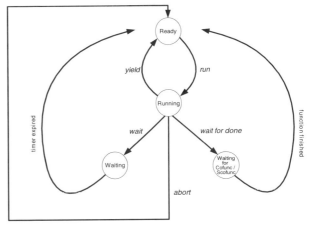

Figure 8.4: Task states for cooperative multitasking.

Chapter 8

When one of the tasks uses the `yield`, `abort`, `waitfor`, or `waitfordone` statements, the task will give up control of the CPU so that other `costate` blocks can run. These control statements affect the state of the running task as follows:

- The `yield` statement is used to immediately give up the CPU to other tasks. Once all other tasks have been given the opportunity to utilize the CPU, the task that called yield will be activated again and will continue processing at the statement immediately following the `yield` statement.

- The `waitfor` keyword is a prefix for other C expressions that return a value. The expression is computed each time the task is given CPU time and, until the expression returns a nonzero result, and then the next statement is executed. Each time the expression after the `waitfor` returns a zero result, the task (*costatement* or *cofunction*) gives up control. When the task runs again, it starts to evaluate the same expression again, and the cycle repeats until the expression returns a nonzero result.

- The `waitfordone` keyword is used only when calling special functions that are marked with the *cofunc*, or *scofunc* attributes. The `waitfordone` keyword can be abbreviated with `wfd` and will wait until the cofunction is done executing.

- The `abort` statement causes the costatement or cofunction to terminate execution. If a costatement is `always_on`, the next time the program runs the costatement, it will restart from the top. If the costatement is not `always_on`, it becomes inactive and will not execute again until turned on again by the program.

A running task can also get blocked while waiting for a semaphore signal or a message. The task then has to wait until the event occurs before it can be made ready to run again. Program 8.8 provides an example where one task has to wait for a signal from another task.

8.7.2 Cofunctions

Cofunctions are similar to functions and the key difference between costates and cofunctions is that cofunctions can receive arguments and can return values. Unlike C functions, cofunctions cannot receive structures and cannot be nested. Cofunctions can call other cofunctions as long as a `cofunction` is only in use once at any point. The scofunc handles this case by only allowing one instance of the `scofunc` to be active at once.

Basically a *cofunc* is run as a separate task, which causes the calling function to be suspended until the *cofunc* completes. The difference between a *cofunc* and a *scofunc* is that a *cofunc* cannot be called multiple times within the same *waitfordone* statement. This is because Dynamic C would attempt to run both functions simultaneously, which is not allowed. A single-user cofunction can be used instead, since they will be called in sequence. In the current walking of the tasks, a `cofunction` or any single instance of an indexed `cofunction` can only be active at a time. The scofuc has a lock in it to insure this property where it is hard for the user to decide when things will be active or not.

A `waitfordone` statement block can be used to launch several functions in parallel. If the tasks can run simultaneously, they are placed inside the `waitfordone` block. Dynamic C will run all the `cofunc` functions in parallel and waits until the longest running one completes

Multitasking Overview

before allowing the task to continue with the next statement. While this can be a useful arrangement, the programmer has to be careful because all the functions will appear to run at the same time.

Program 8.3 illustrates how a `waitfordone` block is used to launch three cofunctions together; all three functions will be called and run at the same time. Assuming that the `get_data()` function takes the longest time to execute, the `waitfordone` block will wait until that happens. Thus, *cofunctions* provide another way to do multitasking in Dynamic C.

Program 8.3: Using a waitfordone block to launch multiple cofunctions.

```
waitfordone
{
        display_text("Waiting…");
        get_data();
        led_control(FLASH_GREEN);
}
```

8.7.3 Scofunctions

What happens when the same cofunction needs to be called multiple times within a `waitfordone` block? Consider the code fragment shown in Program 8.4A:

Program 8.4A: Calling a cofunction multiple times in the same block.

```
waitfordone
{
    display_text("Waiting…");
    get_data();
    display_text("Done!");
}
```

Since all the cofunctions in the block are launched at once, it does not make sense to simultaneously launch the `display_text` cofunction two times with different parameters. Dynamic C does not allow this to happen, since each confunction only has a single data structure to tracks its execution.

Dynamic C provides a method to deal with this situation: the `display_text` cofunction needs to be defined as a "*single user cofunction*," or `scofunc`. Now, after calling `display_text` for the first time, the task will wait for the *scofunction* to complete before it is called again. Using this technique, a single-user cofunction can be called multiple times from within the same `waitfordone` block.

Program 8.4B shows how the two cofunctions in Program 8.4A should be defined:

Program 8.4B: Defining `cofunction` *and* `scofunction`.

```
cofunc void get_data()
{
<code removed for brevity>
}


scofunc void display_text ( char *pText )
{
<code removed for brevity>
}
```

When the `waitfordone` block is encountered, Dynamic C calls the first `display_text` and the `get_data` functions at the same time. Once the first call to the `display_text` function has completed, `display_text` is called again the second time around. Therefore, the `display_text` function must be called sequentially, but the `get_data` function can be run in parallel along with the `display_text` function.

8.8 Dynamic C's Implementation of Preemptive Multitasking

Dynamic C implements preemptive multitasking through two methods:

- It provides the `slice` statement to assign timeslices to tasks
- It supports the μC/OS-II real-time kernel as an add-on module

The two approaches are mutually exclusive—`slice` statements cannot be used together with μC/OS-II. Moreover, TCP/IP function calls cannot be inside of a `slice` statement.

The following syntax is used for defining a `slice` statement:

slice ([context_buffer,] context_buffer_size, time_slice)

[name]{[statement|yield;|abort;|waitfor(expression);]}

- `context_buffer_size`: This value must evaluate to a constant integer. The value specifies the number of bytes for the buffer context_buffer. It needs to be large enough for worst-case stack usage by the user program and interrupt routines
- `time_slice`: The amount of time in ticks for the slice to run. One tick = 1/1024 second
- `name`: When defining a named slice statement, the context buffer needs to be supplied as the first argument

When an unnamed slice statement is defined, this structure is allocated by the compiler.

[statement | yield; | abort; | waitfor(expression);]

The body of a slice statement may contain:

- Regular C statements
- `yield` statements to make an unconditional exit.

Multitasking Overview

- `abort` statements to make an execution jump to the very end of the statement.
- `waitfor` statements to suspend progress of the slice statement pending some condition indicated by the expression.

8.8.1 Dynamic C's Task States for Preemptive Multitasking

The state diagram in Figure 8.5 shows the possible tasks states and the transitions to other states. Only one task can be running at a time, and it can make function calls to change its own state or the state of another task.

Only a running task can stop its own execution altogether; it cannot terminate other tasks directly.

When the kernel creates a task, it allocates a timeslice to it. The timeslices are measured in ticks, where each tick equals 1/2048th of a second. The following events change the state of a running task:

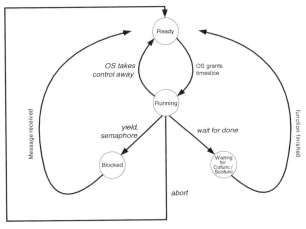

Figure 8.5: Task states for preemptive multitasking.

- When the timeslice is up, a context switch happens and the task moves from a *running* to a *ready* state, while control is granted to the next *ready* task of equal priority.
- If the running task encounters a `waitfordone` statement or block, it waits until the related *cofunctions* or *scofunctions* are executed, before it is ready to run again. If the task runs out of its allocated timeslice, execution continues at the same place when control is returned to the task.
- If a `yield` statement is encountered, the task immediately gives up the CPU to other tasks.
- An `abort` statement would cause a jump to the end of the `slice` block. Therefore, the running task would immediately give up control to the next ready task.

A semaphore signal or a message can also block the running task. The task then has to wait until the event occurs before it can be made ready to run again.

8.8.2 Task Priorities

There is no direct support in Dynamic C for task prioritization. This is achieved by using slices of unequal time length. Although this is not true RTOS prioritization, it does control the amount of time a task could have to the CPU. Consider Program 8.5 that assigns unequal time slices to two tasks:

Program 8.5: Code fragment showing tasks with unequal time slices.

```
task1slice = 1024;
task2slice=512;

slice(4096,task1slice)
{ // task #1
        //...
}

slice(4096,task2slice)
{ // task #2
        //...
}
```

8.8.3 Dynamic C's Constructs for Inter-Process Communication

Inter-process Communication involves semaphores and messages. There is no direct support in Dynamic C for Inter-process Communication, and this has to be implemented by the user. Some examples are shown in Section 8.9.3.

A semaphore is usually used to ensure mutual exclusion or synchronizing processing related to a resource.

8.9 Project 2: Flashing LEDs with Multitasking

Flashing LEDs serves a useful purpose at this stage—it allows us to examine fundamental multitasking concepts through a simple application. Earlier, we have looked at what it takes to flash an LED: determine a duty cycle, and set the right timing elements in place to turn the LEDs on and off when required.

A task that flashes LEDs is implemented as a big infinite loop, where the state of LED is changed, and the cycle repeats.

8.9.1 Flashing LEDs with Cooperative Multitasking

The project will flash one LED at 1Hz and another at 2Hz, both at a 50% duty cycle. We will use the RCM3400 prototyping board for the project. The DS1 LED will blink at 1Hz and the DS2 LED at 2Hz.

We will skip the details of initializing ports, having already covered that in many projects. The RCM3400 requires Port D to be initialized where the two LEDs are connected.

Multitasking Overview

Before we look at the code, we need to get an idea how the overall program will be structured. Let us consider one task whose responsibility it will be to flash DS1, and another task will flash DS2. The tasks will contain infinite loops within, so that the LEDs keep flashing, and the tasks will themselves be enclosed in an infinite loop, so that the tasks keep on executing. The task structure is shown in Figure 8.6A.

```
main()
{
 initPort();   // initialize ports

    while (1)
    {
       costate // task that flashes DS1 LED
       {
          for (;;)
          {
          // body of task
          } // for

       } // costate


       costate // task that flashes DS2 LED
       {
          for (;;)
          {
          // body of task
          } // for

       } // costate

    } // while
} //main
```

Figure 8.6A: Task structure.

> The "big loop" implementation is common where tasks spend life cycling through states.

Program 8.6A shows our first try at the project. We defined two costates, `flash_DS1` and `flash_DS2`, that control the DS1 and DS2 LEDs, respectively. Each `costate` has an infinite loop within and both the costates are enclosed in a `while` loop that makes then run forever.

> The example shows two named costates. Naming costates is not required.

When the program is run, we will observe that only the first costate, `flash_DS1`, will execute. Why? Because once the program starts executing this task, it never gives up CPU control. It violates one of the key tenets of cooperative multitasking, that the running task has to give up control voluntarily, so that other tasks can run. We can observe here that the `flash_DS2` costate never gets a chance to run.

We therefore have to insert a mechanism that will make each task yield control so that the next task can run. One way to do this is to insert a `yield` statement at the end of each task so that it can go through its loop, and then yield control to the other task. While this approach makes both tasks run, the downside is that each task will have to wait for the other task to finish completely. As a result, the DS2 LED will not flash at 2Hz since it will have to wait for the DS1 LED to finish flashing at 1Hz. We therefore need a better mechanism so that each task yields control as soon as it encounters the associated delay.

Chapter 8

Program 8.6A: One task runs, and one doesn't.

```
//mtled1a.c

    costate flash_DS1 always_on   // task that flashes DS1 LED
    {
     for (;;)
        {
            DS1led(ON);
            printf( "DS1 On\n" );
            delay1mS(500);

            DS1led(OFF);
            printf( "DS1 Off\n\n" );
            delay1mS(500);
 } // for
 } // costate

    costate flash_DS2 always_on   // task that flashes DS2 LED
    {
     for (;;)
        {
            DS2led(ON);
            printf( "DS2 On\n" );
            delay1mS(250);

            DS2led(OFF);
            printf( "DS2 Off\n\n" );
            delay1mS(250);
        } // for
    } // costate
```

In modifying the above program, the programmer has to be careful about a few other things:

- If each task gives up control and waits for the other one to finish, there could possibly be a situation where each tasks is waiting for the other one. This condition, deadlock, has to be avoided, otherwise it will hang the application.

- Because each task requires some delay after changing the state of the associated LED, the task calls a delay routine. What happens if both tasks happen to call the delay routine at the same time? There has to be a locking mechanism in place so that both tasks are able to call the same delay routine without stepping on each other's toes.

Multitasking Overview

Program 8.6B includes the modifications to satisfy the above conditions:

Program 8.6B: Two tasks that flash LEDs using cooperative multitasking.

```
//mtled1b.c

    costate flash_DS1 always_on  // task that flashes DS1 LED
    {
        for (;;)
        {
         DS1led(ON);
         waitfor(DelayMs(TIME_DS1_ON));

         DS1led(OFF);
         waitfor(DelayMs(TIME_DS1_OFF));
        } // for
    } // costate

    costate flash_DS2 always_on  // task that flashes DS2 LED
    {
        for (;;)
        {
         DS2led(ON);
         waitfor(DelayMs(TIME_DS2_ON));

         DS2led(OFF);
         waitfor(DelayMs(TIME_DS2_OFF));
        } // for
    } // costate
```

Each costate runs an infinite loop that contains the code to turn an LED on, wait, turn the LED off, and wait again. The LED on and off times are set with the constants `TIME_DSx_ON` and `TIME_DSx_OFF`, respectively, and these constants can be changed to alter the flash rate and duty cycle of the LEDs.

The magic happens in the *waitfor* statements that precede the LED delay. Since the `DelayMs` cofunction keeps returning a "0" until the appropriate time delay is reached, the task gives up control each time the *waitfor* statement encounters the delay function, until the appropriate delay is reached. This allows the other task to run while this task is waiting for the delay. Thus, inserting the *waitfor* statement before each on/off delay allows us to do other things while we wait to flip the LED state.

`DelayMs()` is a `firsttime` function that can only be called from inside of a `costatement`, `cofunction`, or `slice` statement. Its design allows it to be called from multiple costates so that it does not tromp on itself.

Program 8.6C modifies the above program to implement a third task that handles user I/O via the serial port. The program presents the user with a short menu that allows the user to control the LEDs on the prototyping board. Each LED has one of four states: "Off," "On," "Flashing

Chapter 8

Slow", and "Flashing Fast." The following steps should be performed to run this program:

1. Connect the serial cable to jumper block J5 on the RCM3400 proto board
2. Plug the other end of the serial cable into an available serial port
3. Use a terminal emulation program such as HyperTerm
4. Run the program with Dynamic C and wait for the user menu to appear on the screen

> The RCM3400 prototyping board uses jumper block J5 for serial I/O. A suitable cable is available from Rabbit Semiconductor (Part number 540-0009) that will facilitate hooking up a connection to a standard 9 or 25-pin serial connector.

Figure 8.6B shows the user menu that appears on the terminal emulator screen:

```
Time to Control LEDs!

LED DS1 Status: Flash Fast
LED DS2 Status: Flash Slow

User Options:
Press 1 to turn  DS1 OFF
Press 2 to turn  DS1 ON
Press 3 to flash DS1 slow
Press 4 to flash DS1 fast
Press 5 to turn  DS2 OFF
Press 6 to turn  DS2 ON
Press 7 to flash DS2 slow
Press 8 to flash DS2 fast
```

Figure 8.6B: User menu.

The task that manages serial I/O in Program 8.6C presents the menu to the user, and then waits for a key to be pressed. It uses the `waitfordelay` statement to yield program control until a key has been pressed:

```
wfd c = cof_serDgetc(); // yields until successfully getting a character
```

Multitasking Overview

After receiving valid input, the task simply updates the appropriate state variable for each LED. The tasks that drive the LEDs use the state variables to decide the state of each LED. Shown below is a code fragment from the task that controls the DS1 LED:

Program 8.6C: Code fragment from three tasks that share state variables.

```
// mtled1c.c

costate task_DS1 always_on   // task that flashes DS1 LED
{
   for (;;)
         {
           switch (status_DS1)
           {
                 case LED_OFF: {
                       DS1led(OFF);
                       abort; }

                 case LED_SOLID: {
                       DS1led(ON);
                       abort; }

                 case LED_FSLOW: {
                       DS1_time_on = TIME_SLOW_ON;
                       DS1_time_off = TIME_SLOW_OFF;
                       break; }

                 case LED_FFAST: {
                       DS1_time_on = TIME_FAST_ON;
                       DS1_time_off = TIME_FAST_OFF;
                       break; }
           } // case

           DS1led(ON);
           waitfor(DelayMs(DS1_time_on));

           DS1led(OFF);
           waitfor(DelayMs(DS1_time_off));
       } // for
} // costate

costate task_DS2 always_on   // task that flashes DS2 LED
{
<code removed for brevity>
}

costate task_serIO always_on  // task that performs user I/O
{
<code removed for brevity>
}
```

If the LED has to be "Off" or "Solid On," the case statement does its thing and aborts the costate immediately. On the other hand, if the LED has to flash, the task spends the time necessary to turn the LED on and off with the appropriate timing. Since it uses the `waitfor` statement with the `DelayMs` cofunction, the task yields control until the right amount of delay has been reached.

The `for` loop inside the `costatement` is not strictly necessary because the `costatement` will start over from the top after it completes.

This project illustrates another benefit of multitasking—as long as the interface between the tasks is identified and agreed upon, different individuals or different teams could have developed and independently tested the two tasks, and then integrated everything together. As is often the case, the hardware may not be ready in time, but separate tasks can be developed and tested while hardware development is taking place. Moreover, separating the tasks by functionality makes it easy to debug and reuse the code.

8.9.2 Flashing LEDs with Preemptive Multitasking

Let us examine how LEDs can be flashed using preemptive multitasking.

One can write a program very similar to Program 8.6B, except that the `slice` statement replaces the `costate` statement to indicate the use of preemption. Program 8.7 shows one of the tasks that use this mechanism to flash an LED:

Program 8.7: Flashing LEDs with preemptive multitasking.

```
// mtled2.c

slice(TASK_BUFFSIZE,TASK_TIME) task_DS1        // task that controls DS1
{
   for(;;)
      {
          DS1led(ON);
          waitfor(DelayMs(TIME_DS1_ON));

          DS1led(OFF);
          waitfor(DelayMs(TIME_DS1_OFF));
      }
}

slice(TASK_BUFFSIZE,TASK_TIME) task_DS2        // task that controls DS2
{
<code removed for brevity>
}
```

The tasks are given optional names and are assigned a task buffer and time task upon creation. Note that the tasks will not run if the assigned bugger is too small. For instance, if `TASK_BUFFSIZE` is changed from 200 to 100, the tasks will stop running.

Multitasking Overview

Although the program does not use shared variables, this is an appropriate place to describe them. The `shared` keyword allows atomic updates of multibyte variables. For example, if a long variable is being used with preemptive multitasking it is possible that half of the long would be updated with the other half not updated. Integers are already safe because they are updated in an atomic way. For many updates this is sufficient, though there are more complicated multi-variable update situations. This class of operation is important to µC/OS-II as well. µC/OS-II has specific synchronization methods, such as semaphores, to handle this situation. The programmer would have to implement this for slice statements. Interrupts are disabled when the shared variable is being changed.

8.9.3 Using Semaphores

There are other ways to flash LEDs, for example, by using semaphores. In our simple example, we can use a binary semaphore that synchronizes two costates. In Program 8.8A, one task turns the DS1 LED on, while the other task turns it off. The tasks use the semaphores `ledOn` and `ledOff` to signal the LED state to each other, and each tasks use the `waitfor` construct to wait and get synchronized with the other task. In effect, the tasks lock each other out until the LED state needs to change.

Program 8.8A: Using two binary semaphores.

```
// mtled3a.c

<code removed for brevity>

main()
{
char ledOn, ledOff;

initPort();
ledOff = TRUE;
ledOn = FALSE;

  while (1)
  {
    costate turn_on_DS1 always_on // task that turns DS1 LED on
    {
    for (;;)
        {
            waitfor(ledOff);
            ledOff = FALSE;

            DS1led(ON);
            waitfor(DelayMs(TIME_DS1_ON));

            ledOn = TRUE;
(Program 8.8A continued on next page)
```

Chapter 8

Program 8.8A: Using two binary semaphores (continued).

```
        } // for
    } // costate

    costate turn_off_DS1 always_on // task that turns DS1 LED off
    {
    for (;;)
        {
            waitfor(ledOn);
            ledOn = FALSE;

            DS1led(OFF);
            waitfor(DelayMs(TIME_DS1_OFF));

            ledOff = TRUE;
        } // for
    } // costate

  } // while
} //main
```

The same result can be obtained with a single binary semaphore. One has to understand how the `waitfor` construct works to make this happen. Program 8.8B shows the resulting code.

Program 8.8B: Using a single binary semaphore for the same result.

```
// mtled3b.c

<code removed for brevity>

main()
{
char semaphore;

initPort();

semaphore=TRUE;

  while (1)
  {
    costate turn_on_DS1 always_on // task that turns DS1 LED on
    {
    for (;;)
        {
            waitfor(semaphore);

            DS1led(ON);
```
(Program 8.8B continued on next page)

Program 8.8B: Using a single binary semaphore for the same result (continued).

```
            waitfor(DelayMs(TIME_DS1_ON));

            semaphore = FALSE;
        } // for
    } // costate

    costate turn_off_DS1 always_on // task that turns DS1 LED off
    {
    for (;;)
        {
            waitfor(!semaphore);

            DS1led(OFF);
            waitfor(DelayMs(TIME_DS1_OFF));

            semaphore=TRUE;
        } // for
    } // costate

    } // while
} //main
```

Dynamic C does not contain any built-in primitives for using semaphores with slice statements; these need to be created by the programmer.

8.10 Project 3: Using Linux to Display Real Time Data

While this project does not use multitasking on the Rabbit 3000, it is an example of a system-level multitasking where three programs are running concurrently on two CPUs. The Rabbit 3000, of course, is running one, but the Linux box is running two applications for this project, plus the normal complement of background processes. The bash and gnuplot scripts presented here are examples of how a programming problem can be broken up into multiple processes and the overall system is simplified.

The purpose of this project is to demonstrate how to use a Linux platform to receive and display data in real time. Building on the previous Linux project in Chapter 5, we will replace the known voltage input to the RCM3400 with a temperature sensing probe. The probe will be sampled every 5 seconds to record outside air temperature. The RCM3400 will convert the ADC code into degrees and transmit that to the Linux PC. The Linux PC will receive the temperature, archive the data value, and update a graphical plot of the temperature history. Additionally, the RCM3400 will transmit a "time stamp" to the PC along with each temperature reading.

Chapter 8

For this section, we will use equipment setup as in Figure 8.7:

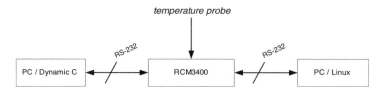

Figure 8.7: Set up to acquire and display real-time data.

The project will:

- sample the temperature probe on the RCM3400
- convert the ADC code into degrees C
- send the current time and temperature to the Linux PC
- archive the received data on the Linux PC
- update a graphical plot of the temperature history every 5 seconds

For temperature measurements, we will use a commercial probe: the Qualimetrics Model 5190 Temperature/Humidity Probe. This device requires 5VDC for power. The Model 5190 returns two linearized 0–1V voltages representing relative humidity 0–100% and temperature in degrees (–40°C to +60°C).

To supply the probe with 5V power, we'll use the RCM3400 development board prototyping area. The temperature signal from the probe will be attached to ADC channel 1, the channel we characterized in the previous project. From our characterization, we know that the RCM3400's ADC channel 1 has very little noise.

8.10.1 Theory: Sampling the Temperature Probe

As in the previous project, we'll use the Dynamic C function anaIn() to return the raw ADC code from the converter. Taking into account the gain setting, we can determine the input voltage as follows:

$$\text{Input Voltage} = \frac{\text{ADC code} * 20}{\text{GAIN} * 2048}$$

According the probe's user manual, the temperature signal ranges from 0 V to 1 V, corresponding to a temperature of –40°C to +60°C, a range of 100°C. Thus 10 mV indicates 1°C above –40°C. We can now convert the input voltage to temperature as follows:

$$\text{Temperature} = \frac{\text{INPUT VOLTAGE}}{10 \text{ mV}} - 40°C$$

8.10.2 Transmitting the Sampled Data to the Linux PC

Once we have the temperature, we use the RCM3400 real-time clock (RTC) to obtain the current time, and format both the time and temperature for transmission to the Linux PC. As we saw in the previous project, we must format our data so that the Linux programs can parse

Multitasking Overview

the data quickly and easily. By reviewing the gnuplot manual, we note that gnuplot can accept time data in a user-configurable format. This allows us more flexibility in how we format the time stamp on the RCM3400.

Dynamic C provides the `getRTC()` function to read the value of the real time clock. The return value from this function is the number of seconds since the first day of 1980. We can use the Dynamic C function `mktm()` to convert this offset into a `struct tm` data structure which breaks out the time into separate year, month, day, hour, minute, and second fields.

Once we have the time and temperature data, building a string to transmit requires only a single call to `sprintf()`, followed by the `serDputs()` call we used in the previous project. For consistency, we then wait for the transmit queue to empty. Finally, we delay for almost 5 seconds and start all over again.

On the Linux PC, we use the same data capture program described in Chapter 5, except that we don't bother to echo the received data to the user via the standard output data stream.

8.10.3 Analyzing the Data Graphically

Gnuplot will be used again to display the incoming data. To make a gnuplot graph with time on the X-axis, we first configure gnuplot to understand our format for date strings. Next, we instruct gnuplot to scale the display in a pleasing manner (scale the time axis automatically, and place a lower bound of 0 on the temperature axis). We then have gnuplot read the data from our archive file. Lastly, we delay for 5 seconds, and have gnuplot replot the data.

8.10.4 Implementation: Capture, Convert, Format, and Transmit the Data

We will re-use the Dynamic C from Chapter 5 for this project. The program begins with the following macros:

- `GAINSET`: represents the ADC gain; a list of alternative gain values is included for convenience
- `DINBUFSIZE`: represents the size in bytes of the serial channel D receive buffer
- `DOUTBUFSIZE`: represents the size in bytes of the serial channel D transmit buffer

Next is the function `sample_ad()`, identical to the one we used in Chapter 5's project.

Next is the function `msDelay()`, copied from the sample programs supplied by Rabbit Semiconductor. This function waits a specified number of milliseconds and then returns. It will be used for coarse loop-timing in our function `main()`.

Our function `main()` begins by declaring its variables. We have five variables:

- `sample` is an integer that holds the most recent ADC code that we've read
- `TempV` and `TempC` are floating point variables that hold the temperature in Volts and degrees C
- `buffer` is a character array used to hold the formatted time and temperature data
- `current` is a `tm` structure that holds the current RTC values broken down for ease of use

Chapter 8

Execution begins by initializing the RCM3400 board by calling `brdInit()`. Next, we initialize the ADC, serial port UART, and the serial port buffers with code used in the previous project. We then begin an infinite for-loop.

Inside the loop we being by sampling the ADC by calling the `sample_ad()` function. If the return value is negative, signifying an error condition, we continue to sample the ADC until the error goes away.

Next, we convert the raw ADC code into a floating point number that indicates the input voltage. The equation from above was rewritten slightly to both remove the division operation as well as to hard-code in the gain constant and resultant input voltage range. Then, we convert the input voltage into a floating point number that represents the temperature in degrees C. We get the current time with calls to `read_rtc()` and `mktm()`.

At this point we start formatting the data for transmission. This requires only a call to `sprintf()` with a defined date and time format, and the computed temperature from above. Lastly, we queue up the string for transmission, delay until the transmission is complete, wait approximately 5 seconds and start the process over.

8.10.5 Capturing Data on the Linux Side

This section discusses the bash shell script that is used to the received date and temperature data on the Linux PC. This script is the same as the script used in Chapter 5's project. The code is presented in Program 8.9:

Program 8.9: Bash shell script used to receive data.

```sh
#!/bin/sh
#
# acquire a batch of data from the serial port
#

COMPORT=/dev/ttyS1
FILEBASE=data.log

# get the time/date
DATE=`date +%j-%T`
FILE=$FILEBASE"-"$DATE

# tell the user what's happening
echo "logging data from $COMPORT to $FILE"
echo "press control-C to exit"

cat $COMPORT | tee $FILE
```

Multitasking Overview

8.10.6 Sampled Temperature Visualization

This section discusses using the gnuplot program to display real time data from a file. The basic premise is to save a script file for gnuplot that has the required commands to display the datafile once, and then repeat itself after a short delay. Program 8.10 lists the script below:

Program 8.10: Using gnuplot to display realtime data.

```
#
# realtime.gnu
#
# A gnuplot script file to display realtime data from a file
#
set timefmt "%Y-%m-%d-%H:%M:%S"
set xdata time
set autoscale x
set xrange [*:*]
set ylabel "Temp C" -4,-11
set autoscale y
set yrange [0:]
plot "data.log-353-17:16:15" using 1:2
pause 5
load "realtime.gnu"
```

The script is initially loaded by gnuplot with the command ' load "realtime.gnu" '.

The script begins by describing the format of our time stamp data, and configuring the x-axis to plot time data. Next, we ask gnuplot to auto-scale the axes, and add a label to the vertical axis. The heart of the script is the plot command, which gives the name of the data file and instructs gnuplot to plot column one of the data file (the time stamp) along the x axis, and column 2 of the data file (the temperature) along the y axis. Finally, the script pauses for 5 seconds and then reloads itself.

The output trend graph is shown in Figure 8.8.

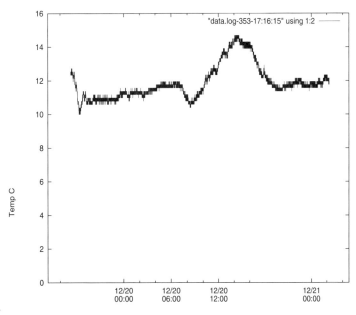

Figure 8.8: Trend graph from gnuplot.

325

Chapter 8

The Linux PC receives a continuous stream of temperature data from the RCM3400. The data is captured in a log file. Gnuplot is used to fire off an updated graphical plot to the monitor every five seconds. The user watching the Linux box gets to see a new updated temperature plot every five seconds.

8.11 Project 4: Designing an Analog Sensor Task

The project will define a general-purpose sensor routine that examines analog sensors. These sensors can retrieve byte values from, say, an ADC, and trigger based on threshold values. The routine can easily be modified to trigger on changing bit status from digital (binary) sensors.

Let us consider some requirements for the sensor task. The user can define the following elements for each analog sensor:

- Enable or disable scanning of a sensor
- Enable or disable alarm generation from a sensor; i.e., whether a trigger from a sensor will cause an alert for the user
- Define hysteresis

The sensor task will be used to monitor temperature values using the thermistor supplied with the RCM3400 prototyping board. It is simple to modify this routine to support other types and a larger number of analog sensors.

The sensor task can constantly monitor sensor status; another task can take action based on triggers, and a third task can update sensor controls based on user input from the above requirements.

In Chapter 9, we will modify this code so that the routine displays sensor values and alerts on a web page, and allows the user to update sensor controls from the web page.

8.11.1 Theory: Elements of a Sensor Task

If temperature rises above or falls below preset thresholds, alarms are generated. There is hysteresis built-in to re-arm sensors if temperature varies between thresholds and associated hysteresis. All the values are in degrees Fahrenheit, although the program can easily be modified to work in Celsius. If a reading goes above an upper threshold an alarm will be generated (if alarms are enabled) and the status bit for that threshold will be set. Further alarms will not be generated until the reading goes below that threshold by the positive hysteresis value. This will also clear the status bit for that threshold. Similar operation occurs for lower thresholds using the negative hysteresis value.

Multitasking Overview

The default values for each threshold are shown in Table 8.1; the temperature defaults are shown in degrees Fahrenheit:

Table 8.1: Default temperature values for the sensor task.

Parameter	°F
Positive Hysteresis	2
Negative Hysteresis	2
Upper Critical Threshold	80
Upper Noncritical Threshold	76
Lower Noncritical Threshold	60
Lower Critical Threshold	55

The four temperature thresholds are illustrated in Figure 8.9:

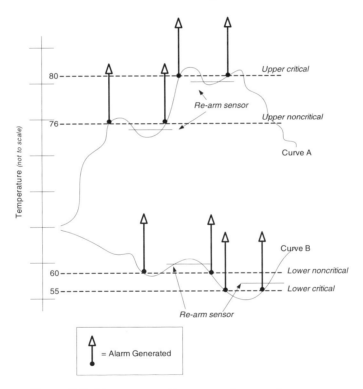

Figure 8.9: Alarm generation over a temperature range.

An alarm is sent out just once for each threshold sensor exceeding a threshold. For instance, if the upper non-critical temperature threshold is set to 65°F, and the temperature rises to 66°F, a "Temperature" alarm will be sent out just once, as the temperature crosses the upper non-critical temperature threshold.

Chapter 8

8.11.2 Theory: Reading Temperature from the RCM3400

The RCM3400 uses an ADS7870 analog-to-digital converter. It features 11 bits of resolution, 8 input channels, and a serial interface. While the part can deliver up to 52k samples per second, the serial interface in the design does not allow sampling at such high rates.

The math needed for reading temperature values is as follows:

Temperature in Celsius, $T_c = T_k - 273.15$

Temperature in Fahrenheit, $T_f = 1.8 * (T_k - 255.37)$

Where T_k is the temperature in Kelvin, and is calculated as follows:

$T_k = (B_t * T_{kstd}) / (T_{kstd} * (\log(\text{fabs}((-D_{raw} * R_s) / (R_{tstd} * (D_{raw} - (D_{max} * \text{Gain})))))) + B_t);$

$$T_K = \frac{\beta_T \cdot T_{KSTD}}{T_{KSTD} \cdot LOG\left(\left|\frac{-D_{RAW} \cdot R_S}{R_{TSTD} \cdot (D_{RAW} - (D_{MAX} \cdot GAIN))}\right|\right)} + \beta_T$$

T_K calculated temperature in kelvin

T_{KSTD} standard temperature in kelvin

β_T thermistor's characteristic beta

R_S stimulus resistor

R_{TSTD} thermistor's resistance at standard temperature

D_{RAW} Measured ADC code (raw data)

D_{MAX} Maximum ADC code

GAIN channel gain in $\frac{v}{v}$

The RCM3400 libraries provide the `anaIn` function that delivers the raw value from an analog channel on the ADS7870. Thus, the raw value for the thermistor is obtained as follows:

D_{raw} = anaIn(channel#, operation_mode, gain_code)

Where *operation_mode* supports single ended, differential, or milliamp inputs. The *gain_code* value provides for a gain of 1 to 20, depending on the input voltage range. More information on this function is provided in the `RCM34xx.lib` library file.

8.11.3 Implementation: Coding the Sensor Task

The project uses a thermistor with the following characteristics:

- Beta = 3965
- Series resistor we used on the RCM3400 prototyping board = 2.5 kΩ
- Resistance at 25°C = 3 kΩ

Multitasking Overview

Before running this program on the RCM3400 prototyping board, the thermistor needs to be connected between the AIN7 and the AGND inputs of the prototyping board.

This project uses the following three tasks to implement sensor scanning with alarms:

- `task_Temp`: continuously scans temperature readings and determines if the temperature crosses any thresholds. If so, it sets up the appropriate variables for the alarm task. This task scans the sensor every 500 milliseconds.
- `task_serIO`: performs user I/O via the stdio window to display temperature thresholds, alarm status, and enable or disable alarm generation.
- `task_Alarm`: quietly sits idle and checks the status once a second to see whether `task_Temp` needs to generate an alarm.

Because of the amount of code required, this program runs out of root space and generates runtime errors such as "Unexpected Interrupt." One way to make this program work is to select the "Separate Instruction and Data Space" option from Dynamic C's project settings.

Program 8.11 shows a code fragment that checks to see whether the temperature has exceeded the upper noncritical or the upper critical thresholds, and then also checks the hysteresis to determine whether the sensor has to be armed again.

Program 8.11: Code fragment for determining temperature alarms.

```
// sensor.c

    Tk = getTemp();                     // acquire temperature
    Tc = Tk - 273.15;                   // convert to celcius
    Tf = 1.8*(Tk - 255.37);             // calculate fahrenheit

    if (Tc >= Threshold_UNC)
        {
                if (Tc >= Threshold_UC)
            {
                AnaTempStatus |= TEMP_AT_UC;
                AnaTempStatus |= TEMP_AT_UNC;

                if ((TempAlarmsEnable & MASK_DISABLE_UC) && SensorArmed)
                    AlarmStatus = TRUE;
            }
            else
            {
                AnaTempStatus |= TEMP_AT_UNC;
                if ((TempAlarmsEnable & MASK_DISABLE_UNC) && SensorArmed)
                    AlarmStatus = TRUE;
            }
        }


        // Check hysteresis to find out whether we should re-arm sensors
        if (Tc < Threshold_UC  - HysterisisP) SensorArmed = TRUE;
        if (Tc < Threshold_UNC - HysterisisP) SensorArmed = TRUE;
```

Chapter 8

The tasks share various variables to control sensor scanning, alarm generation, etc. While Dynamic C allows shared variables with cooperative multitasking, variables cannot be shared with Dynamic C's implementation of preemptive multitasking.

Figure 8.10A shows the user menu for the above program:

```
Sensor Management
-----------------

Temperature is 78.96 F (26.09 C)

User Options:
Press 0 to Display Thresholds
Press A to Enable all Alarms
Press D to Disable all Alarms
Press 1 to Enable Upper Critical Alarm
Press 2 to Enable Upper Non-Critical Alarm
Press 3 to Enable Lower Non-Critical Alarm
Press 4 to Enable Lower Critical Alarm
Press 5 to Disable Upper Critical Alarm
Press 6 to Disable Upper Non-Critical Alarm
Press 7 to Disable Lower Non-Critical Alarm
Press 8 to Disable Lower Critical Alarm
Press 9 to Display Alarm Triggers
```

Figure 8.10A: User menu.

Figure 8.10B shows an alarm condition and the result obtained when a user disables the alarm trigger and then presses the "0" and then "9" keys to display the temperature thresholds and alarm triggers, respectively. The program outputs messages with "Action," "Status," and "Warning" headings to indicate response to user requests, status and warning messages, respectively.

```
    WARNING: *** Upper Non-Critical Alarm ***   Temperature is 79.44 F

    ACTION: Upper Non-Critical Alarm Disabled


    Temperature Thresholds:

    STATUS: Upper Critical: 80.00 F
    STATUS: Upper Non-Critical: 76.00 F
    STATUS: Lower Critical: 60.00 F
    STATUS: Lower Non-Critical: 55.00 F



    STATUS: Upper Critical Alarm Enabled
    STATUS: Upper Non-Critical Alarm Disabled
    STATUS: Lower Non-Critical Alarm Enabled
    STATUS: Lower Critical Alarm Enabled
```

Figure 8.10B: "Action," "status," and "warning" messages.

This project can be modified in many ways, including the following:

- For sake of simplicity, the project uses hard coded values for temperature thresholds. `task_serIO` can easily be modified so that the user can enter these values at will.
- The program is structured so that individual threshold alarms can be enabled and disabled. For instance, the program can be made to trigger alarms on "Upper Critical" and "Lower Non-Critical" thresholds but not others. `task_serIO` can be modified to accommodate modifying variables associated with temperature thresholds.
- The project implements just one sensor. The associated sensor variables can be put into an array so that multiple sensors can be scanned.
- While the project implements an analog sensor and no digital sensors, it is easy to scan binary values of digital sensors and take appropriate action.

8.12 Back to the State Machine from Project 1

In Project 1, we implemented a state machine with a set of rules. Let us revisit the program and examine how easily it can be implemented with multitasking. To keep things really simple, we can code each state as a separate task, and the rest of the program takes care of itself. A code fragment is shown in Program 8.12 that shows the three states in this program.

Program 8.12: Code fragment showing modified state machine from Program 8.1.

```
    RTC3b.c

<code removed for brevity>

    // start of state machine

    state = 1;

while(1)
{
  costate state1 always_on
  {
    for (;;)
      {
          if (state != 1) abort;

          trans_time = 5;
          glBlankScreen();
          glPrintf (0,  0, &fi6x8,  "In State1 for");
          glPrintf (0,  8, &fi6x8,  "%d seconds", trans_time);

          waitfor(DelaySec(trans_time));

          state = 2;                 // signal transition to next state
          yield;
```

(Program 8.12 continued on next page)

Chapter 8

Program 8.12: Code fragment showing modified state machine from Program 8.1 (continued).

```
      } // for
    } // costate

    costate state2 always_on
    {
      for (;;)
        {
            if (state != 2) abort;

            // calculate transition time to next state
            trans_time = random (8,2);
            glBlankScreen();
            glPrintf (0,  0, &fi6x8,  "In State2 for");
            glPrintf (0,  8, &fi6x8,  "%d seconds", trans_time);

            waitfor(DelaySec(trans_time));

            state = 3;                    // signal transition to next state
            yield;

      } // for
    } // costate


    costate state3 always_on
    {
      for (;;)
        {
            if (state != 3) abort;

            // calculate transition time to next state
            trans_time = random (8,2);
            glBlankScreen();
            glPrintf (0,  0, &fi6x8,  "In State3 for");
            glPrintf (0,  8, &fi6x8,  "%d seconds", trans_time);

            waitfor(DelaySec(trans_time));

            state = 1;                    // signal transition to next state
            yield;

      } // for
    } // costate

  } //while
```

8.13 Final Thought

When implemented properly, multitasking can make life easier for the programmer. In order to be productive and effective, the programmer should understand various aspects of the multitasking kernel being used, the capabilities and the pitfalls.

Dynamic C supports both cooperative and preemptive multitasking. Depending on the application and one's own preferences, the programmer can choose to use cooperative or preemptive multitasking, or a combination of the two.

CHAPTER 9

Networking

Networks are ubiquitous, and now exist in places where they did not exist five years ago. Broadband home networks and public WiFi networks are being deployed globally at a great pace. The Internet is the most identifiable form for networking for the lay person—we can now find Internet access in large and small offices, homes, hotel rooms, restaurants, airports, coffee shops, cruise ships and commercial airplanes. We take for granted more and more services that use networking to improve our daily lives. Credit card transactions, email, online banking, e-Commerce, online delivery tracking, and online movie rentals are just a few examples of commonplace services that did not exist a decade ago.

Networked embedded devices are finding uses in diverse areas from building access controls to smart homes to wireless cameras and entertainment appliances. A growing number of industrial devices, as well as consumer and enterprise-grade devices now use embedded web servers for configuration, management, monitoring and diagnostics. Industry analysts are predicting the use of embedded devices in the near future that converge media, entertainment and productivity. Networking is one of the key enablers to that vision.

In this chapter, we will look at networking from the perspective of Rabbit core modules. We will discuss common networking protocols at a high level, examining not how they work or how they are implemented, but how they can be used on Rabbit core modules.

Networking is a very broad field, including local area networks (LANs), wide area networks (WANs), metropolitan area networks (MANs), and wireless technologies. Each of these areas has its own protocols and interfaces. One can get into a lot of detail with networking protocols—DHCP and TCP, for example, have been described in thicker books than this one. Consider the subject of socket programming that we have covered in just one section here—entire books have been written on this subject and the reader is advised to look for more detailed coverage elsewhere. Dynamic C libraries support just a core set of protocols from these technologies, and these are enough for most embedded applications. The goal of this chapter is not to educate the user on networking protocols, but to examine how a networked application can be built using Dynamic C's networking features. Refer to the enclosed CD-ROM for some excellent technical papers on Dynamic C's TCP/IP implementation.

Networking is not limited to Ethernet. RS-485 is a physical interface widely used for building networks. With RS-485, programmers often have to write their own protocol. Although Dynamic C libraries provide strong support for RS-485 and other physical network interfaces, this chapter will focus only on Ethernet-based connectivity.

Networking

We will first examine a number of networking protocols that are supported by Dynamic C, and then build some projects that use some of these protocols. We will also build some applications with C++ and Java that will help us control some of these projects. We will wrap up the chapter with a project that brings it all together—hardware characterization and interfacing, user interface design, and embedded web server programming.

9.1 Dynamic C Support for Networking Protocols

In this section, we will briefly describe some of the networking protocols supported by Dynamic C. The authors assume that readers are familiar with the seven-layer OSI model.

From the programmer's perspective, there isn't much for Dynamic C to do at the presentation and session layers; most of the action happens at the transport and application layers. Embedded applications are most likely to use the layers in the application layer (FTP, HTTP, etc.) or TCP and UDP directly in the transport layer. Most deployed networking uses 4 layers of the OSI; protocols operating at these layers are the ones most likely to be used in applications. Figure 9.1 shows the four layers most relevant to embedded developers.

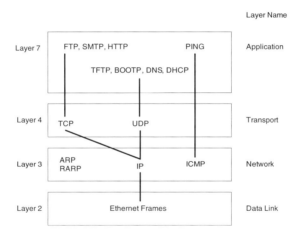

Figure 9.1: The four-layer networking model and related protocols.

9.1.1 Common Networking Protocols

Dynamic C provides support for the following protocols:

IP: The **Internet Protocol** is where the magic starts. The *Data Link* layer deals with switching Ethernet frames, based on MAC addresses, while the *Network* layer uses IP addresses to describe sources and destinations on a network.

ARP: The **Address Resolution Protocol** allows a device to discover a MAC address, given an IP address. This forms a bridge between the TCP/IP protocol suite and any link level address, of which Ethernet is an example.

RARP: The **Reverse ARP** does the opposite of ARP—it provides us with the IP address that is associated with the given MAC address.

ICMP: The **Internet Control Message Protocol** implements various messages not encapsulated by the other TCP/IP protocols. In particular, the well-known "ping" command uses ICMP messages to determine if a network device is reachable.

TCP and UDP are two major transport protocols. TCP is connection oriented, while UDP is connectionless. TCP provides reliable data transfer, while UDP provides best-effort service. These will be described here in some detail, and we will cover some more detail in Section 9.7.

TCP: The **Transmission Control Protocol** is the building block for a host of networking services. The main purpose of TCP is to provide reliable connection-oriented data transfer, and it uses various methods for flow control and error detection to accomplish its mission. Routing and congestion cause the timing of packet arrivals to be non-deterministic, which does not guarantee that packets will arrive in the same sequence in which they were transmitted. TCP uses a sequencing mechanism to line up packets for upper layers in the same order the packets were sent.

As shown in Figure 9.1, the following applications use TCP as the underlying transport:

- FTP: The **File Transfer Protocol** allows us to do just that—transfer files over a network. Internet users often use FTP as a mechanism to download files from a remote host.

- SMTP: The **Simple Mail Transfer Protocol** is used to send and receive email. A number of popular email clients use SMTP for the underlying mail transport.

- HTTP: The **Hypertext Transfer Protocol** is commonly used with browsers. HTTP defines how web pages are formatted and transmitted, and certain commands that the browser must respond to. The actual formatting of the web pages is defined by HTML (**Hypertext Markup Language**).

UDP: Unlike TCP, the **User Datagram Protocol** does not guarantee data reliability. In fact, there is no guarantee whether a packet sent via UDP will get to its destination (that is why the UDP transport is often called a "best effort" datagram service). Moreover, UDP does not reassemble packets to get them lined up in the same order they were delivered. UDP's connectionless nature results in simplicity of implementation code[1] and lower housekeeping overhead. Unlike a TCP connection, which must be synchronized and maintained through the network, UDP requires neither initialization handshake between the source and the destination, nor the networking resources that are tied up in maintaining a reliable connection.

As shown in Figure 9.1, the following applications use UDP as the underlying transport:

- TFTP: The **Trivial File Transfer Protocol** is a simpler version of FTP, and uses UDP to transfer files; it does not use TCP's reliable delivery mechanisms.

- DNS: The **Domain Name System** is a mapping scheme between domain names and their associated IP addresses. For example, every time a browser tries to access http://www.google.com/, the domain name server translates the "google" domain name into its associated IP address: 216.239.39.99, and the browser accesses the

---

[1] In the protocol layer, not necessarily in the user's application.

IP address without going through the DNS translation. If the user types "216.239.39.99" into a browser window, the browser will access the "google" web server without going through a domain name server.

- DHCP: The **Dynamic Host Configuration Protocol** allows dynamic assignment of IP addresses to devices. In the embedded systems context, this means that an embedded system can boot without an IP address, and can negotiate with a DHCP server to receive an IP address. Dynamic assignment of IP addresses is common, since it eases the burden on network administrators to statically assign and manage IP addresses. It is common to have DHCP servers built into routers for home networking.

TCP and UDP ensure integrity of the payload with checksums (up to a certain extent, since the checksum mechanism is not perfect).

TCP is a point-to-point connection protocol, whereas UDP is connectionless and therefore allows for other possibilities, such as broadcast messages.

Some additional applications are of interest to us—we can use the following utilities to debug networked applications, and some others are listed in Section 9.14.

- Telnet: this utility uses TCP to perform remote logins. It is commonly used to log in to networking devices such as routers and switches. For security reasons, some network administrators block telnet access to their devices. Moreover, telnet is not secure because it sends unencrypted data across networks.
- Ping: As Figure 9.1 shows, Ping uses ICMP messages to determine whether a networked device is reachable via its IP address.

9.1.2 Optional Modules Available for Dynamic C

In addition to providing support for the networking protocols listed in Section 9.1.1, Dynamic C supports the following protocols, provided as add-on modules:

- PPP: The **Point-to-Point Protocol** allows a device to perform TCP/IP networking over a serial connection. Most dialup Internet connections use PPP.
- AES: The **Advanced Encryption Standard** is meant as a replacement for the aging DES (**Data Encryption Standard**). While the DES provided a key size of 56 bits, AES supports 128, 192, and 256 bit keys. Although triple DES can be used for added security, it is not as efficient as AES. Dynamic C supports 128, 192, and 256 bit AES encryption and decryption through the `aes_crypt.lib` library. AES is not itself a protocol but is used by other protocols.
- SSL/HTTPS: The **Secure Socket Layer / Secure HTTP** module allows users to have an encrypted web server (HTTPS). This is useful for creating a secure interface to a networked embedded device, and should always be used when the web server is accessible from the Internet, especially when the embedded systems control physical, potentially dangerous, devices.
- SNMP: The **Simple Network Management Protocol** uses a messaging mechanism to manage networking devices. A request / response mechanism is typically deployed

for device management and maintenance of status data. Dynamic C supports SNMP through the `snmp.lib` library.

9.2 Typical Network Setup

Before looking at the Rabbit core module's network connectivity, we will present a "big picture" view of where the core module will exist in a networked development environment. We will highlight both a corporate network and a home network, since a software developer may work in either or both of these environments.

9.2.1 Typical Corporate Network

A corporate network generally has a lot of redundant devices to offer a high degree of availability. The network is used for internal operations, as well as for customer-facing activities such as the corporate web site, eCommerce, and remote connectivity with partners, customers, and employees. Having various network elements in active and standby mode provides for quick failover and recovery. The firewalls secure the internal corporate network against unprivileged access, and there is a "demilitarized zone," the DMZ, that exists outside the corporate firewalls.

The web servers can be on either side of the firewall. If they are on the internal network behind the firewall, port 80 is opened for them to allow web requests to go through the firewall. These details vary, depending on the company's security policies and infrastructure.

Access switches inside this simple corporate set up connect users in the corporate intranet. In addition, application servers in the secure intranet run corporate applications for email, databases, inventory management, etc. The corporate intranet, shown in Figure 9.2, is partitioned into various Virtual Local Area Networks (VLANs) that separate functional access. For example, corporate users and administrators use separate VLANs, while engineers use one called a "Test VLAN" that provides access to the Internet but not to corporate applications. Various other networking elements, for example, that implement storage, content caching and intrusion detection, are not shown here.

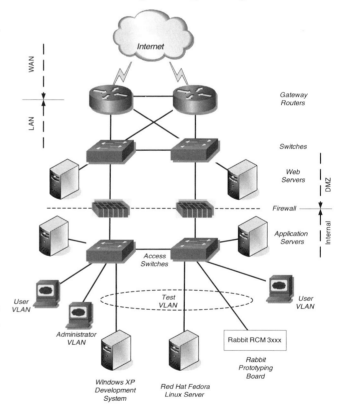

Figure 9.2: Rabbit core module in a corporate network.

Networking

The overall network can use a mix of fiber optic, Gigabit Ethernet, 10/100 Mbit Ethernet, serial, and frame relay technologies in different parts of the network.

During development phase, a Rabbit core module will likely connect to an internal "Test VLAN" that development engineers will use to test networked applications. Dynamic C will run on a Windows® workstation in the same VLAN, and the development engineer may use a Linux machine to test connectivity with the Rabbit core module. The engineer will connect these devices to an access switch that provides 10/100 Mbps connectivity to the engineer's office or lab bench.

It is not necessary to have the development system on the same network as the core module. Since the Rabbit core module is programmed via a serial link, as long as the programmer does not need to test the device over the network from the development machine, the core module can work on a separate network. The core module can be tested via a test system on the test network.

9.2.2 Typical Home Network

Conceptually, the home network consists of a router that connects the home LAN to an external WAN. A cable modem serves as the link between the cable-based broadband service and the router. The Internet service provider (ISP) dynamically assigns an IP address to the router's WAN connection. The router usually implements NAT (Network Address Translation) to allow multiple computers to connect to the internet with only one address provided by the ISP. The router also provides DHCP to assign IP addresses to devices on the home network. A private addressing scheme can be used on the LAN that is separate from that on the WAN. The router connects to a switch that provides layer 2 switching to all the home devices.

Figure 9.3 breaks out the functional pieces of a multifunction home networking device. The router, firewall, Ethernet switch, and even the wireless access point, can be in a single box. Almost everything in the diagram is connected with 10/100 Mbps Ethernet. Unlike the corporate environment, there is no redundancy, and little or no management capability in the networking devices.

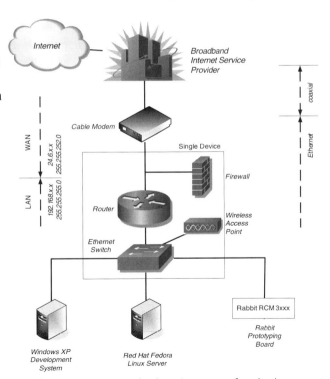

Figure 9.3: Networked environment for the home.

339

9.3 Setting up a Core Module's Network Configuration

Before a Rabbit core module can be used for networking, several decisions have to be made, that include where the core module will fit in the overall network, how it will be addressed, whether the networking configuration will be hard coded, or whether it may change at runtime. While this section will not cover all these issues in detail, it will introduce readers to various aspects of networking that need to be considered when bringing up a core module.

9.3.1 Setting up the IP Address

As is the case with any network device, the Rabbit core module needs to have an IP address that is consistent with the overall network addressing scheme where the module will be used. If this is done incorrectly, various network elements will not recognize the core module and will not respond to queries from the device.

Two methods are commonly used to set up an IP address in networking devices:

- *Static* addressing uses hard-coded or manually-configured IP addresses. These addresses do not change unless manually reconfigured. A module that uses static addressing will work on the specific network segment for which it is configured, but will need reconfiguration to work in other networking segments.

- *Dynamic* addressing uses an external entity, a DHCP server that uses a request / response mechanism to assign IP addresses dynamically. How this works most often is each time a network device power cycles, it requests for an IP address from the DHCP server[2]. After both devices negotiate and agree upon an IP address, the DHCP server leases that IP address and makes an entry in the *DHCP Client Table*. Since the addressing is dynamic, the device can renew its IP address at any time by requesting the DHCP server for a renewal. Dynamic addressing makes it easy to configure network elements and create a "plug and play" environment where devices can be plugged into different network segments without the need for manual reconfiguration. While static addressing is useful for certain applications, such as router ports or web servers, dynamic addressing may be convenient for instances when the device is acting as a client[3].

In previous versions of Dynamic C, the programmer often had to use static IP addressing. With version 7.2, Rabbit Semiconductor improved the DHCP implementation. In previous versions, DHCP used to be blocking (that is, when a DHCP negotiation was happening, no other code on the device could be run), but now it isn't. Whether a programmer uses static or dynamic addressing (or which ones will be supported) is a design-time decision. There are good arguments for either, depending on the purpose of the device. For example, if there are multiple embedded devices running the same firmware and static addressing, they would boot up with the same IP address and cause conflict on the network. In such a scheme, DHCP would be preferable.

---

[2] Not only is the address dynamic, but its duration is time limited (and configurable and negotiable).
[3] When initially "bringing up" target hardware, use of a static address can make life easier, since the IP address is one thing that isn't changing.

Networking

Two steps are required to set up an addressing scheme with Dynamic C:

1. The programmer should create a custom configuration library, say, `custom_config.lib`, based on the `\lib\tcpip\tcp_config.lib` library. The programmer should make all necessary changes to the custom configuration library. In Chapter 6, we covered the details of creating custom libraries and, for this chapter, we will create a library called `custom_config.lib` for custom network configurations.

2. The `TCPCONFIG` macro describes various parameters of the physical interface that will be used in the user program. The `TCPCONFIG` macro works as follows:
   - If the macro has a value of 0, the network configuration is done in the main program, not in any libraries.
   - If the macro has a value less than 100, the network configuration is done by `tcp_config.lib`.
   - If the macro has a value at or higher than 100, the network configuration is done in the custom library, such as `custom_config.lib`.

When using the `TCPCONFIG` macro in a custom library, the programmer must copy the appropriate library code from `tcp_config.lib` to custom library. For example, when using static IP configuration with Ethernet, the programmer will define `TCPCONFIG` to be 100, and copy the appropriate code from `tcp_config.lib` to `custom_config.lib`, so that the static IP address will be defined in `custom_config.lib`. The programmer does not have to replicate all of `tcp_config.lib` in to `custom_config.lib`, but to simply use enough of it to do the custom network configuration.

The top of `custom_config.lib` must contain the following definition:

```
#ifndef CUSTOM_CONFIG_H
#define CUSTOM_CONFIG_H
```

The programmer should set up the `TCPCONFIG` macro according to Table 9.1[4]. The top of `tcp_config.lib` will describe the steps for creating custom networking configurations.

Table 9.1: Table used to set the TCPCONFIG macro.

TCPCONFIG	Ethernet	PPP	DHCP	Runtime	Comments
0					Do not do any configuration in the library; this will be done in the main program
1	Yes	No	No	No	
2	No	Yes	No	No	
3	Yes	No	Yes	No	
4	Yes	Yes	No	No	
5	Yes	No	Yes	No	Like #3, but no optional flags
6	Yes	No	No	Yes	
7	Yes	No	Yes	No	DHCP, with static IP fallback

---

[4] This table is subject to change, since Rabbit Semiconductor is expected to continue development on the libraries.

Chapter 9

Instead of creating a separate library for custom configuration, the other option for the programmer is to modify `tcp_config.lib` directly (to set the IP address, gateway, network mask, and DNS server), and to set the `TCPCONFIG` macro accordingly.

In addition, we can use the `ifconfig()` function to make changes at run-time.

9.3.2 Link Layer Selection

We also need to tell the core module which physical interface we are using. Three interfaces are supported:

- Ethernet: This is the main focus of this chapter, and we will build all our code to support this protocol. The macro `USE_ETHERNET` should be defined to support this protocol, and the macro `USING_ETHERNET` can be queried to find out whether the core module is using Ethernet.
- Point-to-Point Protocol (PPP): This protocol is commonly used over serial ports; it uses encapsulation and transports other protocols over a point-to-point link. For example, a Rabbit core module's serial ports can be enabled to work with PPP. The macro `USE_PPPSERIAL` should be defined to support this protocol and the macro `USING_PPPSERIAL` can be queried to find out whether the core module is using PPP.
- PPP Over Ethernet (PPPOE): With this protocol, an Ethernet frame transports the PPP frame. The macro `USE_PPPOE` should be defined to support this protocol and the macro `USING_PPPOE` can be queried to find out whether the core module is using PPPOE.

We will not cover PPP and PPPOE; readers can find more information about these in the Dynamic C TCP/IP manual.

The link layer selection is part of the setup in `tcp_config.lib`. This should not be done in the application; some of the predefined configurations in `tcp_config.lib` should instead be used.

9.3.3 TCP/IP Definitions at Compile Time

Programmers should use the `#use dcrtcp.lib` directive to choose the networking library.

It is critical to call the `sock_init()` function before proceeding with networking or calling any functions that relate to networking. Moreover, the code should check to insure that the call to `sock_init()` has been successful; a return value of 0 indicates that the call to `sock_init()` was successful.

The network interface takes some time to come up after a call to `sock_init()`. This is especially the case when dynamic addressing is being used, because negotiation with a DHCP server can take time.

A short piece of code will tell us if `sock_init()` has been successful, the interface has come up, and we are ready to proceed:

```
sock_init();
while (ifpending(IF_DEFAULT) == IF_COMING_UP) tcp_tick(NULL);
```

Networking

Once we are able to proceed, we need to make sure that the function `tcp_tick()` gets called periodically. This can be done within the "big loop" of the program or within a separate `costate`.

In certain cases, the following definitions will be useful. These will help us allocate enough TCP and UDP socket buffers, respectively:

```
#define MAX_TCP_SOCKET_BUFFERS    1
#define MAX_UDP_SOCKET_BUFFERS    1

#define TCP_BUF_SIZE              2048
#define UDP_BUF_SIZE              2048
```

These are described in more detail in the Rabbit TCP/IP User Manual. These macros can be modified from the defaults to fit the resource profile of the user application. For example, we might want to increase `TCP_BUF_SIZE` to increase performance, but at the cost of more memory usage. Note that any definitions of these macros must come before the "`#use dcrtcp.lib`" line.

9.3.4 TCP/IP Definitions at Runtime

The `ifconfig()` function is used to make changes at run-time. This function is similar to the `TCPCONFIG` macro, except that it sets network parameters at runtime. In addition, the programmer can use `ifconfig()` to retrieve runtime information. The function allows us to set an arbitrary number of parameters in one call.

A number of other functions are available to look at run-time information:

- `ifup()`: attempts to activates the specified interface.
- `ifdown()`: attempts to deactivate the specified interface.
- `ifstatus()`: returns the status of the specified interface, whether it is up or down.
- `ifpending()`: returns indication of whether the specified interface is up, down, pending up or pending down. This reveals more information than `ifstatus()`, which only indicates the current state (up or down).
- `is_valid_iface()`: returns a Boolean indicator of whether the given interface number is valid for the configuration.

These functions are described in detail in the Rabbit TCP/IP User Manual, provided on the CD-ROM.

9.3.5 Debugging Macros for Networking

Dynamic C provides a numbers of useful macros for debugging networked applications:

- `#define DCRTCP_DEBUG`: turns on network-related debugging
- `#define DCRTCP_VERBOSE`: turn on all verbosity
- `#define DCRTCP_STATS`: turn on statistics counters

The above macros enable debugging and verbosity for functions related to ARP, IP, UDP, TCP, BOOTP, ICMP, DNS, SNMP, PPP, IGMP, etc.

Chapter 9

When the VERBOSE macros are defined, setting the variable debug_on to a number 0 through 6 will enable various levels of TCP-related messages. The higher debug_on is set, the more messages.

9.4 Project 1: Bringing up a Rabbit Core Module for Networking

Here we take our first baby steps. Before learning to do something more exciting over a network, we must first insure that the core module comes up and is accessible via the network. In the following examples, we will bring up the RCM3400 prototyping board. We will verify network connectivity by using the ping command, to make sure we can reach the core module on our network segment.

9.4.1 Configuration for Static Addressing

Program 9.1 brings up the prototyping board with a static IP address. In order to do this, we needed to take the following steps:

Table 9.1 tells us that the TCPCONFIG macro needs to be set to a "1" for static addressing. If we use the default values in TCP_CONFIG.LIB, the board will come up with a private Class A address of "10.10.6.100." Assuming that we are going to make the device work in a Class C private address space, as shown in Figure 9.2A, we will define a static IP address of "192.168.1.50." Therefore, we need to define a custom library and set up the TCPCONFIG macro accordingly.

We will create a custom configuration library, CUSTOM_CONFIG.LIB, and will store it in the same folder as TCP_CONFIG.LIB. We will copy the following definitions from TCP_CONFIG.LIB to CUSTOM_CONFIG.LIB and will modify them to suit the addressing in our environment:

```
#define _PRIMARY_STATIC_IP    "192.168.1.50"
#define _PRIMARY_NETMASK      "255.255.255.0"

#ifndef MY_NAMESERVER
#define MY_NAMESERVER         "192.168.1.1"
#endif

#ifndef MY_GATEWAY
    #define MY_GATEWAY        "192.168.1.1"
#endif
```

The above network addresses will need to be modified if we use the board in other network segments.

Networking

Next, since we are defining our own configuration, we will define a custom value for the `TCPCONFIG` macro. Since values above 100 are read from `CUSTOM_CONFIG.LIB` instead of `TCP_CONFIG.LIB`, we will use a value of 100; this is consistent with static IP configuration from Table 9.1. Except for the line checking for the value of `TCPCONFIG`, everything else is just copied from `TCP_CONFIG.LIB`:

```
#if TCPCONFIG == 100
        #define USE_ETHERNET            1
        #define IFCONFIG_ETH0 \
                    IFS_IPADDR,aton(_PRIMARY_STATIC_IP), \
                    IFS_NETMASK,aton(_PRIMARY_NETMASK), \
                    IFS_UP
#endif
```

Finally, we need to make sure that the master library file `MYLIBS.DIR` has an entry for `CUSTOM_CONFIG.LIB`. Thus, the top two lines of `MYLIBS.DIR` contain the two libraries we have defined so far in the book:

```
CUSTOMLIBS\MYLIB.LIB
LIB\TCPIP\CUSTOM_CONFIG.LIB
```

> Each time Dynamic C gets reinstalled, the `LIB.DIR` file gets rewritten. Therefore, the file needs to be modified each time a Dynamic C upgrade is performed.

Program 9.1 includes the code needed to bring up the RCM3400 prototyping board with static IP addressing. Once the relevant initialization code has been run, the board displays its IP address in the stdio window.

Program 9.1: Configuration for static addressing.

```
// basicStatic.c

#define PORTA_AUX_IO

#define TCPCONFIG 100

#memmap xmem
#use "dcrtcp.lib"


/*******************************/
void main()
{
(Program 9.1 continued on next page)
```

Chapter 9

Program 9.1: Configuration for static addressing (continued).

```
   char buffer[100];

      //debug_on = 5;

      brdInit();

      printf( "\nWaiting to bring up TCP with static addressing...\n");

      sock_init();

      // wait until the interface comes up
      while (ifpending(IF_DEFAULT) == IF_COMING_UP) tcp_tick(NULL);

      /* Print who we are... */
      printf( "My IP address is %s\n\n", inet_ntoa(buffer, gethostid()) );

      while (1)
      {
            tcp_tick(NULL);
      } // while

} // main
```

To state the obvious, we need to make sure that the static IP address is not already in use in the network segment. Two devices on the same network segment, using the same IP address, can cause all kinds of conflicts. Moreover, this can look like the beginning of a network attack to managed switches and intrusion detection systems in a corporate environment and these switch ports may get shut down.

To verify connectivity, we should ping the core module from a workstation to make sure we can reach that IP address on the network. Figure 9.4 shows the output of the `ping` utility:

```
C:\>ping 192.168.1.50

Pinging 192.168.1.50 with 32 bytes of data:

Reply from 192.168.1.50: bytes=32 time=1ms TTL=64
Reply from 192.168.1.50: bytes=32 time<1ms TTL=64
Reply from 192.168.1.50: bytes=32 time<1ms TTL=64
Reply from 192.168.1.50: bytes=32 time<1ms TTL=64

Ping statistics for 192.168.1.50:
    Packets: Sent = 4, Received = 4, Lost = 0 (0% loss),
Approximate round trip times in milli-seconds:
    Minimum = 0ms, Maximum = 1ms, Average = 0ms
```

Figure 9.4: Verifying connectivity to the core module.

9.4.2 Configuration for Dynamic Addressing

In order to support dynamic address allocation through DHCP, we need to change just one macro definition in Program 9.1. The code fragment shown in Program 9.2 does just that:

Program 9.2: Configuration for dynamic addressing.

```
// basicDHCP.c
#define TCPCONFIG 5
```

We do not need to define anything in `CUSTOM_CONFIG.LIB`, since we are using a predefined configuration that will be read from `TCP_CONFIG.LIB`.

Once we compile and run the program, and after the relevant initialization code has been run, the core module will display its IP address in the stdio window. At this point, we can ping the core module to make sure we can reach that IP address on the network.

What happens if there is no DHCP server present? The programmer can set up hard-coded (or configured) "fallback" IP addresses to use in case the Rabbit core module is unable to dynamically receive an IP address. The core module tries to contact the DHCP server several times over a period of about twelve seconds (this can be configured). If there is no response, then it falls back to using a fixed IP address and network mask.

The fallback address should be specified using:

```
ifconfig(IF_DEFAULT, IFS_DHCP_FB_IPADDR, <ipaddr>, IFS_END);
```

See the function description for `ifconfig()` for details.

There is also an `IFS_DHCP_FALLBACK` which tells DHCP whether to allow any fallback, plus `IFG_DHCP_FELLBACK` to test whether the stack is currently using a fallback.

If there is no fallback address, the network port will not be usable, since no host is allowed to have a zero IP address.

9.4.3 A Special Case for Dynamic Addressing

A group of embedded controllers, working together in an environment, can often have the same firmware running in them. This brings us to a special case with dynamic addressing: what happens if all these devices power up and look for a DHCP server, and a DHCP server is not found? Are these devices going to fall back on a default IP address? This will not work, because if they are running the same firmware, they may all default to the same IP address, which will cause conflicts in the network.

When designing for such an environment, the systems designer must consider cases where networked devices have to "look within" for determining an IP address, instead of relying on an external DHCP server. In fact, the Internet Engineering Task Force (IETF[5]) has devoted

---

[5] Look at http://www.ietf.org.

a working group to this area, called "Zero Configuration Networking[6]." Among other things that involve small network connectivity, the working group looks at address resolution without DHCP.

9.5 The Client Server Paradigm

Before we explore network programming in greater depth, it is important to understand the client / server paradigm. This is a common approach to network programming, including the Berkeley Socket API, which we will examine in the next section.

The word "server" may make us think of rack-mounted enterprise-grade machines with multiple CPUs, terabytes of storage and redundant power supplies, running complex applications. A server is completely different in the network paradigm, and, for the most part, is similar to a client. From a networking perspective, the main differences between a client and a server are:

- A client initiates the connection. The client requests services when needed, and the server responds. For the most part, the server is listening for incoming requests and only then does it take action.
- A client generally handles one communication at a time while a server can be communicating with multiple clients. A client may have multiple connections with the same or different servers as part of the communication.

Depending on the connection protocol and the programming interface used, the server and the client need to do certain things in sequence in order to establish communication. For example, we will later examine which calls a TCP client has to make and in what order so that it can establish a connection with a server.

How a client and server communicate depends entirely on the application, and both parties must follow a given set of rules for things to work[7]. For example, a browser on a personal computer knows what to do once a user has logged into a brokerage account, and the server on the other side knows that it will now be accepting encrypted communication that it has to act upon. From an embedded systems perspective, both parties have to use the same protocols and have to know what connections to talk to. Section 9.6 describes a well-known interface that helps us keep the communications in order.

9.5.1 What to be Careful about in Client Server Programming

There are certain special cases that we have to be careful about in client server programming:

- Byte Order: although two systems may send or receive the same data, one architecture may send out the most significant byte first while the other may start conversation with the least significant byte. We need to understand how the client and server communicate so that the data bytes do not get swapped unexpectedly. "Big Endian," which specifies "the most significant byte first," is the general rule for network communication.

---
[6] Look at http://www.zeroconf.org/.
[7] The set of rules is called a protocol.

Ideally, we need to be independent from the platform byte order. This helps make the code portable to platforms with different byte orders, and the code does not depend on the other side to have a specific byte order.

It is often enough to use the standard Berkeley macros `htonl`, `htons`, `ntohl`, and `ntohs` to convert values between host and network byte order:

- Higher Level Protocol Definition: we should be clear about various aspects of the communication, such as who starts communicating, what data needs to be exchanged and the format of the data, how the request and response are structured and what are the timeouts, etc. It is critical to avoid conditions where both the client and the server are indefinitely waiting for a response from the other party.
- Disconnection: both parties should agree upon termination of the communication. In certain cases, it may be enough to assume that communication may be terminated at any time, while in other cases, a given sequence of bytes sent from party to another may indicate that the party wishes to disconnect. Either way, the system should be designed accordingly so that it can recover from expected and unexpected disconnection.

9.6 The Berkeley Sockets Interface

The sockets interface is yet another subject that several books have been written about, and an exhaustive discussion of Berkeley Sockets is outside the scope of this book. We are presenting the sockets interface here only to introduce the reader to the code on the PC side. Note that the Dynamic C TCP/IP API (Application Programming Interface) is not the Berkeley sockets API, although there are some similarities.

The Berkeley Sockets Interface was developed at UC Berkeley in the early 1980s for BSD (Berkeley Software Distribution) Unix. It is commonly referred to as the Sockets API or just Sockets.

The sockets API provides an abstraction from the underlying network stack, where the programmer interfaces with the sockets API, without thinking about how the network stack is implemented on a given hardware platform or operating system. The sockets API also helps with platform independence, because it helps make the application code portable between languages and operating systems. Almost all of the popular languages of today have adopted the Berkeley sockets API; the programs presented in this chapter use this API for Java, C# and C++.

A fundamental concept in the Sockets API is that the destination of a message is a port in the destination machine; the port being a virtual connection spot to plug into or to send a message to. There are "well known port numbers" used by applications in the network stack, and unused port numbers can be dynamically assigned by the application. Therefore, a socket is a combination of an IP address and a port number. Port numbers range from 0 through 65,535, and most operating systems reserve port numbers below 1024. We have to make sure that our applications use port numbers that are allowed by the operating systems we work with. In Section 9.10.1, we will illustrate ports needing to be explicitly opened for I/O since they are blocked by an operating system's built-in firewall.

Chapter 9

While the sockets concept is generally applied to two separate computers on a network, sockets can also be used to communicate between processes running on the same computer. For example, there exists an internal loopback address (127.0.0.1) on most systems that can be used as a software interface.

Another advantage of abstracting the network layer is that the client and server applications do not have to be written in the same language or development platform. For example, we will use the same Rabbit program to work with servers on both the Windows and Linux platforms in various languages. Using an agreed-upon interface and networking protocols makes platform and language abstraction possible.

The Berkeley socket API uses two key types of sockets: Datagram and Stream[8]:

DGRAM (datagram[9]) sockets are used for UDP communication. The following functions work with DGRAM sockets:

- `sendto()`: sends data.
- `recvfrom()`: receives data. This can happen only after a `bind()` operation has been performed.
- `bind()`: attaches the socket to a local port address. This needs to be done before data can be sent or received.

STREAM sockets are used for TCP communications. Since TCP is a connection-oriented protocol, a connection needs to be established before a STREAM socket can send or receive data. There are two ways to create a TCP connection:

- An *active* socket can be created by using the `connect()` function. This requires us to specify the TCP/IP port address of the other party and, once connected (i.e., after the TCP three-way handshake), we can carry out two-way communication. A client usually connects to a server through an active connection.

- A *passive* socket can be created by using the `bind()` and `listen()` functions. Servers often start out with a passive socket, where they wait and listen for an incoming connections. Once that has taken place, making a call to the `accept()` function creates an *active* socket. Multiple active sockets can be created from a single passive socket, since TCP allows multiple connections on a single port, as long as the ip_address:port_number combination is unique. For example, each of the following is a separate connection:

    10.0.0.1:5555 — 10.0.0.2:80

    10.0.0.1:5556 — 10.0.0.2:80

    10.0.0.3:5555 — 10.0.0.2:80

Thinking one level higher, a server creates a socket and "names" it, so that it is identifiable and clients can find it, and then the server waits, listening for service requests. Clients create

---

[8] There is a third type, called *raw*, used for custom protocol development. It bypasses the TCP and UDP layer and works with raw IP packets.

[9] A datagram is to UDP what a packet is to TCP.

Networking

a socket, find a server socket by its name or IP address and port, and then try to connect to the server socket. Once the basic conversation has taken place, both parties can continue with two-way communication.

9.6.1 Making a Socket Connection

We will examine the socket interface both from the server side and the client side. In either case, before a client or server can talk to anyone, it needs to create a socket. For our purposes the most important part of this step is specifying the type of socket, whether TCP or UDP. Applications need to make the `socket()` call first to create the socket.

If the server is creating a socket, it must then use the `bind()` function to attach the socket to an IP address and port number. If using TCP, the server can then get into a listen state and listen for incoming connections. Otherwise, if using UDP, the server can block until it receives a UDP datagram.

If using TCP, the client tries to connect to a server using `connect()` and can then `read()` and `write()` data to that socket. The client does not need to call `bind()`; it gets an arbitrary local port number, which for TCP clients is usually just fine. In case of UDP, the client can simply use `sendto()` and `recvfrom()` to talk to a UDP server.

These operations are summarized in Figures 9.5A and 9.5B.

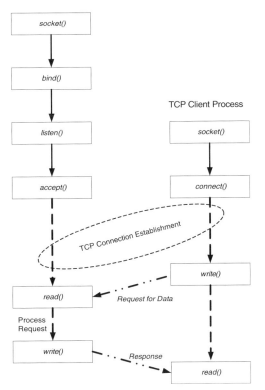

Figure 9.5A: TCP socket operation.

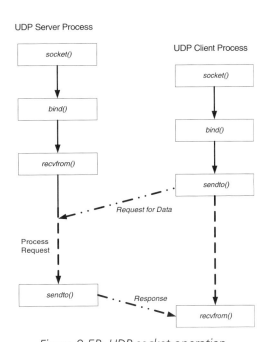

Figure 9.5B: UDP socket operation.

We will keep these socket operations in mind in developing code on the Windows and Linux platforms. We will use the appropriate calls in C++ and Java to use the sockets interface.

9.7 Using TCP vs. UDP in an Embedded Application

UDP is geared towards fast delivery while TCP is focused on guaranteed delivery, albeit at the cost of additional connection and resource overhead. The choice to use TCP or UDP depends on the application.

The additional reliability of TCP means higher implementation overhead of the protocol stack, in terms of code complexity and processing time. This does not mean that using TCP will add complexity and overhead to an application. Depending on what the application is trying to do, using TCP can actually *reduce* application complexity.

Moreover, TCP requires higher networking resources than UDP, since TCP requires an initialization handshake between the sender and receiver, and maintains a connection between the two devices. On the other hand, an application that needs reliable packet delivery will be less complex using TCP than a version building more reliability upon UDP. If the application requires TCP's guarantees, the programmer will need to do much more work to implement error detection and recovery, fragmentation and sequencing with UDP.

A significant factor in choosing TCP or UDP would be whether guaranteed delivery is required or not. A closed loop control system, for example, may not need to use TCP since it will monitor the system's behavior from the response. Moreover, if a system has built-in mechanisms for error detection and correction, it can use UDP instead of TCP.

Another factor is the network distance, which includes the number of hops, congestion, and round trip time (RTT) between the source and destination nodes. For example, if the two endpoints are spread far apart and connected through the Internet, UDP is not likely to work well without "adjustments."

Yet another factor to consider is whether the communicating parties need tight synchronization. For example, since TCP ensures that packet will arrive in order, the receiver can act on them in that sequence. On the other hand, packets may arrive out of order with UDP, and we need to consider what effect that will have on the overall system.

While it may first appear that the simplicity of UDP makes it a good choice for embedded systems, choosing TCP versus UDP depends on the needs of the application. For instance, most embedded system use small messages to communicate, which will fit in small datagrams[10], instead of exchanging many thousands of bytes of data. An application that does not require guaranteed delivery and one where a single datagram can carry all the required data would not need the flow control, buffering, sequencing or reliability offered by TCP.

Let us ponder a bit over the "UDP practical limit of 512 bytes per datagram." The Internet Protocol specification, RFC 791 for IPv4, requires that a host handle an IP datagram size of at least 576 bytes. So the effective size of the data payload would be: 576 less 20 (IP header

---

[10] UDP has a practical limit of 512 bytes per datagram, but it could be 1,472 bytes on local Ethernet. We definitely do not want UDP datagrams larger than the Maximum Transmission Unit (MTU), since that will force fragmentation, and worse, may cause reliability issues.

without options) less 8 (UDP header), which equals 548 bytes. 512 is likely chosen as a nice binary number.

Another significant concept to consider is that TCP is stream-oriented, while UDP is record-oriented (a record being a packet). If one wants the reliability and simplicity of TCP with a record-oriented interface, it can be built quite easily.

Broadcast or multicast applications should use UDP. If these applications start creating connections between each client and server, they can overwhelm the network resources. For example, UDP is often used for real-time streaming. Some packets will be dropped, and that's acceptable. On the other hand, if the application needs to record streamed audio at the highest quality, for example, it should use TCP to maintain a good connection.

If the programmer is trying to build a high degree of reliability in communication by implementing flow control, congestion avoidance, error detection and retransmission, it may be simpler to use TCP and let the lower level routines do the hard work "behind the scenes." Implementing these mechanisms when using UDP will often not be worth it.

Programmers are strongly discouraged from trying to "reinvent TCP" with UDP. The biggest risk is to add congestion to the whole network. Also, the resultant application will be more complex than the equivalent using the built-in TCP stack. This is primarily because the TCP API is well though out, while if a programmer were to reinvent TCP, they will likely create something clumsy and inefficient, unless this is being done with great expertise.

9.8 Important Dynamic C Library Functions for Socket Programming

We will use a number of Dynamic C library functions in our networking applications. Let us examine these from the perspective of the various phases of the connection. The goal here is not to provide too much detail about each function but to introduce the reader to the available functions. To keep things simple, the parameters to the functions are not shown here.

> Parameters to library functions can be examined in Dynamic C by placing the cursor over a function and hitting control-H.

Most of these functions work with both TCP and UDP sockets and the reader has to read about the functions in greater detail or look at sample programs to determine the behavior of each function for use with TCP or UDP communication.

Also note that these functions are current as of Dynamic C 8.61. Future versions may modify these functions or add new ones, and the reader is encouraged to consult the latest Dynamic C documentation available.

9.8.1 Functions used for Connection Initialization or Termination

- `sock_init`: should be called before using other networking functions. It initializes the necessary low level packet drivers. The return value indicates if the initialization was successful.

After this function is called, the network interface will take some time to come up, especially if DHCP is being used. The programs in this chapter use a while loop to wait until the interface has completed initialization or has failed to come up.

- `tcp_listen`: tells the low level packet driver that an incoming session for a particular port will be accepted.
- `tcp_open`: actively creates a session with another machine with the given IP address and port number.
- `tcp_extopen`: this is an extended version of `tcp_open` and also actively creates a session with another machine.
- `sock_close`: closes an open socket. In order to de-allocate resources, it is good practice to close open sockets when the client and server agree to disconnect or when the client or server has established that the other party has gone away for good.

9.8.2 Functions used to Determine Socket Status

- `sock_established`: helps us determine whether a socket is currently active.
- `sock_bytesready`: indicates whether data is ready to be read from a socket.
- `sock_alive`: indicates one of two states of a socket connection: a) reset or fully closed, or b) opening, established, listening, or in the process of closing. `tcp_tick` performs similar checks but involves greater overhead.
- `sock_readable`: indicates whether a socket connection has data available to be read, and how much data is ready to be read.
- `sock_writable`: indicates whether a socket connection can have data written to it, and how much data can be written.

9.8.3 Functions used to Send or Receive Data

These functions work only for TCP communication, but have their counterparts for exchanging UDP datagrams:

- `sock_puts`: In binary mode, sends a string; in ASCII mode, appends a carriage return and linefeed to the string.
- `sock_gets`: It reads a string from a socket and replaces the carriage return or linefeed with a "\0".
- `sock_fastwrite`: Writes as many bytes possible to the socket and returns the number of bytes it wrote. The equivalent UDP functions are `udp_send` and `udp_sendto`.
- `sock_fastread`: Reads the given number of bytes from a socket, or the number of bytes immediately available. The equivalent UDP functions are `udp_recv` and `udp_recvfrom`.
- `sock_awrite`: Does an "all or none" write to a socket, meaning that it writes the given number of bytes to a socket, and, if that amount of data cannot be written, it writes nothing to the socket.

Networking

9.8.4 Blocking vs. Nonblocking Functions

Blocking functions (such as `sock_write()` and `sock_read()`) can block the program while they wait for an event to occur. Nonblocking functions (such as `sock_fastwrite()` and `sock_fastread()`) will either do their task immediately, or indicate to the programmer that they could not (or could only do part of it). Nonblocking functions are more difficult to use, but when an application has other tasks that must be done in a timely fashion, then they must be used (at least in terms of cooperative multitasking). Anything more than a trivial program that uses cooperative multitasking will need to use the nonblocking functions.

> In using these programs, it will be best to configure any firewalls running on the PC to open the ports we need to use. For the sake of consistency, we will use port 8000 for all of our examples on the Rabbit side as well as the PC side, and the firewall may block access to this port on the PC. This will cause some of these programs to not work.

9.9 Project 2: Implementing a Rabbit TCP/IP Server

We will use the RCM3400 prototyping board as a TCP server. A client will establish a connection with the TCP server and will request temperature readings from the server. For each request, the server will send out a sample and a sequence number, followed by a carriage return and a linefeed. The sequence number will ensure that the client will know if it has missed receiving any samples. This precaution is on top of the sequencing mechanism that TCP already implements behind the scenes. This is shown here for illustration purposes only; such precaution, of course, is a waste of bandwidth, unless it is useful to the application.

In order to establish a connection to the server, the client would need to know the server's IP address and port number. While it is easy for us to use dynamic addressing for the server, we will use static addressing so that the client will know for sure the server's IP address. Otherwise, each time the server power cycles, it may come up with a new IP address, and the client may not be able to ever find the server again or that the client code may have to be recompiled again with the server's IP address.

9.9.1 The Server TCP/IP State Machine

In Chapter 8, we looked at the use of "big loop" implementation for multitasking. This is a common technique, especially for cooperative multitasking. We will implement TCP and UDP communication using a combination of state machine programming as well as "big loops" for cooperative multitasking. The TCP Server program implements the TCP state machine shown in Figure 9.6.

The server takes periodic temperature readings, specified by the number of milliseconds between samples in the `SAMPLE_DELAY` constant. Flashing the sample LED requires 50ms each time, and this serves as the minimum delay between samples.

Once a connection is established with a client, the server sends the readings and a sequence number to the client. The client application can use the sequence numbers to determine if any samples were not received.

The server sends the number of samples defined by the `SAMPLE_LIMIT` constant, and then sends a special code, "END". The server also resets the sequence number at this point. The client terminates the connection once it receives the "END" code.

The Rabbit TCP server uses the DS1 and DS2 LEDs on the prototyping board to indicate the following:

- The DS1 LED remains on while the client and server are connected.
- The DS2 LED flashes each time a sample gets sent.

The Rabbit server uses two costatements: one to implement the TCP state machine and the other one to take periodic temperature samples. The `costate` that takes temperature samples keeps an eye on a connection being established, and, once that happens, it starts transmitting the temperature readings and sequence numbers to the client.

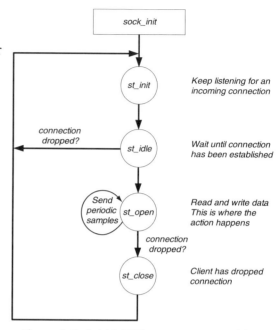

Figure 9.6: Rabbit TCP server state machine.

Program 9.3 shows a code fragment for the Rabbit Server's TCP state machine:

Program 9.3: Rabbit-based TCP server.

```
// ADCservTCP.c

<code removed for brevity>

tcp_state = st_init;
sequence = 0;

for(;;)
{
   tcp_tick(NULL);

   costate
   {
   switch (tcp_state)
   {

   case st_init:
       if (tcp_listen(&serv, 8000, 0, 0, NULL, 0))
```
(Program 9.3 continued on next page)

Program 9.3: Rabbit-based TCP server (continued).

```
            tcp_state = st_idle;
         else
         {
            printf ("\nERROR - Cannot initialize TCP Socket\n");
            exit (99);
         }
         break;

      case st_idle:
         if (sock_established(&serv) || sock_bytesready (&serv) != -1)
         {
            tcp_state = st_open;
            connectled(LEDON);
         }
         break;

      case st_open:
         if (!sock_readable (&serv))
         {
            tcp_state = st_close;
            sock_close(&serv);
         }
         break;

      case st_close:
         // not much to do here; the socket is closed elsewhere

      default:
      } // switch

      if (!sock_alive (&serv))
      {
         tcp_state = st_init;
         connectled(LEDOFF);
      }
   } // costate

} // for infinite loop
```

The costatement that determines whether data needs to be sent to the client is not shown here.

9.9.2 Working with General-Purpose TCP Utilities

Before we write any client code on the PC platform, we must make sure that the Rabbit TCP server runs as planned. This will also help us establish a reference point for the future; if commonly available utilities are able to talk to the Rabbit but our Java or C++ code cannot, we will easily know that we need to start debugging the PC code.

Chapter 9

A simple way to check the server is to just use a telnet connection. The telnet utility is described in Section 9.1.1 and can be invoked with the IP address and port number of the TCP server:

```
C:>telnet 192.168.1.50 8000
```

A more versatile utility, Netcat, can be invoked as a TCP or UDP client or a server. For this project, we will invoke Netcat to connect with the Rabbit TCP server (Netcat is described in Section 9.14.4). A screen output is shown in Figure 9.7:

```
C:\Netcat>nc 192.168.1.50 8000
86.784568, 0
86.784568, 1
86.784568, 2
86.784568, 3

<output removed for brevity>

86.784568, 197
86.772720, 198
86.784568, 199
END
86.784568, 0
86.784568, 1
86.784568, 2
86.784568, 3
^C
C:\Netcat>
```

Figure 9.7: Netcat talking to the Rabbit TCP server.

We can verify that the Rabbit server is sending us the samples and sequence numbers, as well as the "END" code after the maximum number of samples hard coded in the program. Now that we are satisfied that the Rabbit TCP server works as expected, we will use Java and C++ based TCP clients to talk to the Rabbit TCP server.

9.9.3 Working with a Java TCP/IP Client

The core of the Java client is shown in Program 9.4:

Program 9.4: Java-based TCP client.

```
// tcpClient.java

<code removed for brevity>

        socket = new Socket(server, port);

        BufferedReader reader
```
(Program 9.4 continued on next page)

Program 9.4: Java-based TCP client (continued).

```
        = new BufferedReader(
            new InputStreamReader( socket.getInputStream() ) );

    while( (line = reader.readLine() ) != null )
    {
        if(line.equals("END"))
        {
        break;
        }

        System.out.println(line);
    } // while

    socket.close();
```

While it is easy to enter command line arguments for the server's IP address and port number, the Java client uses constants for these values.

The code first creates a socket with the `Socket(server, port)` call, and then creates a stream reader with the `socket.InputStreamReader()` call.

When the work is done, it closes the socket with the `socket.close()` call.

9.9.4 Working with a C++ TCP/IP Client

The TCP client built with Microsoft C++.net has similar functionality as the Java client in the previous section. Although this code is compiled with the .net compiler, it has been built as a Windows32 console application. Similar to the Java program, the TCP client uses constants for the server's IP address and port number; it is easy to change the code so that the user can enter these values manually.

A fragment of the C++ connection code is shown in Program 9.5:

Program 9.5: TCP client built in C++.

```
// TCPClient.cpp

<code removed for brevity>

// Connection Phase
printf("\nLooking for Rabbit TCP Server at %s:%d\n", RABBIT_SERVER,
        htons(RABBIT_PORT));

SOCKET clientSock=Connect(RABBIT_SERVER);

// Send / Receive Phase
Send(clientSock,"\n");
```

(Program 9.5 continued on next page)

Chapter 9

Program 9.5: TCP client built in C++ (continued).

```
while (!done)
{
recvbytes=recv(clientSock, receivedStr, 100, 0);
receivedStr[recvbytes-1]='\0';
printf("\n%s", receivedStr);
//printf(": %d bytes", recvbytes);

Send(clientSock,"\n");
if (!strnicmp(receivedStr,"END",3)) done = 1;
}

// Closing Phase
printf( "\n\nReceived END message from Rabbit Server!!");
printf( "\nTerminating TCP connection...");

closesocket(clientSock);
WSACleanup();
```

The client uses the Winsock library to implement various phases of the communication, such as *connect*, *send/receive*, and *close*. If the C++.net application was built instead of the Windows32 console application, we could have used the various goodies that Windows provides, such as text boxes and scroll bars. Moreover, C++.net would have allowed us to use the try/catch exception handling mechanism that we will use with Java.

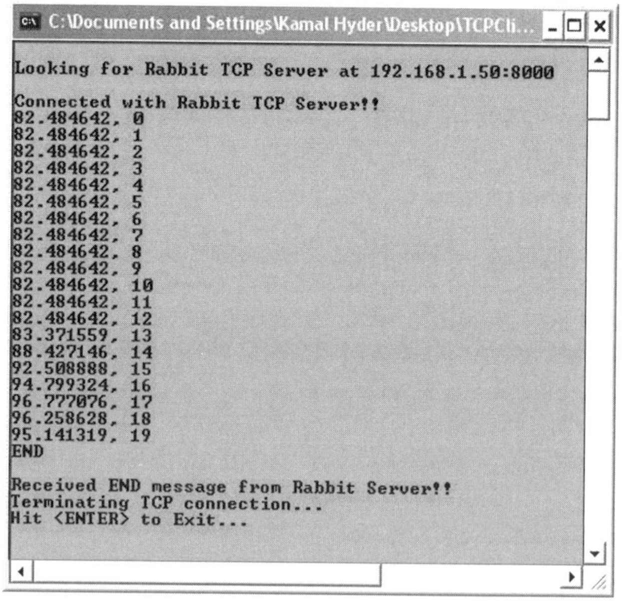

The client displays a couple of status messages in the console window, and then starts to display the temperature data (temperature and sequence number). Once the Rabbit server has reached the sample count, the server sends an "END" string to the client, at which point the client terminates the connection.

Figure 9.8: Output of the TCP client.

Figure 9.8 shows an output from the C++ client. To reduce ambiguity and aid in debugging, the client displays the IP address and port number of the server it is seeking.

While the code was running, we held the thermistor that made the temperature rise. Then we let go of the thermistor, at which point the temperature started falling. The Rabbit server was programmed to send the "END" terminator after twenty samples, and the client program ended as expected.

9.10 Project 3: Implementing a Rabbit TCP/IP Client

In this project, we will program the RCM3400 prototyping board as a TCP client that will read a channel from the analog-to-digital converter (ADC) and send a reading to the server. When connecting to a server, a TCP client typically does an active open, which requires us to specify the IP address and port number of the server. Once connected, the client carries out two-way communication.

In any two-way communication, it is important to define the highest level protocol. Although TCP takes care of various lower level details for this project, we still need to establish how the client will signal start of communication, how the server will request a channel number for the ADC, and how the client will return the appropriate ADC reading. Just as significant is knowing how the connection will be terminated. We have a simple scheme here—the server can either send a channel number in the range 0 through 7 (yes, the Rabbit client does range checking) or the server sends a "quit" and the client terminates the connection. The higher level connection sequence is shown in Figure 9.9A:

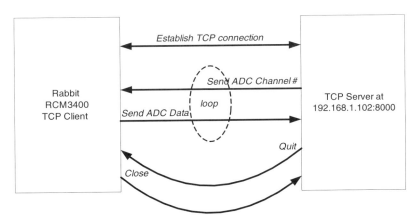

Figure 9.9A: Two-way communication between Rabbit client and server.

The state machine for the Rabbit TCP client is shown in Figure 9.9B. This is similar to the state machine for the Rabbit TCP server, except for a couple of states that wait for connection to be established and send data to the server.

Either party can drop the connection at any time, and we need to account for this possibility.

Chapter 9

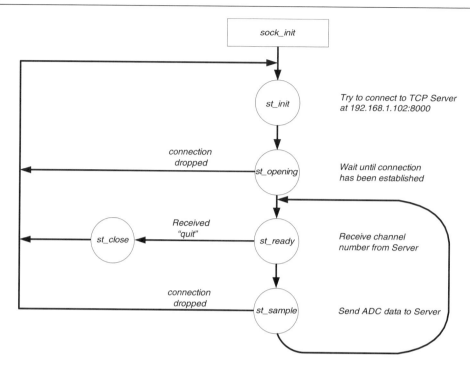

Figure 9.9B: Rabbit TCP client state machine.

The Rabbit client comes up with DHCP and displays its IP address in the stdio window. It then tries to connect with a TCP server at IP address 192.168.1.102 port 8000. Program 9.6 shows the Rabbit Client TCP state machine:

Program 9.6: Rabbit-based TCP client.

```
// ADCclientTCP.c

<code removed for brevity>

switch (tcp_state)
{

case st_init:
   // printf( "In Init State  ");
   connectled(LEDOFF);
   if (tcp_open(&serv, 0, resolve (SERVER), PORT, NULL))
         tcp_state = st_opening;
   else
         exit (EXIT_ON_ERROR);
   break;
case st_opening:
   if (sock_established(&serv) || sock_bytesready (&serv) != -1)
   {
(Program 9.6 continued on next page)
```

Program 9.6: Rabbit-based TCP client (continued).

```
        tcp_state = st_ready;
           connectled(LEDON);
        }
        break;

    case st_ready:
        // check for the socket being readable
        // this is important to check in case the client is trying to
        // close a connection

        if (!sock_readable (&serv))
        {
               tcp_state = st_close;
               sock_close(&serv);
               break;
        }

        if (sock_gets(&serv, buff, sizeof(buff))>0)
        {
               if (((!strcmpi(buff, "q")) || (!strcmpi(buff, "quit"))))
               {
                      tcp_state = st_close;
                      sock_close(&serv);
               }
               else
               {
                      channel = atoi(buff);

                      if ((channel>=0) && (channel <= 7))
                      {
                      printf("\nChannel Number = %d", channel);
                      }
                      else
                      {
                      printf("\nInvalid Channel Number!  Data from \
                             channel 7 will be sent");
                      channel = 7;
                      }

                      tcp_state = st_sample;
               } // else
        } // if
        break;
    case st_sample:
        // printf( "In Sample State  ");
        // not much to do here; the other costate is doing the sampling

    case st_close:
        // printf( "In Close State  ");
        // not much to do here; the socket is closed elsewhere

    default:
    } // case
```

Chapter 9

The Rabbit code implements the TCP state machine in a costatement. It tries to initialize the TCP socket in state `st_init`, and returns an error if it cannot. In `st_opening`, it proceeds to the ready state if the socket is active or if there are bytes ready to be read. It proceeds to `st_sample` once it has received a valid channel number from the TCP server.

A different costatement keeps track of whether we are in the `st_sample` state. If so, the Rabbit code sends a sample and returns to `st_ready` so that it can receive a new channel number from the server.

In development phase of the project, it is good practice for the server and client to print their IP address on the screen. This will avoid confusion and will aid in debugging. When doing a final build, the developer can simply comment out the `printfs` or encapsulate them in a VERBOSE macro. We used a bunch of `printfs` in the code for debugging purpose and then commented them out once we confirmed that the code worked as planned.

9.10.1 Disabling the Windows XP Firewall

At first, we had trouble with this project. While the Linux box connected with the Rabbit client, the Windows PC did not. The machine was running Windows XP Professional, with the Internet Connection Firewall (ICF) enabled. The firewall blocked access to port 8000 for the PC-based servers. We had to choose between disabling the firewall or opening up our ports of choice—we opted to open up port 8000 for both TCP and UDP, and the following procedure can be used to do that on a Windows XP system. Other software firewalls will require different instructions for creating the port 8000 hole.

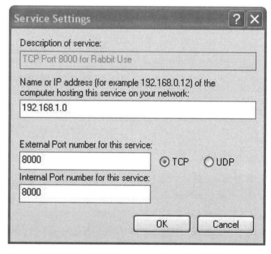

Figure 9.10A: Adding a setting for TCP port 8000 in Windows XP Professional.

From the "Network Connections" folder, right click on the appropriate connection. This will most likely be labeled the "local area network." Choose Properties ⇒ Advanced ⇒ Settings.

From the "Advanced Settings" dialog box, click on "Add" to define a new setting. Figure 9.10A shows the values we entered for opening up TCP port 8000. Similarly, we have to add a new setting for UDP port 8000.

After adding another setting for UDP, the "Advanced Settings" dialog would look similar to the one shown in Figure 9.10B. This confirms that we have opened up port 8000 for both TCP and UDP.

If the user is running a software firewall client on the PC, the appropriate steps should be taken to disable the firewall completely or to "punch a port hole" through it.

Networking

Figure 9.10B: Port 8000 opened up for TCP and UDP.

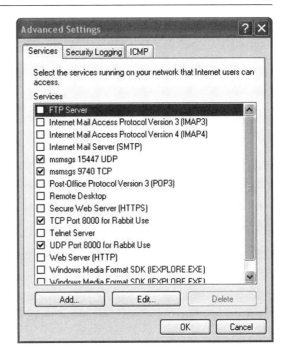

9.10.2 Verifying the Client Code

Simply putting a few `printfs` in the right places in the code can save debugging time. We coded the client to output its IP address in the stdio windows, as well as the IP address and port number of the server it is trying to connect to. Moreover, the client prints the channel number that it receives from the server. Having this information on the screen removes any ambiguity and confusion that may arise when things do not work as expected. The output of the Rabbit client is shown in Figure 9.11A.

```
Welcome to the Rabbit TCP Client!!

My IP address is 192.168.1.103

Trying to connect to TCP Server at 192.168.1.102:8000.


Connected to Server at 192.168.1.102:8000.

The server should enter ADC Channel to read from Client, or QUIT

Channel Number = 4
Channel Number = 7
Bye!
```

Figure 9.11A: Output from Rabbit client.

Chapter 9

The output from the Netcat server is shown in Figure 9.11B. The command line arguments shown for Netcat make it listen at 8000 for incoming connections.

```
C:\Netcat>nc -l -p 8000
4
2.000000
7
1070.000000
q

C:\Netcat>
```

Figure 9.11B: Output from Netcat server.

Now that we have verified that the Rabbit TCP client is able to make a connection and that it is working as planned, we can implement a PC-side server in C# and Java.

9.10.3 Working with a Java TCP/IP Server

A fragment of the Java TCP server is shown in Program 9.7:

Program 9.7: Code fragment that implements the Java server.

```java
// TCPClient.java

<code removed for brevity>

try {
    System.out.println("Input details from keyboard" +
        '\n' +"If you want to exit the connection enter 'q' or 'quit':");

    // The Client reads the standard input from keyboard
    BufferedReader inFrmUsr =
    new BufferedReader(new InputStreamReader(System.in));

    // When some client asks for tcpServSocket,There is a connection
    // established between the tcpServSocket and tcpClientSocket
    Socket connSocket = tcpServSocket.accept();

    // This stream provides process input from the socket
    BufferedReader inFrmClient =
    new BufferedReader(new InputStreamReader(connSocket.
    getInputStream()));

    // This stream provides process output to the socket
    DataOutputStream  outStream =
        new DataOutputStream(connSocket.getOutputStream());
```

(Program 9.7 continued on next page)

Program 9.7: Code fragment that implements the Java server (continued).

```java
        while (bflag)
        {
            // Places a line typed by user in the strInput
            strInput = inFrmUsr.readLine() ;

            outStream.writeBytes(strInput + '\n');

            // Places a line from the client
            clientStr = inFrmClient.readLine();

            System.out.println("From Client: " + clientStr);

            if ((strInput.equalsIgnoreCase("q")) ||
               (strInput.equalsIgnoreCase("quit")))
                    bflag = false;
        }

        // Close the socket
        tcpServSocket.close();
    } // end try

    catch (IOException exp)
    {
        System.out.println("Connection closed between client and server");
    } // end catch
```

Similar to Project 2, the Java code creates `InputStreamReader` instances for the socket and user input. It also creates `DataOutputStream` for output to the Rabbit client and processes input until it receives a "quit" signal.

The try/catch blocks in Java form a useful mechanism for exception handling. The *try* keyword guards a section of the code, and, if a method throws an exception, code execution stops at that point and the *catch* block, the exception handler, is executed. The programmer can insert the appropriate code in the exception handler to inform the user.

Chapter 9

The output of the Java server is shown in Figure 9.12:

```
C:\ java>java TCPServer
TCP Server IP Address : desktop-fast-28/192.168.1.102
Server hostname       : desktop-fast-28
Server listening on port 8000

Input details from keyboard
If you want to exit the connection enter 'q' or 'quit':
5
From Client: 2.000000
9
From Client: 1068.000000
7
From Client: 1068.000000
q
From Client: null
Server exiting...

C:\ \java>
```

Figure 9.12: Output from Java server.

If the channel number entered by the server is outside of the range 0 through 7, the Rabbit client outputs the ADC value from channel 7.

9.10.4 Working with a TCP/IP Server in C#

The C# code runs as a server. The lifecycle of the server is, to listen to incoming request on a specific port. Once the client establishes the connection, both the client and server enter a loop where the client keeps listening for server request and returns a response based on the channel requested. The server, on the other hand, queries the user for channel numbers, which in turn it sends to the client. A fragment of the C# code is shown in Program 9.8:

Program 9.8: Code fragment that implements the C# server.

```
// TCPServer.cs

<code removed for brevity>

Console.Write("Waiting for a connection on port["+port+"] ... ");

// Perform a blocking call to accept requests.
// We could also user server.AcceptSocket() here.

TcpClient client = server.AcceptTcpClient();
Console.WriteLine("Connected!");

data = null;
```
(Program 9.8 continued on next page)

Program 9.8: Code fragment that implements the C# server (continued).

```csharp
// Get a stream object for reading and writing
NetworkStream stream = client.GetStream();
do
{
    //Send the Channel Number from User Input
    Console.WriteLine("Input details from keyboard ['q' or 'quit'
    exits]");
    data = Console.ReadLine();
    data =((data.ToLower()=="q")?"quit":data)+"\n";

    //convert data to bytes
    byte[] msg = System.Text.Encoding.ASCII.GetBytes(data);

    // Send the user request.
    stream.Write(msg, 0, msg.Length);

    //make sure the buffered data is written to underlying device
    stream.Flush();

    Console.WriteLine(String.Format("Sent: {0}", data));

    int i;

    // Receive the data sent by the client.
    if((i = stream.Read(bytes, 0, bytes.Length))!=0)
    {
         // Translate data bytes to a ASCII string.
         data = System.Text.Encoding.ASCII.GetString(bytes, 0, i);
         Console.WriteLine(String.Format("Received: {0}\n", data));
    }
}
while(!data.StartsWith("quit"));
// Shutdown and end connection
Console.WriteLine("Closing connection");
client.Close();
```

9.11 Project 4: Implementing a Rabbit UDP Server

So far, we have focused on TCP communication, and a critical part of that is connection establishment before data can change hands. With UDP, there is no need to establish such a connection—the UDP server just initializes itself to listen on port 8000 and waits until a client sends a datagram to that port. Unlike TCP, communication happens one datagram at a time instead of on a "per connection" basis.

The UDP server sample presented here does essentially the same thing as the TCP samples presented earlier.

Unless the client sends a "C" to close the connection (akin to a "finish" request), the server keeps sending temperature readings and a sequence number with each reading.

Chapter 9

The Rabbit UDP server uses the DS1 and DS2 LEDs on the prototyping board to indicate the following:

- The DS1 LED remains on while the client and server exchange data. Once the server receives a "finish" request, it turns the LED off. The LED will turn on when the server receives the next datagram.
- The DS2 LED flashes each time a sample gets sent.

The state machine for the UDP server is shown in Figure 9.11.

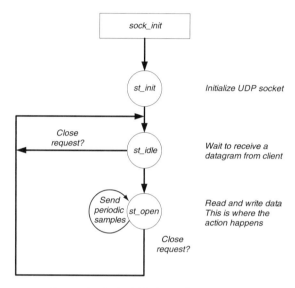

Figure 9.13: Rabbit-based UDP server.

The code for UDP server's state machine in Program 9.9. Not shown is the `costatement` that sends periodic data to the client.

Program 9.9: Rabbit-based UDP server.

```
// ADCservUDP.c

<code removed for brevity>

for(;;)
{
    // It is important to call tcp_tick periodically otherwise
    // the networking subsystem will not work

    tcp_tick(NULL);

    // UDP state machine
```
(Program 9.9 continued on next page)

Program 9.9: Rabbit-based UDP server (continued).

```
costate
{
    switch (udp_state)
    {

     case st_init:
         if (udp_open(&serv, PORT, -1, 0, NULL))
          udp_state = st_idle;
         else
          {
           printf ("\nERROR - Cannot initialize UDP Socket\n");
            exit (EXIT_ON_ERROR);
          }
         break;

     case st_idle:

        // look at data in the buffer for a close request
        buff[0] = 0;

        if ((udp_recvfrom(&serv,buff,sizeof(buff),&cli_ip, &cli_port)) >= 0)
        {
            if (buff[0] != 'C')
            {
            udp_state = st_open;
            connectled(LEDON);
            }
        }
         break;

     case st_open:
         // nothing to do here; the other costate will check for udp_state
         // to determine whether we are in st_open state

        // keep checking for new packet; if so,
        // update client IP address and port number dynamically

        buff[0] = 0;

        // look at data in the buffer for a close request

        if ((udp_recvfrom(&serv,buff,sizeof(buff),&cli_ip, &cli_port)) >= 0)
        {
        if (buff[0] == 'C')
            {
            udp_state = st_idle;
```

(Program 9.9 continued on next page)

Program 9.9: Rabbit-based UDP server (continued).

```
            connectled(LEDOFF);
          }
      }
         break;
      case st_close:

      default:
      } // switch

   } // costate

} // for
```

As is the case with most other programs in this chapter, we use two costatements—one to run the UDP server state machine and the other one to send data to the client, once we are in the right state.

9.11.1 Working with a Java UDP Client

To use the program, the user would launch the Java client and would hit <enter> to send a datagram to the UDP server. Each datagram requests a new temperature sample from the Rabbit. The Java client repeats this sequence ten times and then quits.

The Java UDP client is shown in Program 9.10.

Program 9.10: Java UDP client.

```
// udpClient.java

<code removed for brevity>

    for (;;)
    {
         count++;

         String sentence = inFromUser.readLine();
         sendData = sentence.getBytes();

         sendPacket = new DatagramPacket
                     (sendData, sendData.length, remoteAddr, PORT);

         clientSocket.send(sendPacket);


         receivePacket = new DatagramPacket
                         (receiveData, receiveData.length);
```

(Program 9.10 continued on next page)

Networking

Program 9.10: Java UDP client (continued).

```
            clientSocket.setSoTimeout(2000);
            clientSocket.receive(receivePacket);

            receivedData = new String
                    (receivePacket.getData(), 0, receivePacket.getLength());

            System.out.println("FROM SERVER: " + receivedData);

            if (count == 10)
            {
                    clientSocket.close();
                    break;
            }

    } // for
```

Unlike the TCP examples, the Java code creates `sendPacket` and `receivePacket`, which are new instances of the `DatagramPacket` class. It creates a new client socket each time it exchanges data with the UDP server, and then quits once it has counted up to ten responses from the server.

9.11.2 Working with a C++ UDP Client

The code functions as a client where the client sends packets containing a channel number to the rabbit server. The client then blocks for response on the same endpoint from the server, displaying it to the user once it receives it. A code fragment of the C++ client is shown in Program 9.11:

Program 9.11: Code fragment for C++ UDP client.

```
// cppUDPClient.cpp

<code removed for brevity>

do{
   std::cout << "Input a string :" << std::endl;
   std::cin >> szBuf;
   if(!isalpha(szBuf[0])&& strlen(szBuf)==1 && (szBuf[0]>='0')&&(szBuf[0]<='7'))
            inValidInput=false;
}
while(inValidInput);
nRet = sendto(theSocket,                // Socket
szBuf,                                   // Data buffer
strlen(szBuf),                           // Length of data

(Program 9.11 continued on next page)
```

Program 9.11: Code fragment for C++ UDP client (continued).

```
0,                                    // Flags
(LPSOCKADDR)&saServer,                // Server address
sizeof(struct sockaddr));             // Length of address

<code removed for brevity>

// get acknowledge from server
nFromLen = sizeof(struct sockaddr);
nret = recvfrom(theSocket,            // Socket
    outBuf,                           // Receive buffer
    sizeof(outBuf),                   // Length of receive buffer
    0,                                // Flags
    (LPSOCKADDR)&saServer,            // Buffer to receive sender's
    address&nFromLen);                // Length of address buffer
```

9.12 Project 5: Web Enabling the Sensor Routine

This project will revisit the analog sensor routine from Project 4 in Chapter 8, and we will use several methods to build a browser interface for the sensor routine. Before we start, it will be useful to consider the advantages of standard interfaces versus what we have done so far with regard to customized PC applications.

9.12.1 Introduction to Browser-based Control

Most of the projects developed so far in this chapter have one thing in common: they rely on a Windows or Linux-based C++ or Java program to communicate with the Rabbit prototyping board. There is a better way to do this—to use a standard browser interface. There are many reasons why we should build such an interface into our application:

- While we can develop applications on the PC in various languages, including C, C++, Java and Perl, doing so takes time, and each application and the user interface have to be debugged and tested. Alternatively, we can use a standard browser interface without having to build and test a browser.

- Our C++ and Java applications need to be delivered to the end user and installed on the end system. This may be a tedious and time consuming process, prone to errors. Building and delivering application software to customers means we have to manage the process at our end, while customers have to work with their IT department's policies to be able to install the software. Browsers are now ubiquitous. No custom software needs to be installed on the PC to work with our browser-enabled embedded application.

- Regardless of how good an interface we develop, the end user has to get comfortable with it and will often have to be trained to use it. Most users, however, are comfortable with their favorite browser and require no training—we will be hard pressed to find a desktop without a browser or a computer user who is not familiar with one. Operating system vendors have spent years of effort and hundreds of millions of dollars on making the browsers user friendly and we can take advantage of their investment.

- If our application needs to run on multiple operating systems, say, Apple OS X™, Microsoft Windows™, and Linux, we will need to develop, distribute and maintain various versions of the programs, and provide the appropriate version to each customer. This can quickly get complicated, and the complexity is further compounded by our need to support different versions of the same operating system. A standard browser interface provides system independence—we develop code for the browser interface, not for the operating system.

- Since our focus is building embedded applications, those are the resources we must build and develop, instead of developing end user interfaces. An organization often has to ask where they add the most value and what they need to focus on to have the highest competitive advantage.

Most manufacturers understand these reasons and have thus started enabling their embedded devices with browser interfaces. Almost all home networking devices, for example, are configured through browsers instead of a command line interface (CLI). Even enterprise-grade networking devices and office printer and copiers now use browser-based configuration instead of the CLI.

This section will not offer a detailed discussion of the browser or associated languages and mechanisms such as HTML, CGI, Java Applets, Get and Post methods, etc. This information is commonly available and excellent books have been written on these subjects. We assume the reader knows these technologies and we will instead focus on applying these concepts to the Rabbit platform.

In the following examples, the Rabbit core module will act as an HTTP server, and the user will be able to simply point to it by entering the IP address of the Rabbit server in the browser's URL window. To aid consistency, we will use a static IP address of 192.168.1.50 for the Rabbit core modules.

There will be two iterations of the project: first, we will build a simple static web page that will display date, time and temperature on a web page. Next, we will build a dynamic web page that will allow the user to set temperature thresholds and enable or disable alarms.

To develop browser interfaces to projects in this chapter, we will take advantage of Rabbit-Web, a software module introduced by Rabbit Semiconductor to ease the chore of developing web interfaces. Using RabbitWeb greatly reduces the amount of code we need to write.

9.12.2 Introduction to RabbitWeb

RabbitWeb aims to reduce the need to write CGI (Common Gateway Interface) functions for web interfaces. CGI functions are essentially C functions that are called when a web browser requests a specific page from the web server. These C functions are responsible for handling the request from the browser and for generating the response. This allows for great flexibility, since the developer has the power of the C language for generating content dynamically. A price is paid for this power, however: the difficulty and length of development using CGIs is often high. Furthermore, experimenting with the layout of the generated content is difficult, since CGI programming usually results in having portions of the web page scattered across C code. Additionally, developing good error-checking and error-reporting CGIs is quite difficult.

RabbitWeb allows the developer to register C variables for direct use with the web server. These C variables can be displayed in web pages and can be updated through the web server (with all of the difficult details handled by the web server). Variables can have associated error-checking expressions, called "guard expressions." Any updated variables must pass their guards, or the updates will not be applied. RabbitWeb allows these errors in user input to be reported back to the user. Additionally, variable updates can be restricted to specific users and types of authentication, thus allowing developers to secure their web interfaces. When a variable has been updated, an associated "update callback function" can be triggered to alert the rest of the user program.

One of the key concepts behind RabbitWeb is that it allows the developer to make presentation decisions within HTML, rather than forcing those decisions to be made with C code. Since HTML is inherently concerned with presentation, it would be more natural to make presentation decisions within HTML. This is the same concept that has driven the popularity of web scripting languages such as PHP and ASP.

CGI programming will always allow the most flexibility, but RabbitWeb eliminates the need to write CGI functions in most common cases. In our case, it will allow us to develop a rather complex interface to the sprinkler system in Project 6 with surprisingly little effort.

A lot more detail on RabbitWeb is included in the accompanying CD-ROM.

9.12.3 A Static Web Server Built with RabbitWeb

The first iteration of the project will display date, time and temperature on a web page. We will develop two separate files for this application:

`sensor1.zhtml` is the HTML code that gets delivered to or presented to the browser.

`sensorweb1.c` is the code run in the Rabbit HTTP server. As we will see, this code will embed the `sensor1.zhtml` file within the Rabbit executable, using pragma `#ximport`.

The `main()` program from `sensorweb1.c` is shown in Program 9.12A:

Program 9.12A: main() section of `sensorweb1.c`.

```
// sensorweb1.c

#define PORTA_AUX_IO

#define TCPCONFIG       1       // DHCP not recommended for the web server
#define USE_RABBITWEB   1

#use "dcrtcp.lib"
```
(Program 9.12A continued on next page)

Program 9.12A: main() section of `sensorweb1.c` *(continued).*

```
#use "http.lib"

<code removed for brevity>

main()
{
char    buffer[100];
float Tk,       // calculated temperature kelvins
    Tf;         // converted to farenheit

    brdInit();

    //first-time call of this function will take 1 second to charge up cap
    //use single-ended and gain of 1
    Tk = getTemp();
    web_temp = 1.8*(Tk - 255.37);            //calculate fahrenheit

    sock_init();
    http_init();

    /* Print who we are... */
    printf("\nWelcome to the Web-Based Temperature Sensor\n");
    printf("\nMy IP address is %s\n\n", inet_ntoa(buffer, gethostid()) );

while (1)
    {
    http_handler();
    handle_web_vars();
    } // infinite loop

} // main
```

Being a web server, the program uses static instead of dynamic addressing. Moreover, setting the `USE_RABBITWEB` macro to "1" allows us to use RabbitWeb.

As usual, we will call `brdInit()` to bring up the hardware. We have seen the temperature-related function `getTemp()` in Chapter 8 and will not describe it here.

We will call `sock_init()` to initiate the TCP engine and then `http_init()` to get the web server started. These two calls are crucial otherwise the system will simply not work. When the core module comes up, it will print its IP address on the screen.

There isn't much else to do after that—the program operates in an infinite loop where it calls another key function, `http_handler()`, which needs to be called periodically. It performs the various housekeeping tasks to keep the web server running. We can insert our own handler, such as `handle_web_vars()`, to update the appropriate web variables. In Section 9.12.4, we will add another handler to the infinite loop.

Chapter 9

A screen capture of the static web page is shown in Figure 9.14.

The screen capture tells us that the date, time, and temperature variables need to be passed from the C program to the web page. Being a static web server, the browser window will simply display these values and will not allow the user to change any of these settings. As shown in Program 9.12B, RabbitWeb provides a simple mechanism to define such variables and pass them to the web page. In Section 9.12.4, we will get more creative with variables and use "guard expressions" to perform error checking on user values.

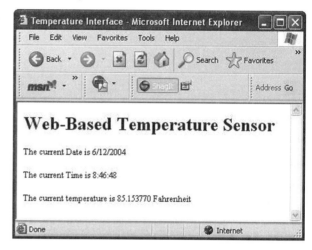

Figure 9.14: Screenshot of the static interface.

Program 9.12B: Passing C variables to the web page.

```
// sensorweb1.c

<code removed for brevity>

float web_temp;
#web web_temp

int date_month, date_day, date_year;
int time_hour, time_min, time_sec;

#web date_month
#web date_day
#web date_year
#web time_hour
#web time_min
#web time_sec
```

We should also examine how the variables get presented to the browser. The `sensor1.zhtml` page is shown in Program 9.12C. RabbitWeb uses the `<?z echo($variable) ?>` tag to convert variable values into their equivalent HTML values. The rest of the program is pure HTML.

Program 9.12C: Displaying variables in HTML.

```
<HTML><HEAD><TITLE>Temperature Interface</TITLE></HEAD>
<BODY><H1>Web-Based Temperature Sensor</H1>

<P>The current Date is
```
(Program 9.12C continued on next page)

Networking

Program 9.12C: Displaying variables in HTML (continued).

```
<?z echo($date_month) ?>/<?z echo($date_day) ?>/<?z echo($date_year) ?>
</P>

<P>The current Time is
<?z echo($time_hour) ?>:<?z echo($time_min) ?>:<?z echo($time_sec) ?>
</P>

<P>The current temperature is
<?z echo($web_temp) ?>
</P>
</BODY></HTML>
```

The final item to examine is how the C program loads the HTML page into flash. This is described in detail in the next section. The project will load three different web pages, one requiring authenticated access.

9.12.4 A RabbitWeb Dynamic Web Server with Authentication

This project picks up from Project 4 in Chapter 8. In that chapter, we implemented an analog sensor task that would take periodic temperature readings, compare them against preset thresholds, and trigger alarms accordingly. We will now use RabbitWeb to enable that project for a web interface.

The program converts readings from a thermistor into temperature values and compares them to four temperature thresholds: Upper Critical, Upper Non-Critical, Lower Non-Critical, and Lower Critical. It allows for scanning of each threshold, and shows on the screen whether one of the threshold alarms has been triggered.

Before coding such a project, it is important to consider various elements of the user interface. We need to know which variables the program will expose to the user, and which variables the user will be able to change and how (for example, whether it uses check boxes, radio buttons, etc.). We decided to implement three web pages for this project:

- The "Monitor" page displays the date, time, current temperature, and the status of each sensor. It needs read-only access to variables that deal with date and time, the four temperature thresholds, whether scanning has been enabled for each threshold, and if a threshold has generated an alarm. This page has been implemented with file `sensor2_monitor.zhtml`.

- The "Set Time" page allows us to change the date and time on the core module. For additional security, this page is protected with first time authentication. The user name is "task" and the password is "mangler." Once the user has authenticated, the protected web page is available until the browser is closed and a new browser session is initiated. This page has been implemented with file `sensor2_settime.zhtml`. RabbitWeb performs additional verification on the entered date and time values.

- The "Set Temperature Thresholds" page allows us to change values for each threshold and whether the system will scan for that threshold. This page has been implemented

Chapter 9

with file `sensor2_settemp.zhtml`. RabbitWeb does some rule checking on the entered temperature thresholds.

The main code module for this program is `sensorweb2.c`, which has a scanning routine very similar to the one presented in Chapter 8. We will describe the code here not from the perspective of the scanning routine, but how the sensor routine is integrated into RabbitWeb. More importantly, we will examine how RabbitWeb provides for a two-way interface between the C program and associated ZHTML pages.

The "Monitor" page is shown in Figure 9.15.

Let us first examine how the web pages are described at compile time in the C program. Since we will use two images to indicate whether an alarm condition exists or not, we will load the images into flash as well:

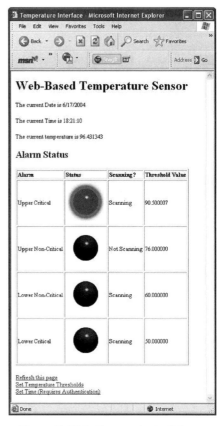

Figure 9.15A: Screenshot of the "monitor" page.

```
// sensorweb2.c

<code removed for brevity>

#ximport "sensor2_monitor.zhtml"    monitor_zhtml
#ximport "sensor2_settime.zhtml"    settime_zhtml
#ximport "sensor2_settemp.zhtml"    settemp_zhtml
#ximport "ledset.gif"               ledon_gif
#ximport "ledreset.gif"             ledoff_gif
```

Next, we will associate the #ximported files with the web server:

```
SSPEC_RESOURCETABLE_START
    SSPEC_RESOURCE_XMEMFILE("/", monitor_zhtml),
    SSPEC_RESOURCE_XMEMFILE("/index.zhtml", monitor_zhtml),
    SSPEC_RESOURCE_XMEMFILE("/settemp.zhtml", settemp_zhtml),
    SSPEC_RESOURCE_XMEMFILE("/admin/index.zhtml", settime_zhtml),
    SSPEC_RESOURCE_XMEMFILE("/ledon.gif", ledon_gif),
    SSPEC_RESOURCE_XMEMFILE("/ledoff.gif", ledoff_gif)
SSPEC_RESOURCETABLE_END
```

We created a special "`admin`" directory above which stores files that require authentication. RabbitWeb uses the concept of groups, where each group can be granted separate access privileges through authentication. We also defined an "`admin`" group for authenticated access to the Rabbit clock:

```
#web_groups admin
sspec_addrule("/admin", "Admin", admin, admin, SERVER_ANY, SERVER_AUTH_
BASIC, NULL);
```

The following two lines create an "`admin`" user and add it to the admin group. The user ID is "task" and the password is "mangler":

```
userid = sauth_adduser("task", "mangler", SERVER_ANY);
sauth_setusermask(userid, admin, NULL);
```

We will now examine how program variables are set up so that their values can be displayed on and altered by the web pages described above. For instance, we know that none of the web pages should alter the value of the temperature variable, so we should declare it "read only" by all groups:

```
float web_temp;
#web web_temp groups=all(ro)
```

As described in Section 9.12.2, RabbitWeb allows us to use "guard expressions" to watch against invalid user data. We have implemented the following rules with guard expressions:

- Time and Date entry: the rules for time and date are well known... these include the treatment of the number of days in each month, consideration for leap years, as well the range for the year the system operates in. For example, it is easy to set up a guard expression to enforce that minutes and seconds are within the "0 to 59" range:

    ```
    #web time.minute ($time.minute >= 0 && $time.minute < 60)
    #web time.second ($time.second >= 0 && $time.second < 60)
    ```

    The "date" variables (year, month and day) are treated in an interesting manner, since we need to verify for leap years and the number of days per month. The syntax of the guard expression allows us to call the function `checkday()` each time the user attempts to change one of the date (day, month and year) variables:

    ```
    #web time.year   (checkday($time.day, $time.month, $time.year))
    #web time.month  (checkday($time.day, $time.month, $time.year))
    #web time.day    (checkday($time.day, $time.month, $time.year))
    ```

    The `checkday()` function does the sanity checking and returns an error in case of invalid input. This leaves us with two `#web` lines for both `time.month` and `time.year`, and that is perfectly fine. RabbitWeb will check all of them and mark the variable as invalid if any of them fail.

Chapter 9

- Similarly, the `checktemp()` function is called each time the user attempts to change a temperature threshold. This function verifies that the four temperature thresholds meet the following conditions:
  - All temperature thresholds must in the range `TEMP_MAX` and `TEMP_MIN`. These are hard coded values and can be changed at compile time to accommodate for the temperature probe being used and the environment where the system will be deployed.
  - Upper Critical Threshold must exceed Upper Non-Critical Threshold,
  - Upper Non-Critical Threshold must exceed Lower Non-Critical Threshold, and
  - Lower Non-Critical Threshold must exceed Lower Critical Threshold

If `checkday()` or `checktemp()` return an error, the webpage uses RabbitWeb tags and the `error()` function to flag the user. For example, the following code in `sensor2_settime.zhtml` informs the user, in red text, that invalid date or time data has been entered:

```
<?z if (error()) { ?>
   <FONT color="#bb0000">
   The date and time you entered was invalid.  Please try correcting the
   error and setting the clock again.<P>
   </FONT>
<?z } ?>
```

Using the `error()` function, once RabbitWeb has flagged a user entry as an error, it does not update the associated variables.

Another interesting aspect of RabbitWeb is how vales are passed back from the HTML page to the C program. As an example, Let us look at the radio buttons in `sensor2_settemp.zhtml` that enable or disable scanning of the upper critical temperature threshold:

```
<INPUT type="radio" name="<?z varname($alarm.Scan_UC) ?>" value="1"
   <?z if ($alarm.Scan_UC == 1) { ?> CHECKED<?z } ?>
<INPUT type="radio" name="<?z varname($alarm.Scan_UC) ?>" value="0"
   <?z if ($alarm.Scan_UC == 0) { ?> CHECKED<?z } ?>
   >
```

These lines treat the radio button in its checked and unchecked state, respectively. The web variable, `$alarm.Scan_UC`, is passed back to the C program in runtime. In Section 9.13.6, we will look at how RabbitWeb treats an array of checkboxes.

Networking

Figure 9.15B shows a snapshot of the "Set Time" page.

Note the list box for the "month" field; the selection is defined in the C code and the HTML follows these values.

The temperature thresholds are defined as shown in Figure 9.15C. Entered values for thresholds are validated and the appropriate variables are updated. Similarly, the scanning function for each sensor can be altered and the values passed back to the C program.

In keeping with our practice of recommending improvements to projects, let us offer some suggestions for enhancing this program:

- Instead of hard coding usernames and passwords, define defaults for these values and allow the user to change them.
- The system can allow the user to change positive and negative hysteresis and sensing time for the analog sensor. The code presented here uses hard coded values.
- The system can send out an email for each temperature threshold that triggers an alarm.
- The HTML could be reworked and optimized for "surfing" from a cell phone browser.

Figure 9.15B: Screenshot of the "set time" page.

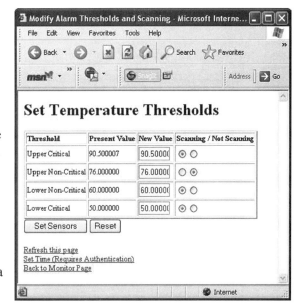

Figure 9.15C: Screenshot of the "set temperature thresholds" page.

9.13 Project 6: Building an Ethernet-Connected Sprinkler Controller

In this section, we bring together all of the tools discussed thus far to create a fully web enabled embedded system. We have selected a sprinkler controller as the project, both because the scope of the project is manageable in the context of a chapter, and because this particular control problem will benefit greatly from an intuitive browser-based interface.

9.13.1 Using Best Practices for Design: Defining the Requirements

When defining a project, the first step is identifying requirements. There are a number of formal approaches to capturing requirements. Most have fancy academic names like Structured Analysis Employing Hierarchical Decomposition, but "common sense" will be our approach.

When generating requirements, it is important to keep in mind the difference between a feature and a requirement. For example, a gold-plated welded-chain steering-wheel is a feature. A requirement that might imply a need for such a device is "the vehicle interior must look cool."

Another example is that a keypad and LCD module as a user interface is a feature. A requirement that might imply a need for an LCD and keypad is "a field service technician must be able to communicate with the system."

The distinction between the requirement and the features that fulfill the requirement are important. Consider the requirement "a field service technician must be able to communicate with the system." Several methods exist that might fulfill the requirement besides just slapping an LCD and keypad on the front of the box. For example a serial port or USB port might do the trick if we know that field service technicians routinely carry laptops or PalmPilots™.

Here are the system level requirements we settled on for the sprinkler controller

1. Must be able to be controlled remotely via a PC or laptop using commonly available software. No additional software need be installed on the PC or laptop.
2. Must be capable of replacing the existing controller in Bob Perrin's front yard.
3. Must be capable of deciding to reduce water when it is raining. We will refer to this as the "rainy day" rule.
4. Must be capable of determining if a day is particularly hot and decide if watering beyond the "normally scheduled water application" is needed. We will refer to this as the "hot day" rule.
5. Must have an intuitive user interface. Strictly speaking, this is a difficult one, since measuring and testing this requirement can be tricky.
6. The system must be inexpensive.

Now that we have the system level requirements, we can generate hardware and software features that fulfill the system requirements.

Networking

Features

1. (to Requirement 1) To be "remotely" controllable, the system will be designed with an Ethernet interface.
2. (to Requirement 1) To be configurable using commonly available software, the system will be designed to support a remote browser interface (HTTP interface).
3. (to Requirement 2) Bob's current system has four 24 VAC valves that can be actuated one at a time. The new system will be able to drive the same valves.
4. (to Requirements 2, 6) Bob has two Opto 22 solid state relays and two twelve volt electromechanical relays in his junk pile—to reduce cost, the four channels will be implemented with these devices as the valve drivers.
5. (to Requirements 3,6) To determine if the day is wet, the system will include a moisture probe. Bob has a Wescor S-460 leaf wetness probe in his collection of surplus sensors that will allow the system to determine if it is raining or has recently rained. The S-460 will require an analog input. To reduce system cost, the system will use the Wescor probe.
6. (to Requirements 4, 6) To determine if a day is "hot," the system will need to measure temperature. Bob has a PT100 temperature probe in his collection of junk, to reduce system cost, the system will use a PT100 temperature probe. This will require an analog input. We will use an RCM3400 because it has analog inputs.
7. (to Requirement 5) We will have to concoct a simple web-based interface for setting a watering schedule.
8. (to Requirements 3, 4, 5) The software that reads the sensors will need to have some pre-determined thresholds for what is "rainy / wet" and what is "hot." The user will not be able to change these hard coded thresholds—as that flexibility is just one more level of complication for the user.

Our six requirements generated eight features. Once we have the feature list, we can go about the business of defining the system.

9.13.2 Using Best Practices for Design: Sprinkler Controller Hardware

In California, automated sprinkler systems are commonplace. Not so for other parts of the world. Before delving into the nitty-gritty of the electronics, let's zoom back and look at Bob's existing irrigation system.

Figure 9-16A shows the four valves in Bob Perrin's front yard. The web enabled sprinkler controller will ultimately allow easy control of these devices.

Each valve is actuated by a 24 VAC solenoid. Figure 9-16A shows the wires that carry current to the solenoids. The wires enter the garage wall behind the sprinkler valves.

Figure 9.16A: Four electrically operated sprinkler valves control drippers, sprayers and two circuits of pop-up sprinklers.

Chapter 9

A 4-station commercially available sprinkler controller directs the behavior of the four valves shown in Figure 9-16A. Figure 9.16B shows the controller and its cryptic user interface. This controller is mounted inside the garage on the wall shown behind the valves in Figure 9.16A. As is common with new homes in California, this setup came with the house.

The user interface shown in Figure 9.16B is infuriating. Getting a program into this little box is tantamount to booting an old PDP with toggle switches. The knob and buttons shown in Figure 9.16B have names that might as well be in Cyrillic. What is the ABC button for? What is the difference between the "Manual Station(s)" and the "Manual Cycle" setting?

Figure 9.16B: A commercially available sprinkler controller.

While the hardware shown in this project will not be in a nice little molded plastic box, the user interface will allow the user to set up a watering schedule intuitively over Ethernet (and with a trip to the computer store, **wireless** Ethernet). Once the sprinkler controller works, we can always spend the time and money to spiff up the package. "Pretty packaging" was not a requirement set forth in the previous section.

One of the requirements for this project was "cost effectiveness." Most of the components for this project were salvaged from the techno-junk piles in the authors' garages.

Figure 9.16C shows the accumulation of hardware assembled for the project. The only purchased component ended up being the test valve (bottom center). The moisture sensor (top right) and the PT100 resistive temperature device (RTD) (bottom right) were left over from previous projects.

Since the valves in Figure 9.16A were unavailable for bench top experimentation (an addition constraint placed on the project by Bob's wife), the authors were forced to purchase an inexpensive 24-volt valve. The new valve served as a test load.

For convenience, banana plugs were installed on the valve solenoid to allow quick connection of the RCM3400-based sprinkler controller.

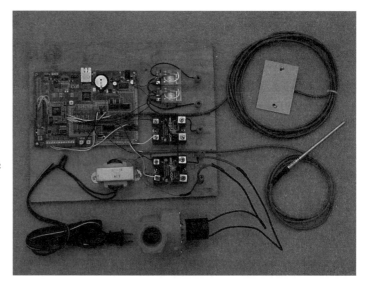

Figure 9.16C: The RCM3400 development board allows quick integration of a wide variety of sensors and actuators.

Networking

The Rabbit Semiconductor RCM3400 development board was selected for the sprinkler controller project. The RCM3400 sports eight analog inputs (we need two), an Ethernet port, plenty of digital I/O (we need to control four valves), and a real-time battery-backed clock.

Figure 9.16D shows a closer view of the actual RCM3400 development board. To interface a moisture sensor, temperature probe and high-current driver channels, we only needed to add a few parts in the prototyping area.

The valve drivers were derived from the discussion of BJT drivers and relays in Chapter 5. Figure 9.16E shows the schematic for the valve drivers. Notice a metal oxide varistor (MOV) placed on all relay contacts to protect the devices against the inductive fly-back from the sprinkler valve solenoids—more on this latter.

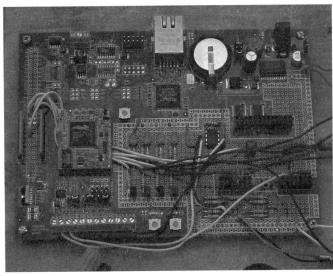

Figure 9.16D: The RCM3400 development board is a feature rich, easily configurable tool.

The two electromechanical relays required a fly-back diode across their coil. Without this diode, the NPN transistors driving the relays would be short lived. The 1N914 diodes were soldered directly across the relay coil's contacts and cannot be seen in Figure 9.16D.

The solid-state relays (SSRs) do not generate any inductive fly-back and their driver circuit contains no fly-back diode.

The RCM3400 development board has many features to show off the RCM3400's ability to read switches, drive LEDs, talk to an LCD and Keypad, interface to Ethernet and read analog signals. The downside of having all this demo hardware is that most of the RCM3400 digital I/O lines are pre-assigned to I/O features on the development board.

Our sprinkler controller must control four valves. We had to choose four I/O lines on the RCM3400 development board that were either unused or could be modified for use in our design.

After looking at the RCM3400 development board's schematic, we decided that Port D bits 4 to 7 were going to drive our solenoids. First, we had to disconnect the devices on the RCM3400 development board that were wired to Port D's upper nibble.

PD6 and PD7 were dedicated to driving two active low LEDs (DS1 and DS2) on the RCM3400 development board. We disconnected these LEDs by simply removing their current limiting resistors R47 and R57. This freed up PD6 and PD7 for controlling valves.

PD4 and PD5 were used by the RCM3400 development board for switch input. There were two pull-up resistors (R43 and R44) connected to these bits. We removed these resistors from the development board.

Chapter 9

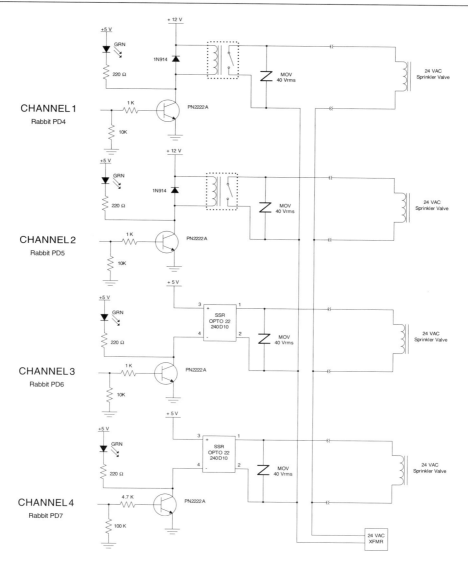

Figure 9.16E: Two SSRs and two electromechanical relays drive four valves. The software driving the channels cannot tell the difference between the two types of relays.

The fact that PD4 and PD5 were used as inputs on the development board and we wanted to use them as outputs meant that we had to change the initialization settings. We could either modify Rcm34xx.LIB in Dynamic C's /LIB/RCM3400 directory or just re-initialize the port after the brdInit() function is called. We chose to do the latter and reinitialized Port D direction registers. For readers who would rather fix the library, we have included a modified Rcm34xx.LIB library on the CD-ROM.

While modifying Rcm34xx.LIB is expedient, if we ever want to use our Dynamic C installation with another stock RCM3400 development board, Rcm34xx.LIB will have to be restored.

Networking

By default, the `brdInit()` function also initializes the output bits. We initialized PD4 through PD7 to be LOW; the valve drivers shown in Figure 9.16E are active high.

The code fragment in Program 9.13 reinitializes the Port D direction registers:

Program 9.13: Setting up Port D to work with the valve drivers.

```
//-----------------------------------------------------------------
// Port D configuration
//
// PD0    RS485 transmit enable    Output Low (disabled)
// PD1    ADC Busy line      Input      Low (ADC device driven)
// PD2    ADC device select        Output High (active low)
// PD3    Not used                 Input         Pulled-up core
// PD4    Valve #1                 Output Active High
// PD5    Valve #2                 Output Active High
// PD6    Valve #3                 Output Active High
// PD7    Valve #4                 Output Active High
//-----------------------------------------------------------------
WrPortI(PDCR, &PDCRShadow, 0x00);     //clear all bits to pclk/2
WrPortI(PDFR, &PDFRShadow, 0x00);     //clear all bits to normal function
WrPortI(PDDCR, &PDDCRShadow, 0x00);   //clear all bits to drive high and
//low
WrPortI(PDDR, &PDDRShadow, 0x0c);     //set bits 3,2,0, clear bits
//7,6,5,4,1
WrPortI(PDDDR, &PDDDRShadow, 0xf5);   //set bits 7,6,5,4 2,0 to output and
//clear 3,1 to input
```

In the brief time between power being applied to the RCM3400 and the execution of the `brdInit()` function there will be period while PD4..7 will be HIGH-Z. To ensure that the valves are kept OFF during this pre-firmware-initialization period, 10 K pull-down resistors were added to the BJT drivers—shown in Figure 9.16E.

The 24 VAC transformer can only reliably power one sprinkler solenoid at a time. There is no hardware assuring mutually exclusive solenoid actuation. For simplicity and cost we put the burden of mutually exclusivity on the Rabbit firmware.

Each driver channel has a MOV across the relay contacts to suppress the inductive flyback caused by the sprinkler solenoid.

Chapter 5 discussed MOV operation and showed how MOVs can be used to suppress high-voltage transients on power and data communication lines. Inductive flyback is simply a form of high-voltage transient and an MOV can be used effectively to protect a switch.

In Chapter 5 we discussed RC snubber circuits for flyback suppression. A MOV is another tool commonly used for controlling flyback in AC systems. Each approach has pros and cons.

Some systems use both an RC snubber and an MOV. The RC snubber can reduce the harmonic content associated with the leading edge of the flyback while the MOV will clamp prolonged voltage excursion to a safe level. This keeps the snubber's capacitor reasonably sized.

Chapter 9

Since we were not overly worried about radiated emissions, we opted to forgo the RC snubber and simply use the MOV.

To verify the effectiveness of the MOV for flyback suppression, a Tektronics digital storage oscilloscope (DSO) was used to observe the operation of the relay drivers with and without an MOV. We used an electromechanical relay channel for these experiments.

Figures 9.16F and 9.16G show how the circuit operated without a MOV installed.

Channel two (CH2) probed the Rabbit port pin used to drive the BJT and subsequently the valve control relay. CH2's vertical sensitivity is shown as 200 mV/division. The channel had a 10x probe attached, so the actual sensitivity at the probe tip was 2 volts/division. When the CH2 trace is "high" the relay is actuated.

The upper trace, Channel 1 (CH1), shows the voltage across the relay switch contacts. When CH1 is flat, the relay contacts are closed and the valve solenoid is actuated. CH1 vertical sensitivity was set to 50 volts/division.

Figure 9.16F illustrates the large voltage spikes that are generated by inductive flyback.

The negative going spike that occurred when the solenoid was turned off is greater than 300 volts and went completely off the measurement scale.

Figure 9.16G illustrates that inductive flyback occurs both when the solenoid is energized and when it is de-energized.

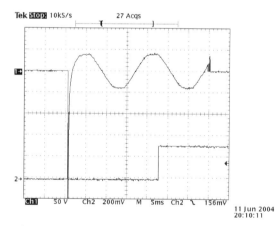

Figure 9.16F: Without a snubber inductive flyback will generate high voltages. The negative going spike on the left is greater than 300 volts.

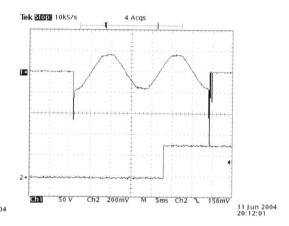

Figure 9.16G: Flyback occurs when turning on or turning off the solenoid.

Figure 9.16H shows how the circuit behaves after an MOV was installed across the relay contacts (as shown in the final circuit configuration—see Figure 9.16E). The inductive spikes are clamped to ± 60 volts. This is within the contact rating of the relays.

Now that we can control all four solenoids, we turn our attention to the sensors.

The finished system will support a "Hot Day" rule that, if enabled, will increase the amount of water dispensed on hot days. This means the Rabbit must be able to read the outside temperature.

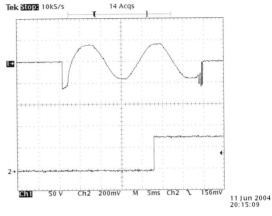

Figure 9.16H: With a MOV installed, the flyback excursions are clamped to ±60 volts.

The RCM3400 development kit came with a fragile little thermistor. This was briefly considered for the project and discarded as too mechanically fragile. Besides, Rabbit has all the functions and sample programs worked out for measuring thermistors. Where's the challenge in that?

We settled on using a PT100 RTD. These devices are a platinum wire or film, usually mounted on a ceramic substrate and secured inside a protective stainless steel probe. The PT100 is shown in Figure 9.16C (bottom right). PT100s are widely used in the biophysical, meteorological and industrial controls industries.

PT100 probes come in a few standard flavors. All PT100's have a nominal 100 ohm resistance at 0 °C. However, the different flavors of probe are built with different grades of platinum and have different temperature coefficients.

The PT100 we had was surplus and we did not know the exact temperature coefficient of the probe. This meant we needed to characterize the probe to get accurate measurements.

If we must go to the trouble of characterizing a probe, we might as well just wait until we have the entire measurement channel wired up and characterize the entire channel. The work is the same. And whatever gain and offset errors may exist in the measurement channel will be calibrated out.

Figure 9.16I shows the circuit we used to interface the PT100 to the RCM3400.

We experimented with values for the resistors in the bridge section and gain settings for the instrumentation amplifier (the AD620). Precision designs often attempt to limit the current in the PT100 to 50 micro-amps or so. This requires the high and low side resistors in the bridge to be about 50K. It also requires gain on the order of 10^3 v/v.

When we tried this, we noticed that not only did we not have the sensitivity that we wanted, but also that high gain was amplifying unwanted noise.

We could have placed some capacitors in the circuit to reduce the unwanted noise, but increasing the signal to noise ratio and dropping the gain was a better solution.

Chapter 9

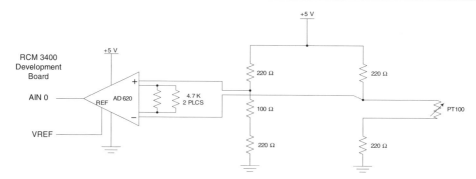

Figure 9.16I: The PT100 interface is simple, but effective.

In the end, we settled on the values shown in Figure 9.16I. The stimulus current is about 10 mA. The AD620's gain is set around 22 v/v. Even without filters, there is no significant noise.

The 10 mA of stimulus current is three full orders of magnitude greater than the 50 microamps found in precision designs. The physical size of the PT100, and the imprecise application (determining if it is "hot" outside) means the self heating associated with the 10 mA stimulus current is negligible.

When using amplifiers in a single supply configuration a designer generally must keep the amplifier's inputs away from the supplies by a volt or more. This of course varies from part to part. The bridge configuration in Figure 9.16I keeps the AD620's inputs between 2 and 3 volts.

Some amplifiers do not have output stages that will drive fully rail-to-rail. And even if they do, sometimes the linearity of the amplifier isn't that great as the output approaches a rail.

The AD620, like many instrumentation amplifiers, has a REF pin. This adds an offset to the amplifier's output. We tied this to the VREF output from the RCM3400. This allows the amplifier to produce voltages between about 2 and 4 volts over the temperature range in which we are interested.

Once the hardware was completed, we used a small test program to sample the RCM3400 channel and display the results using `printf()`. For brevity, we will omit this code, but it is very similar to that used in both Chapters 5 and 8 for reading AIN0.

With the program running, we immersed the PT100 in ice water and recorded the voltage reported by the RCM3400. We also used a commercially available digital thermometer to measure the temperature of the ice water. We repeated the experiment with hot water. The results are shown here.

Environment to which the PT100 was exposed	RCM3400 reported voltage	Temperature reported by digital thermometer
Ice Water	2.115	34.5 °F
Hot Water	4.015	138.5 °F

Networking

From this information, we can derive a function that will accept the measured voltage and return a temperature. Over the limited range of interest, the PT100 is known to be fairly linear. We will thus derive a simple straight-line (linear) function of the form shown here.

$$y = m \cdot x + b$$
$$f(x) = m \cdot x + b$$
$$temperature(volts) = sensitivity \cdot volts + offset$$

Using the numbers from the table above, we derived the following values,

$$sensitivity = 54.2 \frac{°F}{volt}$$
$$offset = -80.133°F$$

This allows the RCM3400 to compute the temperature of the PT100's environment from the voltage measured.

Now that we can measure temperature, we can turn our attention to the last environmental condition that we had—a requirement to sense rain.

Since the authors had a few surplus Wescor S-460 Leaf Wetness sensors lying about, one of these biophysical sensors was pressed into service as a rain detector. The top right section of Figure 9.16C shows the Wescor sensor.

The Wescor S-460 consists of a circuit board containing two gold plated electrodes in the form of interdigitated combs covered with latex-based "secret sauce" paint. The sensor requires about a 4 VAC stimulus voltage. The data loggers commonly used with this sensor supply AC at 300 hertz.

The requirement for an AC stimulus voltage is probably to reduce the likelihood of unwanted plating of the electrode when a DC bias is applied over an extended period of time.

The S-460 datasheet says that when dry the sensor has an impedance of about 3 mega-ohms. When water is on the surface, the impedance drops to about 1 K.

The Rabbit 3000 has a PWM module that can be used to generate a square-wave with about a 50% duty cycle and a 300 Hz frequency. If we were concerned with building a precision instrument, perhaps we might have considered building an oscillator with an output that was more spectrally pure than a square-wave. But, we are just interested in detecting the presence or absence of rain.

Figure 9.16J shows the interface circuit used to drive and measure the Wescor S-460 Leaf Wetness sensor.

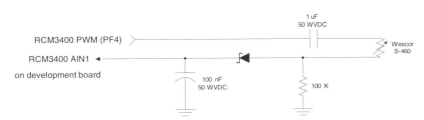

Figure 9.16J: The moisture sensor requires an AC stimulus—the Rabbit PWM feature and a single 1uF capacitor provide it.

Chapter 9

The 1-microfarad capacitor decouples the DC bias of square-wave. This will prevent undesired plating of the electrodes.

The Wescor S-460 is placed in a voltage divider with a 100K resistor. As the impedance of the Wescor sensor changes, the signal at the center of the voltage divider will increase or decrease.

The diode and 100 nanofarad capacitor form a detector and low-pass filter. What's not shown in Figure 9.16J is the RCM3400 development board's analog-input's impedance, about 200K. This 200K is in parallel with the 100 nanofarad capacitor and allows the capacitor to discharge over time.

The RCM3400 actually has an input impedance of 6–7 mega-ohms, but the development board places a voltage divider between the RCM3400 and the screw-terminal connector on the development board (see Figure 9.16D). The voltage divider on the RCM3400 development board, shown in Figure 9.16K, decreases the input impedance to around 200K. The circuit in Figure 9.16J takes advantage of the decreased impedance to bleed the 100 nanofarad capacitor.

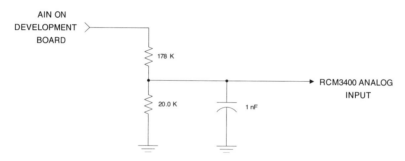

Figure 9.16K: The RCM3400 development board has a bit of extra signal conditioning that extends the RCM3400's analog input range.

The voltage divider on the development board serves a couple of purposes. First, it protects the RCM3400 from over voltages. Second, the voltage divider extends the range of voltages the RCM3400 can measure.

The development board also places a 1 nanofarad capacitor on the center of the voltage divider. This forms a 1-pole anti-alias filter and removes unwanted high-frequency noise.

The RCM3400 measures about 0.2 volts from the rain sensor circuit when the Wescor S-460 is dry and in a 34% relative humidity (RH) environment. The RH was checked with an inexpensive consumer-grade hygrometer[11].

The RCM3400 measures about 0.8 volts when the S-460 has water on the sensor's painted surface.

Picking a 0.675-volt threshold for the "is it a rainy day" rule seems to work satisfactory.

The Wescor S-460 and PT100 were mounted to the eves of Bob's house. There, they are exposed to rain, wind and sun. We expect a five-year minimum service period for the S-460

---

[11] A hygrometer measures atmospheric relative humidity, while a hydrometer measures specific gravity.

Networking

and a much longer period for the PT100. The sensors are high enough that they are not affected by the water spray from the sprinklers.

Now that the RCM3400 can read temperature, detect rain and actuate four valves (one at a time), the system is ready to be web-enabled. Then Bob will be able to control his front yard irrigation system from his laptop—just like surfing the web.

9.13.3 Using Best Practices for Design: Software Requirements

The software required for the sprinkler controller is in large part dependant on the browser-based user interface (UI) we define. The UI must be easy to use, yet versatile enough that watering schedules can be quickly altered based on local weather conditions.

Once the user has selected the watering schedule, we will update an event table. The event table simply contains the list of start and stop times for each channel. The RCM3400 will constantly monitor the event table and engage sprinklers based on the event table and the Real Time Clock (RTC).

The software will also need to take into account the temperature and soil moisture content. If the user opts to allow the watering schedule to be affected by temperature and/or moisture content, the RCM3400 firmware must alter the event table accordingly.

We will allow two environmental modifications from which the user may allow one or both to affect the schedule.

The first rule is the "Hot Day" rule. This rule says—if the temperature is detected to be over 100 °F then each watering interval will be increased by five minutes.

The second rule is the "Rainy Day" rule. This rule says—if the day is wet then decrease each watering interval by five minutes. We could have decided to disable watering entirely, but if the moisture sensor becomes defective and erroneously reports a "rainy" day, we do not want to starve the plants entirely of water.

For simplicity, we will limit the times that a channel can be turned on. For example, channel #1 can be turned on only at the top of each hour. Channels 2, 3 and 4 are allowed start times a quarter of an hour apart.

We will define a standard watering interval as the time a sprinkler is turned on if neither the Hot Day rule nor the Rainy Day rule is applied. We will fix the standard watering interval at 10 minutes.

This means a sprinkler will be on for 15 minutes on a hot day and only on for 5 minutes on a Rainy day.

Next, we can sketch out a user interface that will provide these features.

9.13.4 Using Best Practices for Design: Formalizing the User Interface

Our web interface to the sprinkler system will allow the user to set up the sprinkler schedule, set the clock on the device, and get the current status of the sprinkler system. Each of these three functions will be contained on separate web pages.

The sprinkler schedule page consists largely of a 4x24 table that allows the user to select the times that each of the four sprinklers will be active. If a box is checked, then that sprinkler

Chapter 9

(indicated by the column) will be active starting at the given time for a default of 10 minutes. At the bottom of the page is an additional table that allows the user to enable "rainy day" and "hot day" rules. On a rainy day, the sprinkler duration is reduced to 5 minutes, but only if this rule is active (i.e., checked). On a hot day, the duration is increased to 15 minutes. The sprinkler schedule page also includes links to see the current system status and to set the clock.

The status page is a small pop-up window. It contains a table that shows whether each sprinkler is "ON" or "OFF." It also indicates whether the conditions are currently "hot" and/or "rainy."

The clock page allows the user to set the real-time clock. The interface needs to check the input for correctness. For example, it should take into account the number of days in the month, even for leap years versus non-leap years.

We needed to come up with a clever design for the user interface, since sprinkler systems want only one sprinkler valve to be on at a time, to deliver the highest water pressure. We designed a user interface that allows the user to select no more than one valve at a time. Figure 9.17 shows the HTML mock up of the interface; each checked box below will engage a sprinkler for a 10-minute interval, and no more than one sprinkler valve can turn on at the same time.

Welcome to the Automated Sprinkler System

Sprinkler 1		Sprinkler 2		Sprinkler 3		Sprinkler 4	
12:00 AM	☐	12:15 AM	☐	12:30 AM	☐	12:45 AM	☐
1:00	☐	1:15	☐	1:30	☐	1:45	☐
2:00	☐	2:15	☐	2:30	☐	2:45	☐
3:00	☐	3:15	☐	3:30	☐	3:45	☐
4:00	☐	4:15	☐	4:30	☐	4:45	☐
5:00	☐	5:15	☐	5:30	☐	5:45	☐
6:00	☐	6:15	☐	6:30	☐	6:45	☐
7:00	☐	7:15	☐	7:30	☐	7:45	☐
8:00	☐	8:15	☐	8:30	☐	8:45	☐
9:00	☐	9:15	☐	9:30	☐	9:45	☐
10:00	☐	10:15	☐	10:30	☐	10:45	☐
11:00	☐	11:15	☐	11:30	☐	11:45	☐
12:00 PM	☐	12:15 PM	☐	12:30 PM	☐	12:45 PM	☐
1:00	☐	1:15	☐	1:30	☐	1:45	☐
2:00	☐	2:15	☐	2:30	☐	2:45	☐
3:00	☐	3:15	☐	3:30	☐	3:45	☐
4:00	☐	4:15	☐	4:30	☐	4:45	☐
5:00	☐	5:15	☐	5:30	☐	5:45	☐
6:00	☐	6:15	☐	6:30	☐	6:45	☐
7:00	☐	7:15	☐	7:30	☐	7:45	☐
8:00	☐	8:15	☐	8:30	☐	8:45	☐
9:00	☐	9:15	☐	9:30	☐	9:45	☐
10:00	☐	10:15	☐	10:30	☐	10:45	☐
11:00	☐	11:15	☐	11:30	☐	11:45	☐

Hot Day Rule		**Rainy Day Rule**	
Temperature >100F then Water More (15 minute intervals)	☐	Soil Wet Water Less (5 minute intervals)	☐

Figure 9.17: A simple user interface allows check boxes to allow users to quickly modify the watering schedule.

In the next section, we will describe how the interface was implemented with RabbitWeb.

While the HTML mock up looked good on the printed page, once we looked at it in a browser window, we found it to be non-intuitive—in early trials, some users had difficulty associating a check box to its time of day. We then modified the HTML code to make the interface more intuitive. This constituted a refinement of a feature, not a change of a requirement.

9.13.5 The Web Interface: C Code

The code for this project resides in the file `sprinkler.c`. The web pages are in the files `sprinkler.zhtml` (for scheduling), `time.zhtml` (for setting the time), and `status.zhtml` (for getting system status). We will neither list nor describe every line of code here, but will generally explain how the code behind the interface was written.

The first things to note are the following lines:

```
// sprinkler.c

#define USE_RABBITWEB 1
#define RWEB_POST_MAXVARS 128
#define RWEB_POST_MAXBUFFER 4096
```

Defining `USE_RABBITWEB` to 1 simply allows the developer to use the RabbitWeb extensions. Otherwise, the extensions would not be compiled into the web server. Since this interface uses many variables (recall the 4x24 table), most of which can be updated all at once, we need to raise the number of variables that the RabbitWeb extensions can handle. This is controlled by the `RWEB_POST_MAXVARS` macro (defaults to 64). Additionally, since so many variable updates can be sent to the web server, we need to increase the size of the buffer used to store those variables. `RWEB_POST_MAXBUFFER` (defaults to 2048 bytes) defines the number of bytes we need. This number can be approximated by inspecting a network trace of a request and noting the number of bytes in the request. Section 9.14 mentions some useful utilities for examining network traffic.

```
#use "dcrtcp.lib"
#use "http.lib"
```

These lines simply include the TCP/IP stack and the web (HTTP) server.

```
#ximport "sprinkler.zhtml" sprinkler_zhtml
#ximport "status.zhtml" status_zhtml
#ximport "time.zhtml" time_zhtml
```

The interface files (written in HTML with RabbitWeb extensions) must be imported into the user program. `#ximport` does this, and provides long pointers to the contents of each of the files (`sprinkler_zthml`, `status_zhtml`, and `time_zhtml`).

Chapter 9

```
SSPEC_MIMETABLE_START
    SSPEC_MIME_FUNC(".zhtml", "text/html", zhtml_handler),
    SSPEC_MIME(".html", "text/html")
SSPEC_MIMETABLE_END
```

The MIME table allows certain filename extensions to be associated with specific MIME types (which is how a web browser determines the type of content in a file), as well as associating web server handlers for specific types of content. Note the ".zhtml" entry above: the zhtml_handler parameter means that files ending in ".zhtml" will first be processed by the zhtml_handler before being sent out. This lets the web server know that it needs to search for specific commands with those files (known as ZHTML commands) that allows for dynamic content. This will become more apparent when we look at some of these ".zhtml" files.

```
SSPEC_RESOURCETABLE_START
    SSPEC_RESOURCE_XMEMFILE("/", sprinkler_zhtml),
    SSPEC_RESOURCE_XMEMFILE("/index.zhtml", sprinkler_zhtml),
    SSPEC_RESOURCE_XMEMFILE("/status.zhtml", status_zhtml),
    SSPEC_RESOURCE_XMEMFILE("/time.zhtml", time_zhtml),
SSPEC_RESOURCETABLE_END
```

Each of the imported files is associated with specific names. When the server gets a request for one of these names, it servers the associated file.

A number of variables are declared in the program, such as a schedule[][] array and the hotrule and rainyrule variables. There is nothing special about these declarations, but the "#web" lines that follow them are very special. We dealt with the #web lines in the previous section; the "hotrule" line is not much different:

```
int hotrule;      // 1 if the hot temperatures rule is enabled, 0 otherwise

#web hotrule ($hotrule == 0 || $hotrule == 1)
```

This line registers the variable hotrule for use with the RabbitWeb extensions. This variable can be displayed and updated through the web interface. The expression at the end of the line is the error-checking, or "guard," expression. Each update of the variable's value must pass this expression (i.e., the expression must be true) for the update to be applied. Note the "$" symbols in front of each instance of hotrule. This indicates that the proposed updated value of the variable should be used in the guard.

RabbitWeb also allows arrays to be registered, as in the following:

```
#web schedule[@][@] ($schedule[@[0]][@[1]] == 0 || \
                    $schedule[@[0]][@[1]] == 1)
```

The "@" symbols are used to wildcard the array indices. That is, this statement applies to schedule[0][0], schedule[0][1], schedule[2][13], etc. In the guard, the "@[0]" and "@[1]" expressions refer to the first wildcard and the second wildcard, respectively.

It is also possible to make variables read-only:

```
#web hours[@] groups=all(ro)
#web minutes[@] groups=all(ro)
```

The `hours[]` and `minutes[]` arrays are used to construct the time columns in the scheduling table. They are used for display only. Hence, they are made read-only by the "groups=all(ro)" expressions. This syntax means that these variables are read-only (ro) for all groups. RabbitWeb allows access to variables to be restricted by user group, so these authorization expressions can be more sophisticated than is presented here.

Here is another interesting #web line:

```
#web time.month select("January" = 1, "February", "March", "April", \
            "May", "June", "July", "August", "September", \
            "October", "November", "December")
```

This registration of the `time.month` variable (note that RabbitWeb also allows use of structures) indicates that it is a "selection variable." A selection variable is an integer with specific string values associated with each integral value. Conceptually, it is like an `enum` in C. This construct supports the use of pull-down menus, as we will see later.

```
#web time.day (checkday($time.day, $time.month, $time.year))
```

And here is one last `#web` statement. Not only can we give the guard expression inline, we can also call a separate function to perform the error check. The `checkday()` function makes sure that the number given for the day is valid for the given month and year. Since it even takes leap years into account, then giving this entire expression inline would be cumbersome.

When the time is updated through the clock interface, how does the program know to apply this update to the real-time clock? The answer lies in the following line:

```
#web_update time.year,time.month,time.day,time.hour,time.minute,\
            time.second updatertc
```

This line means that when one of the "time" members is successfully updated, then the function `updatertc()` will be called. `updatertc()` applies the time change to the real-time clock.

There is not much else to study in the C code. `http_init()` initializes the web server. `tcp_reserveport(80)` provides a performance improvement to the web server (port 80) by allowing it to queue requests that it cannot currently service (rather than refusing them outright). And `http_handler()`, which is called from the main loop, drives the web server.

We implemented three `costates` for the project:

- The `mainHandler` costatement drives the HTTP server with the important `http_handler()` call. It also updates the time structure from the real-time clock, so that the time can be displayed properly in the web interface. This costatement runs every 50 milliseconds.

- The `sensorHandler` costatement drives the routine that averages samples from the rain and temperature sensor. The routine determines the "hot day" and "rainy day" status and the appropriate watering interval. This costatement runs once a minute and, if required, could update the sensors even less frequently.
- The `sprinklerHandler` costatement keeps track of the current time and the sprinkler table from the user interface, and determines whether a sprinkler has to be turned on. Since the hardware requirement called for turning on only one sprinkler at a time, the code makes sure all sprinklers are off before it turns one on. It uses `printfs` as a debugging aid to inform the user on the stdio window when a sprinkler gets turned on and for how long. This costatement runs every fifteen seconds; it would do just as well if it ran once every thirty seconds.

9.13.6 The Web Interface: The ZHTML Scripting Language

This section does not purport to be an HTML primer. It will only cover the features that are unique to the ZHTML scripting language. There are many fine books and tutorials available on HTML; www.w3.org contains the definitive resources in the form of various standards.

Special commands in .zhtml files allow the developer to create dynamic content. These commands are encased in "<?z" and "?>" tags. The `zhtml_handler()` mentioned earlier searches for these tags within .zhtml files. When it finds a command within those tags, it performs the appropriate action. One important consideration when writing ZHTML is that each ZHTML command must be in separate "<?z" and "?>" tags. That is, we cannot include multiple ZHTML commands in a single set of tags.

```
status.zhtml
```

We will begin by looking at the sprinkler system status page, `status.zhtml`. The first ZHTML command is a `for` loop:

```
<?z for ($A = 0; $A < count($status, 0); $A++) { ?>
```

`for` loops work just like in C. ZHTML allows temporary loop variables with the names $A, $B, $C, etc. to be used in `for` loops. The `count($status, 0)` expression evaluates to the number of elements in the first dimension of the `$status[]` array. This line, then, begins a loop that iterates over the members of `$status`. This will be used to build a table that indicates whether each sprinkler is "ON" or "OFF".

```
<?z if ($status[$A] == 0) { ?>
   <FONT color="#bb0000">OFF</FONT>
<?z } ?>
```

ZHTML also allows `if` statements. We check the value of `$status[$A]`, and if it is 0, then we display an "OFF" message in red. ZHTML allows simple integral comparisons for `if` expressions. The other `if` statements in the rest of `status.zhtml` are similar to the one above.

Networking

```
time.zhtml
```

Now let us look at the `time.zthml` file, which contains the interface to set the date and time of the sprinkler controller.

```
<?z if (error()) { ?>
   <FONT color="#bb0000">
   The date and time you entered was invalid.  Please try correcting the
   error and setting the clock again.<P>
   </FONT>
<?z } ?>
```

In addition to integral comparisons, ZHTML has a few special expressions for `if` statements. The `error()` expression evaluates to true if we are displaying this page as a result of an error in input. We use this to tell the user to try entering the date and time again.

```
<?z if (updating()) { ?>
   <?z if (!error()) { ?>
      <FONT color="#007700">The clock was set successfully.</FONT><P>
   <?z } ?>
<?z } ?>
```

The `updating()` expression evaluates to true if we are displaying this page as the result of an attempted variable update. These nested `if` statements will display a "success" message if we have updated without error.

```
<FORM action="/time.zhtml" method="POST">
```

This is not a ZHTML command, but rather the beginning of an HTML form. We note it here because of the action and method parameters. The action parameter indicates that after the received variable updates from the form have been processed, the "`/time.zhtml`" page will be displayed (or in this case, redisplayed). For the web server to process the variables, the method parameter must always be "POST".

```
<SELECT name="time.month">
<?z for ($A = 0; $A < count($time.month); $A++) { ?>
   <OPTION
   <?z if (selected($time.month, $A)) { ?>
      SELECTED
   <?z } ?>
   >
   <?z print_opt($time.month, $A) ?>
<?z } ?>
</SELECT>
```

The above code generates the pull-down menu for the month. Note that the name option to the SELECT tag is "`time.month`", without the "`$`". The variable name must always

Chapter 9

match, but must not include the "$" sign when given as parameters to a form element. "`count($time.month)`" returns the number of options in the selection variable, `time.month`. "`selected($time.month, $A)`" is another special `if` statement expression. It returns whether or not the $Ath option is selected. The "`print_opt($time.month, $A)`" command outputs the string corresponding to the $Ath option. The current month is selected by default for the user.

```
<INPUT type="text" name="time.day" size="2" maxlength="2"
       value="<?z print($time.day) ?>">
```

The `INPUT` HTML tag creates a text input field (indicated by the `type` attribute). The default value is the current day, given by the "`print($time.day)`" command. `print()` outputs the current value of the given variable using a default format specifier.

```
<INPUT type="text" name="time.hour" size="2" maxlength="2"
       value="<?z printf("%02d", $time.hour) ?>">
```

This line creates a field for the hour. Notice the `printf()` ZHTML statement. `printf()` allows us to provide a format specifier. Note that `printf()` only allows a single variable, unlike `printf()` in C.

The remainder of `time.zhtml` is either similar to what we have already studied, or contains only standard HTML.

```
sprinkler.zhtml
```

Although `sprinkler.zhtml` is the longest ZHTML page, we have already covered almost all of the ZHTML features that it uses. One interesting technique used on this page has nothing to do with ZHTML. Rather, it uses JavaScript:

```
<SCRIPT language="javascript">
<!--
function openStatus()
{
   NewWindow = window.open('status.zhtml', 'status',
'width=400,height=200');
}
-->
</SCRIPT>
<A HREF="javascript:openStatus()">Check the sprinkler status</A>
```

The above JavaScript function opens a new window with a width of 400 pixels and a height of 200 pixels. This new window will display the "`status.zhtml`" page. The "A HREF" tag contains the link to the JavaScript `openStatus()` function that we defined.

We display the hours in the table in a slightly different manner than before:

```
<?z echo($hours[$A]) ?>
```

Networking

The `echo()` command is actually the same as the `print()` command noted earlier.

Normally, checkboxes in HTML only send a value to the web server when the checkbox is selected. If the checkbox is not selected, then no value for the corresponding variable is sent to the web server. This is problematic for RabbitWeb. It cannot distinguish between a variable that has not been sent because it is not selected, and a variable that has not been sent because it is not contained in the form. Hence, if a user deselects a checkbox, then the associated variable is not changed by RabbitWeb simply because it does not get an update for that variable. The following code implements a workaround:

```
<INPUT type="hidden" name="<?z varname($schedule[$B][$A]) ?>" value="0">
<INPUT type="checkbox" name="<?z varname($schedule[$B][$A]) ?>"
       value="1"
   <?z if ($schedule[$B][$A] == 1) { ?>
      CHECKED
   <?z } ?>
>
```

The "hidden" INPUT tag above creates a variable that is always sent to the web server. Hence, the web server will always receive a value of "0" for each of `$schedule[$B][$A]`. However, if the checkbox for `$schedule[$B][$A]` is selected, then it will also receive a value of "1". The web server will accept the last value received as the definitive value. Therefore, if the checkbox is not selected, the variable will be given the value "0"; if it is selected, it will be given the value "1".

Note the `varname()` command within the name field to the INPUT tags above. Array variables, when given in name fields, must be modified so that they can be sent in the update to the web server. The `varname()` command encodes the variable name so that it can be sent. `varname()` can be used for all variables, not just array variables; however, it is necessary for array variables.

The remainder of `sprinkler.zhtml` is similar to material that we have already covered.

9.13.7 Project Screenshots

Figure 9.18A shows the status screen that presents sprinkler and sensor status to the user. We simulated rain by putting some drops of water on the rain sensor, and turned on the "rainy day" rule. The rain sensor, with the thresholds set at 0.650 to 0.675 volts, worked as expected.

Before connecting the sprinkler controller to real sprinklers, we used `printfs` in the code to simulate sprinkler operation. Figure 9.18B shows the output on the `stdio` screen that corresponds to sprinkler settings from Figure 9.18C.

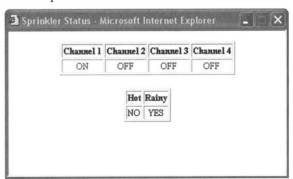

Figure 9.18A: Sprinkler status page.

Chapter 9

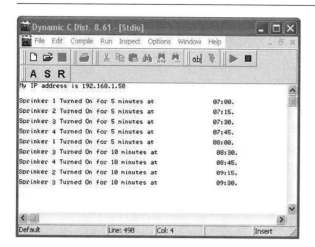

Figure 9.18B: Status output on the stdio window.

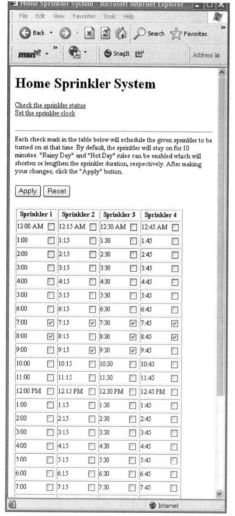

Figure 9.18C: Web interface for the sprinkler system.

Figure 9.18B shows that the simulation started with moisture on the rain sensor, and the "rainy day" rule made the sprinklers run for five minutes. As the moisture on the sensor dried up in the next ninety minutes, although the "rainy day" rule was still in affect, the sprinklers reverted to the standard watering interval of ten minutes, as defined in the project specification.

Figure 9.18C shows the web interface for the sprinkler system. It does not look too different from Figure 9.17, and provides links for setting the clock and displaying sensor status.

9.13.8 Final Thought

We have successfully met the requirements for our web-enabled sprinkler controller. Now we consider other requirements that might be interesting.

What are the consequences of someone hacking into the sprinkler controller? Would a prankster be successful in killing the lawn by forcing the sprinkler controller to over or under-water while the homeowner is on vacation?

If the home network upon which the sprinkler controller resides is not suitably protected it is conceivable a miscreant could maliciously alter the watering times. This could result in a brown or dead lawn if the homeowner is inattentive or absent. It could also result in a huge waste of water.

To avoid this situation, we might consider some of the following approaches.

- Secure Socket Layer (SSL)-based authentication might be required of a user before the user could modify the system settings.
- Additional safeguards such as hard-coding a minimum and maximum allowable watering interval for each valve.
- We could enforce that each sprinkler is not turned on for more than a total of one hour per day.
- We could add separate layers of authentication to the interface, to only allow certain users to modify the setup, while other users can only view the setup.

Anytime someone talks about security, an oft-overlooked element is simply that of physical security. All of the valves for the sprinkler system are simply sitting in a front yard. An individual intent on doing harm, need only walk up to the exposed valves and disconnect a few wires, unscrew the solenoids, or manually turn on the valves. If the homeowner is on vacation, the same harm is done as "hacking into" the sprinkler system.

In a case like this, the extra effort of locking the "electronic" door is probably not worth the engineering.

Another issue to consider is that of reliability. What does it mean for the system to be reliable? In a home appliance, it means the system will do what it is told and when it is told. In this case, it means that as long as the system has power, the controller will accurately keep track of time and execute the watering instructions given to it.

To improve reliability, we could improve the packaging so the device isn't easily damaged. We already have a small battery on the RCM3400 development board that will back up the RTC, but the watering schedule is only stored in volatile RAM. We could ensure a copy of the current watering schedule is stored in the FLASH. Upon re-boot after a power outage, the system could then reload the most recent watering schedule.

Often, people like to talk about "availability" when discussing web-based systems. Generally this term is applied to mean "accessible and operable". This can be tied to reliability, but often marketers and engineers draw a distinction.

Consider a firewall. A firewall may have excellent availability, but may have poor reliability. For example, a firewall may never crash (being highly available), but it may not stop unauthorized access to a network (not be reliable).

In the case of a simple home appliance, the end user will see "availability" and "reliability" together into "its-a-working-or-not-ability," thus rendering the distinction between availability and reliability mute. The end user will only care that the device appears to both water the lawn on queue, and be available for configuration via browser at any time.

Safety is another issue to consider. We have not fused anything. At a minimum, the AC transformer should be fused. This would protect the system from an accidental fire should the wires going to a solenoid short. This is an imperative in a residential setting.

Clearly, this project has a long road to becoming a commercial product. However, it does illustrate the powerful features of the Rabbit hardware and Dynamic C's RabbitWeb tools. The entire system involved very little hardware to interface to the sensors and AC solenoids. The software is small and easy to understand and modify.

As far as the authors know, there is no other combination of hardware and software on the market other than Rabbit and Dynamic C with RabbitWeb that will allow a project of this scope to be assembled as quickly, simply or compactly.

9.14 Some Useful (and Free!) Networking Utilities

We should not take code libraries at face value—things do not sometimes work as planned, and, in the networking world, it is important to be able to independently observe network traffic while debugging. The following tools will be handy in debugging applications.

9.14.1 Ping

This useful utility is found on almost all operating systems, and its simplicity and ubiquity make it a popular application. The key function of Ping is to quickly establish whether a device on the network is reachable. Moreover, because Ping also reports the length of time between the outgoing request and the response, it can help us determine network conditions. Curious programmers should go one step further and explore how Ping uses ICMP messages to perform its function.

Figure 9.19A shows how we used Ping in a Windows XP command window to reach the default gateway in Figure 9.3.

```
C:\>ping 192.168.1.1

Pinging 192.168.1.1 with 32 bytes of data:

Reply from 192.168.1.1: bytes=32 time=1ms TTL=150
Reply from 192.168.1.1: bytes=32 time<1ms TTL=150
Reply from 192.168.1.1: bytes=32 time<1ms TTL=150
Reply from 192.168.1.1: bytes=32 time<1ms TTL=150

Ping statistics for 192.168.1.1:
    Packets: Sent = 4, Received = 4, Lost = 0 (0% loss),
Approximate round trip times in milli-seconds:
    Minimum = 0ms, Maximum = 1ms, Average = 0ms
```

Figure 9.19A: Pinging with Windows XP.

Networking

Ping also resolves fully qualified domain names with the DNS server to check for connectivity with remote hosts. For example, Figure 9.19B shows how we can try to ping Rabbit Semiconductor's URL.

```
C:\>ping www.rabbitsemiconductor.com

Pinging www.rabbitsemiconductor.com [216.167.101.65] with 32 bytes of
data:

Reply from 216.167.101.65: bytes=32 time=83ms TTL=238
Reply from 216.167.101.65: bytes=32 time=82ms TTL=238
Reply from 216.167.101.65: bytes=32 time=83ms TTL=238
Reply from 216.167.101.65: bytes=32 time=82ms TTL=238

Ping statistics for 216.167.101.65:
    Packets: Sent = 4, Received = 4, Lost = 0 (0% loss),
Approximate round trip times in milli-seconds:
    Minimum = 82ms, Maximum = 83ms, Average = 82ms
```

Figure 9.19B: Pinging a URL.

The utility resolves the URL and displays the resulting IP address that it is trying to reach. We can look for packet loss and average response times as well.

9.14.2 Traceroute

While Ping helps us establish whether we can send a packet to a network device, we can use traceroute to find out to trace the path of the packet and discover how the packet gets to its destination. We can examine each line of the response, which gives us the IP address of a router along the path.

If we discover that a network host can no longer be pinged, we can use traceroute to determine how far we can get to the destination before we lose connectivity.

Figure 9.20 shows the results in a Windows XP environment; we can examine the path taken by a packet to Rabbit Semiconductor's URL:

```
C:\>tracert www.rabbitsemiconductor.com

Tracing route to www.rabbitsemiconductor.com [216.167.101.65]
over a maximum of 30 hops:

  1    10 ms    11 ms    12 ms  10.149.184.1
  2    15 ms    17 ms    11 ms  12.244.101.113
  3    11 ms    22 ms    13 ms  12.244.67.169
  4    13 ms    13 ms    17 ms  12.244.67.14
  5    27 ms    23 ms    27 ms  12.244.72.206
  6    23 ms    22 ms    22 ms  gbr2-p50.sffca.ip.att.net [12.123.13.62]
  7    41 ms    27 ms    23 ms  tbr1-p012702.sffca.ip.att.net
 [12.122.11.69]
  8    22 ms    27 ms    23 ms  ggr2-p300.sffca.ip.att.net
 [12.123.13.190]
  9    23 ms    24 ms    24 ms  p16-0-1-1.r20.plalca01.us.bb.verio.net
 [129.250.9.73]
 10    84 ms    86 ms    99 ms  p16-0-1-3.r21.asbnva01.us.bb.verio.net
 [129.250.2.193]
 11    88 ms    95 ms    84 ms  p64-0-0-0.r20.asbnva01.us.bb.verio.net
 [129.250.2.34]
 12    91 ms    88 ms    88 ms  p16-5-0-0.r01.mclnva02.us.bb.verio.net
 [129.250.2.180]
 13    91 ms   113 ms    84 ms  p4-9-0.a00.alxnva02.us.da.verio.net
 [129.250.17.58]
 14    92 ms   103 ms    86 ms  ge-5-0.a01.alxnva02.us.da.verio.net
 [129.250.61.10]
 15    85 ms    88 ms    90 ms  ge0031.ed2.wdc.dn.net [216.167.88.124]
 16    83 ms   105 ms    82 ms  rabbitsemiconductor.com [216.167.101.65]

Trace complete.
```

Figure 9.20: Tracing the route to a destination.

9.14.3 Ethereal

This open source protocol analyzer is popular with developers and is a useful troubleshooting aid. We can capture live packets as they traverse the network and the utility, with its knowledge of hundreds of protocols, helps us easily decipher what it is seeing on the network. In a Windows environment, it requires us to install the WinPcap packet capture driver.

This utility can be downloaded from http://www.ethereal.com. The URL also contains the mirror to the WinPcap packet capture driver.

Networking

9.14.4 Netcat

Netcat started out as a useful debugging tool on the Unix platform and has made its way to the Windows environment. For our purposes, this open source tool serves as an invaluable "reference" TCP or UDP server or client that handles inbound or outbound connections. If the Rabbit code that we write connects with Netcat, we know it works.

Netcat for Windows can be downloaded from http://www.atstake.com/research/tools/network_utilities/. A version of netcat for the Cygwin environment is also available.

9.14.5 Online Tools

A number of tools are available online as well for searching through the Internet. These tools can help us Ping destinations on the Internet, explore DNS and registration records. Some of the tools are available at:

http://network-tools.com/

http://samspade.org/t/.

9.15 Final Thought

In this chapter, we have merely scratched the surface of what is possible with a networked embedded system. Any networked embedded system will likely not work in isolation. The designer has to think from a *systems* perspective and which networked element can provide what services. For example, in a given industry vertical, an embedded system may serve a monitoring, diagnostic, or billing function, but it will have to report to a system at a higher layer. In such an environment, the designer has to consider systems-wide issues involving access, performance, reliability, availability and security.

As our world becomes more connected, people will come to expect the devices in their daily lives to be both user friendly and interconnected. The full featured TCP/IP stack from Rabbit along with other more proprietary tools, such as RabbitWeb, allow a system designer to web enable a design with a minimum investment in time and money.

As with any technology, tools used to mold the technology into products are as important as the technology. Rabbit's highly integrated embedded processor is a remarkable technology. Ethernet along with the various protocols we have discussed is also a very useful technology. C-compilers, assemblers and debuggers and a full-featured TCP/IP stack are the tools used to marry these technologies and create products.

Dynamic C has been used throughout this book for developing the Rabbit side code. As we have also used Java, C# and C++ under both Windows™ and Linux for the PC side, engineers have tool choices on the Rabbit side as well.

For all good processors, third party tools exist. Rabbit is no exception. Softools has created an ANSI C compiler, an assembler and linker bundled under an IDE. The tools offer a traditional development environment with multiple source files.

Rabbit Semiconductor has licensed its feature rich TCP/IP stack to Softools, thereby giving developers not only great hardware, but an ability to choose the tool of their preference.

CHAPTER 10

Softools—The Third Party Tool

Until now, we have used Rabbit Semiconductor's Dynamic C for all of the projects and examples. While Dynamic C is an excellent tool, other tools exist for supporting Rabbit Semiconductor's microprocessors. There are some free tools available in the public domain. There are also proprietary tools that are tied to third party hardware but based on Rabbit microprocessors.

By far the most professional of the third party tools is the tool set offered from a company named Softools Inc. These tools grow out of roots that run two decades deep in C compiler and assembler development.

Users of this toolset have access to an ANSI C compiler, macro assembler, linker, project-manager, source code debugger and other tools all tied together in a nice IDE. This chapter will explore the Softools environment and tool suite in detail.

For interested readers, some of the sample programs from previous chapters that can be ported from Dynamic C to an ANSI C compiler have been ported to the Softools WinIDE and can be found at http://rabbitbook.softools.com. The complete set of Softools Rabbit WinIDE tools can be downloaded and installed from http://www.softools.com/rabbitdownload. The only caveat is that without a license file, the downloaded tools will only compile to RAM and are time-limited for evaluation but are otherwise the full version of the tools.

10.1 Who is Softools?

Softools, Inc., an embedded tools developer, was founded in 1989 with the goal of making better tools for the ZiLOG Z280 than were available at the time. The tools were expanded to support the 8085, Z80 and Z180 microprocessors. Softools improved on the compiler techniques of the day, allowing Z80 and Z180 programs to be written as if the processor had a linear address space for code while an advanced linker laid out the program, adding banking support without source code changes.

In 1991, Softools released its ANSI C compiler for the Z80, Z180, Z280 and 8085 processors. These tools utilized the same advanced memory management techniques, which allowed programs limited in size only by the amount of EPROM or flash memory available in the target.

In 1994 Softools released the lowest cost in-circuit emulator (ICE) available for the Z180 and a standalone DOS-based source debugger.

In 1995, Softools added far pointer and data support to the Z180 C compiler. Now programmers could access the full 20-bit (1MB) address space with code and data. Companies who might have been forced to change microprocessors were now creating programs that were 300 and 400 kilobytes long.

In 1999 Softools released the WinIDE™, which tied the tools together in a modern Windows®-based IDE. A new global optimizer was also introduced with the WinIDE.

Softools began developing the Rabbit 2000 version of the WinIDE in early 2000. This package would inherit the tried and true assembler, ANSI C compiler (updated with a few nice features of the latest 1999 C standard), the global optimizer and a new source debugger.

Softools expertise has been strictly focused on the Z80 derivative processors for over a decade (almost two decades counting its founder's previous experience). They have perpetually improved their tool set. Over the years, users have suggested features and enhancements that have found their way into the tool-chain. The current Rabbit WinIDE tools benefit from this experience and have inherited the elegant features and clever optimizations Softools has developed over the last 15 years.

The Rabbit WinIDE tools are summarized in the following sections, showing the most common features. Of course, extensive documentation of the Softools suite is beyond the scope of this book. The tools come with a complete set of online manuals.

10.2 The Rabbit WinIDE

Like most Windows programs, the WinIDE has many familiar menus and functions. These control various parts of the WinIDE.

The first component of an IDE is a source code text file editor. The editor works great for editing and debugging. The WinIDE also works with the editor of the user's choice. This is referred to as an "External editor."

The WinIDE can open project files in an external editor, including positioning on a source line. The upside is that the user can remain in familiar territory with their favorite editor and not learn a new editor. The downside is that an external editor can never know about the program, so the programmer cannot set breakpoints or use other debugger functions in the external editor.

The WinIDE does have an override so the programmer can force a project file to be opened in the WinIDE editor to allow setting breakpoints or using one of the debugger's functions.

10.2.1 Project Files

The heart of a WinIDE managed program is the project that controls it. Although one can open and edit a file outside of a project, almost all operations involve a project source file. A project contains the files that constitute the program. It also saves the options for the assembler and compiler as well as segment information and linker settings.

When a project is created, the user decides if it will create a program (the linker is used to create the final program) or if it will create a library (the librarian will be run to build a library). We should mention the difference about LIB files—Dynamic C LIB files are text files containing source code while WinIDE LIB files are binary files and can only be used with the WinIDE linker SLINK.

Chapter 10

If a program is large, or composed of separate units which can be tested and "put away," it is advisable to create libraries of these items, and keep the project for application source files only. If a library needs to be updated, it's not difficult to switch projects, make the edits and build the library and switch back. The WinIDE will notice the LIB file is new and will force at a minimum that the program be linked to pick up the library changes.

A project can be as small as one source file (and the requisite C startup file and SCRabbit.lib runtime library) or as large as several hundred files. Only program files which are edited by the programmer are assembled or compiled. There is little overhead over using multiple small source files over a few larger ones. Consider that the smaller source files will compile faster, easily making up for any extra time required by the WinIDE to process a large number of project files.

A project is normally created using the "New Project" menu choice or toolbar button. This opens the New Project Wizard which steps the user quickly through several pages of settings. Two issues which do need mentioning are the stack size prompt and the addresses for RAM.

The default stack size is 512, which may be adequate for many programs. However, TCP/IP applications require a much larger stack. 3,072 (0C000h or 3k) is a minimum stack size for these programs.

The New Project Wizard page requesting input of addresses for Flash and RAM is critical for creating a program that runs in RAM and flash. The starting address of flash is entered into the dialog and defaults to 0. Only in rare cases would this be changed—the Rabbit begins execution at physical address 0x00000 upon boot.

The physical end of RAM is entered and defaults to 1FFFFh. This value is a source of confusion for two reasons. First, all the examples use this value to ensure they will run on all Rabbit core module based systems as all systems have at least 128k of RAM installed.

The logical end of RAM is entered and defaults to DFFFh. This wouldn't be changed except in special cases.

Second, the WinIDE debugs in RAM at the address range normally used by flash memory, so this address appears to be wrong. But it's not. The value 1FFFFh is the end of 128k. The Softools linker has a unique feature of being able to set the *last* address of a segment and also to have segments before this grow downward in memory. If the

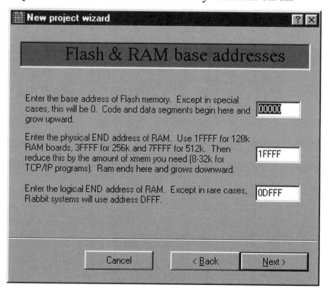

Figure 10.1: New project wizard: Flash & RAM addresses.

Softools—The Third Party Tool

Rabbit controller has 256k of SRAM, setting this to 3FFFFh is good, and if the controller has 512k, 7FFFFh is a good choice for this.

There is one exception to these values—a TCP/IP program requires extended memory (Softools has implemented xmem functions so the full featured Z-World TCP/IP stack discussed in Chapter 9 will be compatible). Extended memory is taken from the end of SRAM.

The amount of extended memory required by a TCP/IP program varies from 8k to 32k or even more. For controllers with 256k or 512k, there is probably enough memory to start with 32k and reduce this later if necessary. A 128k SRAM equipped controller can start with 16k with a note that one may have to reduce extended memory requirements in the TCP/IP program settings.

Therefore, for TCP/IP programs, the end address of RAM might be reduced as follows: 128k – 1CFFFh; 256k – 37FFF; 512k – 77FFFh to allow for 12k, 32k and 32k of extended memory, respectively.

The project window supports the common functions for working with projects found in most tools. Files may be added to, or removed from, projects. Files maybe renamed.

Assembly and C files that are added to a project inherit the (global) options set for the project which are found in the Options menu for Assembler and Compiler. If files are being added that require a specific setting or symbol definition, this can be set in the options before adding the files. A file in the project can have its own private options set by choosing Options in the context menu for the selected file. When the global options are changed, files in the project are updated with the final options unless the files have private options. In that case, a list of these files will be shown, allowing the programmer to not change or override private options. Normally, one keeps all private options when changing the global options.

The WinIDE uses the terms "build" and "make." Build means to assemble and compile everything as if no OBJ files are present. Make means only assembling and compiling what has changed, which could mean doing nothing at all if nothing in the project has changed.

10.2.2 Models with Projects

Project files support models. A model is a variant or derivative of a program build. Each variant is built with the same set of source files but can be changed with different assemble-time or compile-time definitions.

The term model came about because of the idea that different models of a product can be built with the same source code. For example, a *Debug* and a *Release* model might be built. Different models can be used to target languages, e.g., *English, Spanish, German, Italian*, and so on.

The WinIDE maintains the output files of each model and can make or build one model or all models with one command. One requirement is that all the same source files be used in the project for all models. However, source files can be completely excluded at build time with conditional compilation, allowing models (e.g., *Lite, Developer*, and *Deluxe*) to be created where some contain features that other models omit.

Chapter 10

10.2.3 Project Source File Properties

The WinIDE has a properties window, which works with the project files. Assembly and C project source files which have been built (i.e., they have an OBJ file) have properties which can be viewed in the properties window.

Information about a source file's include files, such as their size and modification dates and times, can be displayed. Information commonly buried in map files such as a particular source file's segment usage can also be found.

If, for example, a source file contains the C function *getMotorSpeed* and this function is 200 bytes long and uses 4 static long variables, the segments list in the properties window would show segment CODE with C8 (hex) 200 (decimal) bytes and BSS with 10 (hex) 16 (decimal) bytes. This segment list can be helpful to ensure segments used by the source file have been allocated as intended.

10.2.4 Project Message Window

The WinIDE logs errors from the assembler and compiler. Messages which have a corresponding source file line number can be double clicked to jump to the source line (in the external editor or WinIDE editor as configured). Pressing **F1** on an error will also open the full description from the assembler or C compiler manual for the message. The full set of messages in the message window can also be sent to a WinIDE editor window for searching, saving, or emailing for technical support.

10.2.5 WinIDE's SASMRabbit Assembler

The assembler used in the Rabbit WinIDE is an advanced, full featured multipass assembler. Some interesting features include access to symbols, logical and physical addresses; creating multiple segments for specific placement in the final program, alignment, and the ability to take *any* expression with any operator and operand and defer evaluation until link time.

The programmer can resolve things at link time that many assemblers require code and time to do at runtime because they lack this capability. For example, if the programmer wants to set the Rabbit SEGSIZE start for DATASEG and set DATSEG to the start of user defined segment TABLE, we could do the following:

Program 10.1: Setting SEGSIZE and DATASEG.

```
ioi ld     a, (SEGSIZE)          ; Get SEGSIZE
    and    a, 0F0h               ; Keep STACKSEG start
    or     a, sgst TABLE >> 12   ; Upper nibble start boundary
ioi ld     (SEGSIZE), a          ; Set SEGSIZE
    ld     a, (sgat TABLE >> 12) - (sgst TABLE >> 12)
ioi ld     (DATASEG), a          ; Set DATASEG
```

For this discussion, let us assume segment TABLE will be linked at logical address C000h and physical address 70000h. The third instruction takes the upper nibble of the logical start of segment TABLE and adds that to the lower nibble of STACKSEG. sgst gets the logical start of a segment while sgat gets the physical start address. This ORs 0Ch into SEGSIZE.

Softools—The Third Party Tool

The fifth instruction is one step more complicated. Since the Rabbit MMU adds the logical nibble to the final address, it must be removed from the MMU DATASEG register.

The first part is the physical address of "TABLE shift right 12," which is 70h. The second part is the 0Ch used above. This loads 64h (70h – 0Ch) into DATASEG. If the program loads a word from TABLE+1230h, which is 0D123h, the 0Dh selects DATASEG, which adds DATASEG, 64h, to 0Dh, getting 71h, for a final runtime address of 71123h.

One might argue that we could simply load these values immediately in the instructions with constants. The advantage here is that, at link time, we can move segment TABLE to any logical address (e.g., 9000h or 5000h) and any physical address and the code above works without being changed or assembled.

The linker's job is to resolve symbols and locate segments. We should be able to do so without changing the source code.

Unfortunately, most development tools can't do complex expression evaluation at link time, and require source code changes when changing the memory layout of the program.

10.2.6 SAMRabbit Macros

Microsoft's M80 8080/Z80 macro assembler from the mid 1980's set a de-facto standard for macro support. Yes, even Microsoft once had its fingers in the Z80 and 8-bit tool market. Softools' SASMRabbit is compatible with M80 and its macro syntax.

Macros can be used for many tedious tasks. They can also be used for clever things, and to take it to the extreme, even for things like calculating checksums of program strings to detect program tampering.

Here's a practical example that creates a 256-byte lookup table with the parity for the byte where the byte would be used as the index into the table. If the byte taken from the table is zero, then the byte has even parity, and conversely is one for odd parity bytes. This has a direct application on the Rabbit 3000 as parity for the serial ports must be calculated in software.

Program 10.2 shows a macro to generate the parity table and the fastest code to generate parity for a byte in register A:

Program 10.2: Generating a parity table.

```
genParityTable    .macro
byte         =     0
    .rept   256
par =        0
mask         =     1

    .rept   8
    .if     byte & mask
par =        par+1
    .endif
mask         =     mask << 1
```
(Program 10.2 continued on next page)

Program 10.2: Generating a parity table (continued).

```
        .endm

        .if     par & 1
                .db     1
        .else
                .db     0
        .endif

  byte          =       byte+1
        .endm
        .endm
;
; Use macro to build 256 byte even parity table
;
parity:    genParityTable
; Lookup parity for byte in A, set Z flag for even parity
;
        bool    hl              ; Zero H
        ld      l, a            ; Byte to HL
        ld      de, parity
        add     hl, de          ; Index into table
        bit     0,(hl)          ; Test bit 0, set Z flag if even parity
```

10.2.7 CMACROS

The WinIDE ships with CMACROS.inc which contains a number of macros for writing assembly code which is accessible by C code. These complex macros allow creating assembly functions with optional arguments with names, and including an optional stack frame with named locals. The names defined are to be used with the indexed addressing mode using register IX for access to local variables. More details regarding CMACROs can be found in the WinIDE documentation.

10.2.8 WinIDE's SCRabbit C Compiler

The C compiler included in the WinIDE is a globally optimizing ANSI C[1] compiler, adhering mostly to the C90 standard with a few features of the C99 standard. The most notable C99 feature is variable arguments in macro definitions, which by far is the most desirable change to the C preprocessor made since the C90 standard. It enables creating a macro to remove printf, or include something similar to printf that has one and optionally many arguments.

The downside of the C99 standard coming out so late is that many compilers do not support any, or are only supporting some C99 features. Using C99 features could make programs not portable to another compiler that has not implemented the used features.

---

[1] SCRabbit rounds bit fields up to byte-sized entities and issues a warning. It does not support true bit fields for a couple of reasons. They cannot be implemented efficiently and they are not always portable between different processors or implementations of C.

10.3 SCRabbit Optimizer

Using the microprocessor's instruction set to the fullest and generating small and fast code should be two of the goals of every compiler designer.

The Softools compiler takes advantage of most of the new Rabbit instructions not available in the Z80 or Z180 instruction sets. This change alone showed a significant improvement compiling code for the Rabbit versus the Z180. The primary reason, among several, is due to the better 16-bit addressing modes of the Rabbit. For example, the Rabbit 3000 can assign one local `int` in C to another `int` in 4 bytes and 20 clock cycles (in one of 2 addressing modes), while the Z180 requires 12 bytes and 58 clock cycles (with only once choice of addressing mode). Because an `int` and `unsigned int` are often used in C programs, the change to both the code size and speed becomes a substantial 66% improvement.

The goal for small and fast code was achieved by optimizing the compiled code. There are several optimizations for reducing memory references as these are much slower than register operations. These reduce code size and increase speed.

The optimizer removes common code, which will reduce code size but does not speed it up.

Many other optimizations are done by a sophisticated peephole optimizer which also reduces code size and increases the speed of the compiled program. The optimizer in SCRabbit is global, meaning an entire function is optimized allowing it to remove code and improve code over a large part of the program.

Were it not for the high-speed PCs of today, the current optimizations used by the WinIDE could take many minutes to compile a large file. As computers have gotten more powerful, optimizing compilers have gotten better.

The optimizer does not restrict the ability to include inline assembly code. In fact, inline assembly code which uses C to load registers can be optimized. Once literal assembler instructions are parsed, the optimizer is disabled.

The compiler also supports two compiler options which are not mutually exclusive. Optimize for size and optimize for speed. Surprisingly, these have nothing to do with the global optimizer, but instead change the heuristics and code generation choices for the code generators. The global optimizer works on the output of the compiler regardless of how these two settings are specified.

10.3.1 Error Checking and Warnings

SCRabbit contains a number of diagnostics that generate errors when Softools considers the infraction to possibly cause a program to not run as intended. Some are considered undefined behavior by the C standard, which allowed Softools to do as it wanted.

SCRabbit contains over 150 diagnostics which generate warnings for suspicious expressions or sub-expressions, for things which might affect portability of the program, for things which might not be efficient, and for things which could cause problems but which are deemed common in "normal" code and to not cause them to be escalated to errors.

Chapter 10

A number of users who moved code to their compiler have mentioned that the compiler found outright bugs in their programs before it was even run. Though a *Lint* utility is always a good thing to use, Softools has done what they could to warn the user about many things that just might not be right.

One interesting warning that deserves some attention because it's important on an 8-bit system like the Rabbit is "Applied ANSI `char` promotion to `int`." This doesn't mean anything is wrong or bad. In the following valid program fragment it's going to occur:

Program 10.3: `char` *promotion to* `int`.

```
int     x;
char    a, b;
x = a + b;
```

The warning is caused because a and b must be promoted to `int` before adding them and storing the result into *x*. The warning might alert the programmer to save code because if a and b, when added, never result in a value over 255, *x* could be declared as a `char`.

This would save the promotion from char to int two times in this trivial example. Imagine a program that does this a thousand times. The warning is even more important for signed chars, since the promotion to signed `int` takes a little more code. If the expression or sub-expression can be evaluated entirely in `char` width, the warning will not occur (nor will the promotions).

Some programmers think that the standard states that all chars in expressions must be promoted to ints. The fact is, if the result fits in a `char`, or more accurately, the result is the same whether it had been promoted or not promoted, the promotion clearly is not necessary.

In cases where SCRabbit cannot be sure it can suppress this promotion, it will promote and issue the warning, which will appear to be wrong (on the compiler's part). These are cases where Softools had to take the conservative path and the compiler will only skip the promotion when it can positively be determined not to affect the result.

10.4 SCRabbit Segments

SCRabbit generates code, far code, constants, strings, initialized variables and uninitialized variables into different segments. This allows the separation of these items, which further allows them to be located at link time where they are required to be placed. Segments and memory layout and linking are covered in more detail later but segment names will be defined here to make the following sections clearer.

Table 10.1 summarizes all of the segments used by a Rabbit WinIDE program. Note that not all segments will necessarily appear in a program.

Table 10.1: Segments used by a WinIDE program.

Segments used by SCRabbit		
Segment name	Use	Type
CODE	Program code	Near
CONST	Constants and C strings – variables declared with `const`	Near
DATA	Constants that are part of initialized C variables	Far
FARCODE	Program code	Far
FARCONST	Constants – variables declared with `const` and `far`	Far
BSS [2]	Uninitialized C variables	Near
BSSZ	C variables initialized to 0	Near
IDATA	C variables initialized from segment DATA	Near
STACK	Program stack	Near

10.5 SCRabbit #pragmas

SCRabbit uses a number of `#pragmas` to control the behavior of the compiler, or to do specific tasks in as portable a method as possible. Two of the most common and useful SCRabbit `#pragmas` will now be examined.

10.5.1 #pragma SEG

This `#pragma` is used to change segments used by the compiler. There are two possibilities. The first is to map a compiler-defined segment (see Table 10.1) to another compiler-defined segment. The second is to map a compiler-defined segment to a user-defined segment. The former allows us to combine two segments, like putting strings in the code segment. The latter allows us to place data in any named segment.

Once this pragma is specified, all compiler generated accesses to the old segment will take place in the new segment. The syntax is:

```
#pragma SEG (oldSegment, newSegment )
```

---

[2] BSS is used strictly for historical reasons dating back to FORTRAN Assembly and Unix. It comes from the term "Block Started by Symbol" and was used for uninitialized variables as it is now. Note that TEXT was and perhaps still is used for code storage, but it can be confusing since we all think of text quite differently.

For example `#pragma SEG (CONST, FARCODE)`[3] moves constants into the far code segment. This could be useful for a large table that can be accessed by a far function. By moving CONST to FARCODE, the function can access constants in its own space and eliminate using up CONST memory which resides in near (root) space. The danger in doing something like this is we must not pass a pointer to this const data to another far function. Because CONST is now mapped into FARCODE, it will not be available to every function in FARCODE.

A second example is to define one or more structures that are uninitialized and can be stored into battery-backed memory. If the battery-backed memory is stored in another physical SRAM, the segment would also be specified at link time to fall into the address space of the battery backed SRAM.

Program 10.4: Placing a structure in a segment.

```
#pragma SEG (BSS, BATT_BACKED)
struct {
    char    name[20];
    char    age;
    long    phone;
} employees[100];
```

This stores 100 records of employees with their name, age, and 7 digit phone number. The compiler will place these in segment BATT_BACKED.

Only segment BSS is cleared on program start, so BATT_BACKED would always retain its values. Of course, the program must have conditional code to initialize the array the first time it is run, or via operator input.

There is more than one approach that can be used to determine when battery-backed RAM is valid or when it needs to be reset to the "factory defaults," perhaps with some notification to the operator. Notification to the operator is important—the reason battery backed RAM isn't valid could be an indication of a hardware problem or simply indicates a dead RAM backup battery needs replacing.

To check for valid battery-backed RAM, many programmers use one or more C variables set to known values, or sometimes an initialized array is set to a known value. The variable is checked on startup to determine if RAM has been preserved. The robustness of this check determines how well the program will handle power fails or glitches. If a few bytes of battery backed RAM should change, but the power on test variables do not change, the system will continue to run with these changed values still in battery backed memory. The degree to which this affects the program could range from not at all, just cosmetic showing in a display, to catastrophically if calibration or control data has been corrupted.

A robust method is to checksum, or better still, CRC the entire battery backed area. If battery backed items are contained in their own segment, the segment starting address and size can be

---

[3] Segment names are not case sensitive but are shown here in uppercase to suggest that they are special names in the compiler.

used to know how much RAM to CRC. If the data is kept in a C struct, the `sizeof` operator can be used.

When the application program writes to battery backed RAM, it should check the CRC first to ensure it matches the power on CRC. There isn't much point in writing a value and updating the CRC if the memory area already has a bad value (this could have occurred accidentally by the application). If the CRC matches first, the values in RAM can be changed and the CRC recalculated and stored. However, a power failure or reset during this procedure is bad—the CRC on power up will not match and the program's defaults would have to be restored.

Optimally, two regions should be used with an indicator (preferably a single byte) in a fixed battery backed location indicating which region is current. Access to a single area is much more efficient, but not as reliable as using two regions. To update a variable, the current region should be copied to the unused region. After the CRC check, update, and final CRC procedure is completed on the copied region, the indicator would be changed to indicate that the just-copied and updated region is current. This method guarantees that one of the two regions is always accurate.

Resetting RAM using operator input is not difficult to implement and is a good idea to provide for the user. It is a mechanism that the user can manually invoke to "clean the slate" and have the program start anew. Most electronic equipment and even cheap toy games have these pinhole-accessible reset buttons.

10.5.2 #pragma init_file

This `#pragma` is used to initialize an array with the contents of a file. Softools added this to their compiler primarily for including web pages and images into Ethernet applications. The C language had no provision for doing this. Dynamic C's similar construct is #ximport.

SCRabbit is different in that it does not modify the file's image with the file's size. We can use `sizeof` to get the size of the array, which is the size of the file used with `init_file`. So that SCRabbit will know the name of the array, it requires that the array be an *extern* array declaration before use as follows:

```
extern const char far  index[];
#pragma init_file (index, "web pages\index.html" )
```

The array `index`, which has been declared as `far` and `const` since it can be stored in far space and wouldn't be changed (or specifically would be stored in flash), will be initialized with the contents of the file `index.html`. `sizeof index` will result in the value equivalent to index.html's file size.

TCP/IP and HTTP examples in the WinIDE installation use this `pragma` many times to be able to store web pages and images.

10.6 Near and Far Functions

The Rabbit processor's innovative MMU supports near and far functions. SCRabbit creates far functions by default, so generally one need not be concerned about near and far code. We bring it up because there is a WinIDE compiler option (which is on by default) to create all plain functions (near or far is not specified) as far. There is rarely a reason to change far functions to near functions. This is because near functions take up room in the precious 56k of near (root) space and should be used only in two cases:

1. Interrupt service routines. All function pointers are 16-bit—SCRabbit creates the linkage to allow the programmer to use a far function addresses which are referenced with 16-bit pointers. This linkage destroys the IY register and therefore **must not** be used for an interrupt service routine.[4] By default, when an interrupt modifier is used on a function definition the function is created as a near function.
2. Functions that need to use the XPC register for general memory access. XPC is used to map far functions into the XPC logical address space. Changing XPC in a far function would crash the program.

10.6.1 Far Data and Pointers

SCRabbit supports far data and far pointers. Far data can be stored anywhere in the one megabyte address space and can be accessed directly as all C data can. A far pointer can be used to access something far by its address.

In practice, far data are defined as aggregate data definitions, i.e. arrays, structs or a mixture of both of these. Smaller data constructs shouldn't be stored as far. Doing so only wastes code and incurs execution overhead.

A far pointer is 24 bits but is stored as 32 bits to take advantage of the faster Rabbit 16-bit loads and stores. Syntactically, the `far` qualifier is just like the `const` qualifier.

Declaring far pointers can be tricky. If a pointer points to something qualified as far, it is a far pointer. Otherwise, the pointer is stored far. Rarely should one declare a pointer to be stored far. The following declarations should clear this up:

```
char far *farPointer;
far char *farPointer;
```

These far pointer declarations are identical. The placement of the far is decided by the programmer's preferred style. The pointer "points to" a far `char` and therefore is a far pointer.

```
char * far pointerFar;
```

This is a pointer "stored far" that points to a near character. The overhead to load the pointer (stored far) is significant compared to loading the 16-bit value directly. Of course, loading a far pointer means loading 32-bits, but doing this from near space is a lot more efficient.

We'll finish this with a bad example:

```
char far * far farPointerFar;
```

---

[4] One exception is a timer callback, which can be a far function because the function making the callback has taken care of this issue.

This uses far to qualify two things—the pointer and the `char` pointed to. Although the compiler will honor this and generate correct code, it is very inefficient to dereference this pointer—it must generate a far load to load the 32-bit far pointer and another far load to get the pointed to data.

10.7 Inline Assembly

The WinIDE's C compiler supports inline assembly. SCRabbit is optimizing and certainly generates efficient enough code. That is, efficient enough to write a program, including interrupt service routines entirely in C. This does not preclude the programmer from using assembly code. There are a few reasons to support it within a C source file:

1. It allows SCRabbit to inline all Rabbit port I/O including shadowed I/O. This has been reported to show a ten fold improvement in I/O intensive applications over not inlining I/O.
2. An interrupt which occurs at a high frequency should probably be written in assembly.
3. Porting code already written in Z80, Z180 or Rabbit assembly code. This is especially true when the C function contains both C and assembly code.
4. Creating code with exact timing or code that counts processor cycles. As mentioned in Chapter 6, one should never rely on the speed of compiled C code in timing critical applications.

Remember that a carefully coded and cycle counted assembly language sequence may not function correctly in the presence of interrupts. A common example is a serial port or SPI port implemented by "bit-banging" the serial data. An interrupt occurring while the 10 or 11 bits are being sent or received will most likely corrupt the byte being processed.

10.8 Library Support

The traditional standard C libraries are included in the WinIDE. This includes the familiar string and memory functions, arithmetic and trigonometric functions, formatted stream I/O, time and date and other functions.

Equally important to the standard library functions is support for the processor architecture and peripherals. There are also macros for inline generation of code for changing or restoring the interrupt priority (IP).

Peripheral support begins as simple and efficient inline macros for configuring or resetting the watchdog timer. There is support for creating the interrupt vector table and setting interrupt vectors. The library takes care of ensuring that the debugger isn't adversely effected by creating or changing interrupt vectors.

SCRabbit library support extends to the periodic timer, where it can be used to call a user specified function a set number of times per second. The realtime clock can be read and set.

Parallel port I/O is accessed through Rabbit I/O fast inline instructions.

The four (Rabbit 2000) or six (Rabbit 3000) serial ports are also supported. These are interrupt driven for transmit and receive using built-in efficient queues. The bulk of the library's serial code is shared by all four or six serial ports. Even the interrupt service routine is shared. Seven bit and eight bit parity are supported, as are Xon/Xoff and hardware handshaking.

Other Rabbit 3000 peripherals are supported by a port of a library included in Dynamic C.

10.9 WinIDE's SLINK Linker

The WinIDE's linker SLINK creates the final program which is going to be executed in RAM or programmed into flash. A linker groups the separately assembled and compiled files into one program image. It assigns the logical and physical addresses of all segments in the program, and then all of the underlying references within a module and between modules. It writes the output files and gathers and groups the debugging information.

A linker allows the use of libraries. For example, the programmer can write an LCD graphics library and use it in multiple projects in binary format (with or without retaining source debugging information). If the library is updated, all projects can be updated by relinking the new library.

SLINK takes all of the OBJ files and groups all of the segments. The order of segments is defined in the WinIDE "Locate modules and segments" dialog. The order of code or data in a segment is taken from the order of modules. Rarely does (or should) the order of modules matter, except for a case of the program's startup module (in a WinIDE program it is cstart.asm) which must be first.

After SLINK has processed all of the OBJ files and libraries, it places the segments at their final addresses based on the order of the segments and any address information specified for the program. SLINK writes the program and generates a MAP file.

The MAP file contains the information provided to the linker to assign addresses to segments and modules. A common question is "How large is my program?" or "How much code or constants are used?" The MAP file always contains these answers.

10.9.1 Locating Modules and Segments

Locating modules and segments is done in a dialog box. This might not be the most intuitive dialog in an otherwise logical development tool.

Figure 10.2 shows the dialog for the TCP/IP SSI demo program that comes with the Softools suite.

Figure 10.2: Linker locate dialog for the TCP/IP SSI demo.

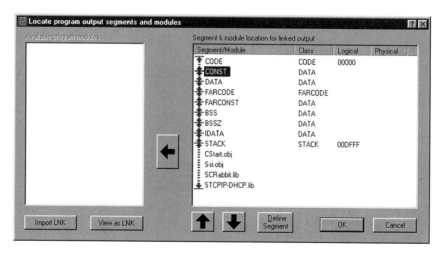

Softools—The Third Party Tool

The segments in the program are listed first. Their order is dependent on each of the segment's settings. A segment can be given a load address, as CODE has been with an address of zero.

Segments CONST and DATA are set to follow the previous segment (in this configuration segments are concatenated). FARCODE and FARCONST follow DATA, but being far segments, they are not using logical space even though they may appear to be in logical space.

The reason for this is as CODE and DATA grow, FARCODE and FARCONST can be pushed right up out of logical space without any ill effect. This is because far segments do not use or even have logical addresses.

Next are BSS, BSSZ and IDATA but we'll skip these and move to the last segment, STACK. The reason for this is that segment STACK is given an end address. This unique feature allows specifying the last address of a segment and the segment will start such that it ends at this address.[5] This is done to allow memory segments to grow downward in the address space.

Segments IDATA, BSSZ and BSS are set to precede the following segment (ultimately preceding STACK). When BSS has grown down to reach the end of segment CONST, the 56k of logical (near or root) space has been used.

Segment STACK's ending address is the key to linking a program. The WinIDE is designed to debug programs in RAM. Reasons for this decision are covered later in this chapter. To accommodate the least common denominator, the "New Project Wizard" and all WinIDE examples set the STACK ending address at 1FFFFh (the end of 128k). This allows all demos and projects initially created to run without problems, albeit not necessarily in the optimal configuration for the system.

Extended memory is taken from the end of STACK to the end of physical memory.[6] TCP/IP programs typically require 8 to 12k of extended memory. Therefore this address would be reduced by 2000h or 3000h. If the target system has 256k of RAM, the end address of STACK would be set to 3FFFFh minus the amount of extended memory. For a 512k RAM board, the end jumps to 7FFFFh (again, minus extended memory). The end address of STACK was covered previously in the "Project file" section.

Softools recommends developing on a 512k SRAM based Rabbit board as this maximizes the size of the program that can be debugged. When running in flash, the WinIDE supports the whole one megabyte address space.

10.9.2 Linktime Program CRC

Softools has a nice solution for ensuring the firmware in a product is the same as when the product was built. SLINK will generate a CRC at link time and write it into the output files during the output stage. This is automatic (i.e., requires no user intervention) and also

---

[5] In reality, it could start at a lower address than the expected address because of alignment. Alignment is always to a lower address when a segment's end address is specified.

[6] This is not done at link time but is done at runtime. The end of stack to the runtime determined end of memory is the size of the extended memory pool. By doing this at runtime, the program can have substantially more extended memory when deployed in flash than when it is being debugged in RAM.

Chapter 10

prevents tampering and therefore is well suited for industries that require this feature such as the FDA and gaming industries.

The SLINK 16-bit CRC is significantly more robust than a simple checksum. The programmer has full control over what is included in the CRC. The area to be CRCed need not be contiguous and may cover multiple segments. If the firmware is modified post-link with a serial number or MAC address, this can be contained in a segment which is not going to be CRCed. This allows the application to still be CRCed while retaining the ability to modify a small part of the application for specific purposes.

10.10 Debugging in the WinIDE

The WinIDE supports many debugging features to allow the programmer to find and fix bugs. Some estimates are that half the time spent working on an application is spent debugging.

This was one reason for Softools' decision to support debugging in RAM. It allows faster compile-load-debug cycles and allows much more flexibility in debugging—even patching an instruction to quickly test a bug fix, or to force a condition in the program that is otherwise difficult to test.

Some debugging features are available before a program is even loaded. The Rabbit's memory can be inspected or changed, and some simple tests can be performed by entering mnemonics right into the Assembly window.

The majority of debugging in the WinIDE is done after a program is loaded into memory. This enables the programmer to access symbol names and to be able to debug in source files that are part of the program.

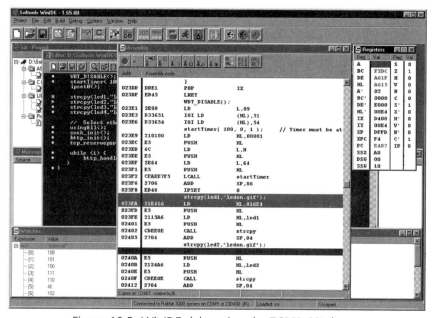

Figure 10.3: WinIDE debugging the TCP/IP SSI demo.

10.10.1 The Debugger Assembly Window

The Debugger Assembly Window shows a disassembly of memory with opcode bytes and also Rabbit mnemonics with symbol substitution. Source code lines are included if a source line matches a displayed disassembly address.

There are a number of debugging features available in this window using function keys or the context menu. Most common is single stepping the code, which allows stepping by disassembled line (not following calls, jumps and returns) or by instruction (following calls, jumps and returns).

If the cursor is positioned on an instruction, a breakpoint can be set on that instruction.

There is also a command to run to a displayed line (a.k.a. "Run to Cursor"). The cursor position can be changed by paging up and down or by using the `Goto` command to jump to any address which may be comprised of a simple constant or a full expression including symbols and C operators. A disassembly of a range of memory can be written to a text file.

10.10.2 The Debugger Registers Window

The Debugger Register Window shows the Rabbit register set (primary and alternate) and processor flags (also primary and alternate). The register pane also shows the **XPC** register and the three **MMU** registers as these work directly with logical addressing in the Rabbit and are frequently referenced. The flag pane also shows the value of **IP**.

Functions available in this window include changing registers as well as setting and clearing flags.

10.10.3 The Debugger Data Window

The Debugger Data Window shows a memory region in a default format of bytes. The format can be changed to words, double words (longs) and floats. Memory can be changed.

This window also supports a few other functions including filling and jumping to an address. Two powerful functions include loading HEX and BIN files to an address and saving a range of memory to a HEX or a BIN file.

10.10.4 The Debugger Stack Window

The Debugger Stack Window shows a memory region in word format at the address of SP. The Rabbit can push words and also 3-byte quantities (for long calls) onto the stack. Viewing the processor stack in word format won't be nearly as helpful in cases where long calls are being made to a function.

10.10.5 The Debugger Watch Window

The Debugger Watch Window is probably the most useful window when debugging C code. This window shows global, static and local variables in up to three formats—decimal, hexadecimal, and character constant.

The best feature is that structs and arrays can be watched, and can be expanded into their constituent elements. Any base type being watched can also be changed (e.g., a member of an array can be changed but not an entire array).

In addition to watches being displayed in the Watch window, tooltip watches are supported by holding the mouse over any variable or highlighted expression. This is quite a time saver when quickly checking a variable without leaving the source code window.

10.10.6 The Debugger Printf Window

The Debugger Printf Window is an output window that can be used at runtime by a program to log or display messages. The programmer can simply add printfs to his code to display variables or messages and they will appear in the Printf window. There is no special debugger code to handle this—the WinIDE can be closed and the output can still be displayed in HyperTerminal or another serial communications program.

10.10.7 The Debugger I/O Window

The Debugger I/O Window can be used to watch the values of I/O ports, or to write a value to a port. There is an option to repeat the I/O write more than once.

The I/O window understands that Rabbit has many shadowed I/O registers[7]. When an I/O port is write-only and the debugger finds a shadow register definition for that address, it will show the shadow register value instead of reading the actual I/O port. If the value of the port is written, the debugger will update the shadow as well as writing to the I/O port.

10.10.8 Debugging Philosophies

In a development environment for a processor such as the Rabbit, there are two choices for debugging a user's application. Each has its advantages and disadvantages. One choice is to load and debug the user's program completely in RAM. The flash memory is not used when debugging (but can be used by the program itself if it requires writing data to flash). The other choice is to load and debug the user's program in flash memory. Tables 10.2A and 10.2B show the major advantages and disadvantages of both debugging approaches:

Table 10.2A: Debugging in flash memory.

Debugging in Flash Memory	
Advantages	**Disadvantages**
<ul><li>Program memory mapping and allocation is the same as the final deployed program.</li><li>All of the controller's RAM is available when debugging.</li><li>Errant pointers won't corrupt the program.</li></ul>	<ul><li>Program takes longer to load since flash must be erased and programmed on every program load.[8]</li><li>Arbitrary breakpoints cannot be easily supported. If supported, they cannot be quickly set and cleared.</li><li>To debug the program, additional code needs to be compiled into the program adding to its size and significantly slowing it down.</li></ul>

---

[7] Shadowed because they are write-only registers. In order to read them, the value written must be saved in memory (a shadow location) and used for the read operation. We have worked a lot with shadow registers in Chapter 6 and Chapter 7.

[8] We did not mention that flash has a finite number of erase and write cycles. For development purposes, this really is not a disadvantage and should not be a cause for concern.

Softools—The Third Party Tool

Table 10.2B: Debugging in RAM.

Debugging in RAM	
Advantages	**Disadvantages**
■ Program loading is very quick. ■ Arbitrary breakpoints can be set, including at assembly instructions. ■ The program being debugged is the same code that will be deployed. No code changes or recompiling needs to be done. ■ The program can be patched on the fly for testing code paths or testing simple bug fixes.	■ RAM must be used for the program code leaving less for the program's use. ■ Memory map is not the same as it is for the deployed program. ■ The development controller should have as much RAM as possible, making it different from the deployed controller which can use less RAM. ■ Less usable RAM is available to the program when debugging than is available to a deployed program. ■ Errant pointers can overwrite the code space and crash the program.[9]

Softools' WinIDE debugger was implemented to debug in RAM. The primary reason is the benefit of the quick edit, compile, load, and debug cycles. Softools feels this saves the developer time through the life of the development cycle, which can be quite substantial for large programs. One benefit is this will save money, which can be substantial as the program development and maintenance time grows from weeks to months to even years.

The ability to set arbitrary breakpoints makes for much more flexible debugging, both in C source and in assembly code.

Many times while debugging, the programmer will get to a point in the C code, or even assembly code, and want to stop at a specific instruction. RAM debugging allows this, while with flash debugging this is impossible to do.

The other reason debugging in RAM was chosen is that Softools feels *all* of the disadvantages listed above for debugging in RAM can be eliminated by the dedicated and committed developer. The program's flash memory on the development controller can be replaced with an SRAM of equal size. Sadly, they are *not* pin compatible, which makes this a tedious and non-trivial modification. Nonetheless, it is possible, at the cost of some time and money.

512k is the maximum RAM provided on most controllers, so debugging in RAM does put a limit on code and data at 512k total. This could become a significant problem as a program grows. The "all RAM" option would solve the problem when a program exceeds total RAM.[10]

---

[9] The Rabbit 3000A, covered in Appendix A, supports write protecting memory regions which would eliminate this disadvantage if implemented.

[10] At the time of this writing, Softools is working on debugging in flash for the cases where applications do exceed 512k or the user must debug with the full controller memory available to the program being debugged.

Chapter 10

Often, debugging in RAM requires the user to link the program differently to move the RAM used by the program from its debug location to its flash runtime location. The WinIDE creates programs that debug in RAM and run in flash without this limitation, that is, unless the program uses far memory (segment FARBSS). Because far memory has different physical addresses between RAM debugging and flash execution, a change to the physical address of segment FARBSS must be made. Softools has experimented successfully with a memory layout that does *not* require a program with FARBSS to be linked differently in RAM and flash by setting the Memory Bank Control Registers specially at runtime.

10.11 Memory Layout

The Rabbit's MMU is very flexible and the added Memory Bank Control Registers (MBCRs) offer yet another layer of versatility. The MBCRs control the chip select of each 256k region of the Rabbit 1 MB address space. A Rabbit-based controller doesn't really have a fixed memory layout, and software can map external memory devices (in 256 KB blocks) into any physical address range desired. With other processors, unless external mapping hardware is used, flash and RAM are at fixed addresses.

A Rabbit system may use the MBCRs to page memory in and out of the address space, without resorting to the MMU segment mapping. Of course the MMU can be used to manage the memory in the pages.

An external four-bit latch (hint: four bits of Rabbit PIO) can be used to map 4MB of page-swapped memory into one of the four 256k quadrants. Extensive data logging is possible with the sixteen 256k pages thus provided. Except for page swapping at 256k boundaries, this memory is fully accessible in C using far pointers. If the last (fourth) quadrant is used and the objects are stored that are a size that evenly divides into 256k, the pointer wrap becomes a trivial and efficient test in C for bit 20 becoming set. When a wrap occurs, the upper bits of the pointer can be used directly to update the 256k page select hardware or PIO bits.

The MBCRs and MMU allow the WinIDE to create programs that run in RAM and flash, sometimes without linking differently for each mode. The WinIDE memory maps shown in Figure 10.4 illustrate the differences for a typical program when debugging in RAM and when running in flash.

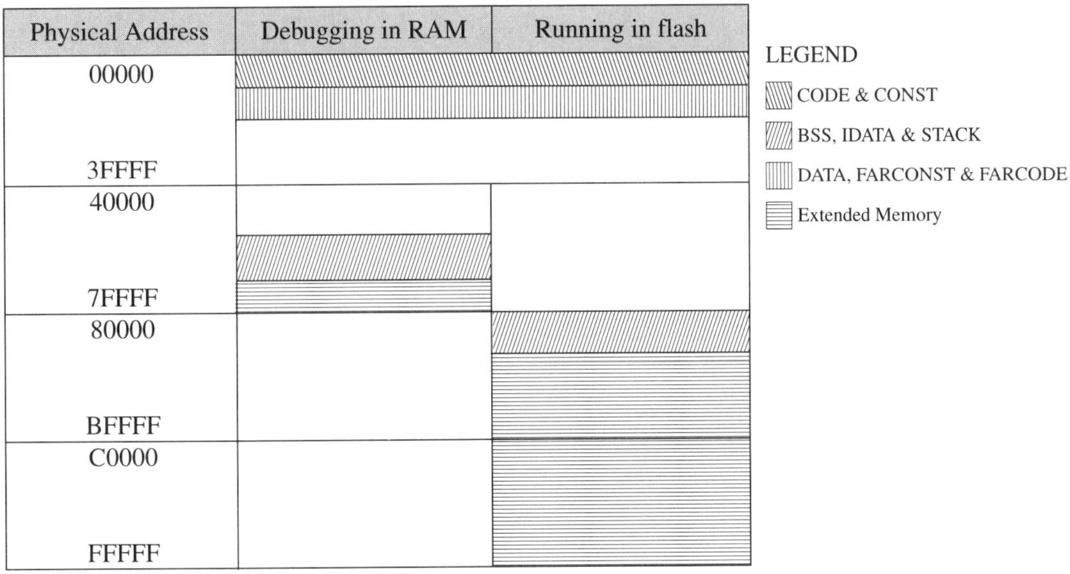

Softools—The Third Party Tool

Physical Address	Debugging in RAM	Running in flash
00000		
3FFFF		
40000		
7FFFF		
80000		
BFFFF		
C0000		
FFFFF		

LEGEND
- CODE & CONST
- BSS, IDATA & STACK
- DATA, FARCONST & FARCODE
- Extended Memory

Figure 10.4: Differences between debugging in RAM and running in flash.

10.11.1 Near and Far, a.k.a. Logical and Physical

When working in the WinIDE, the programmer has to be aware of how Rabbit memory is addressed and how the MMU operates. This is because the processor's native addresses are logical while the linker has to locate segments physically[11]. The majority of Rabbit instructions use logical addresses and the MMU translates them to physical addresses.

The linker assists cstart in configuring the MMU and MBCRs so that the program's logical addresses access the correct physical addresses. A logical address is a regular pointer in C, which is a near pointer—one that fits in a single Rabbit register. A far pointer is a physical address—one that doesn't fit in a single Rabbit register. With one exception, a far pointer can't be used to simply access bytes and words of memory as a near pointer can. For simpler debugging and easier memory addressing, far pointers are stored as actual physical addresses.

Logical addresses are 16 bits in size. If a program used only 16-bit pointers and addresses, it would have a maximum address space of 64k (e.g., the original Z80, although many paging solutions were devised to allow access to larger code and data with the Z80). The Rabbit MMU saves the hardware designer numerous external parts that would be required to implement paging.

The MMU also saves the timing problems that would pay unannounced visits to these designs utilizing faster clocks.

The MMU divides the 64k space into 4 regions to map logical addresses to physical addresses. The 8k XPC region is reserved for far code in Rabbit systems. Although XPC can be "borrowed" by code to access data, the code that does this must be near code.

---

[11] It might help this discussion and what follows to remember that logical addresses are near addresses and physical addresses are far addresses.

Chapter 10

In the WinIDE, the remaining 56k of logical space is used for segments CODE, CONST, BSS, IDATA, and STACK shown in Figure 10.4. If this 56k is exceeded for any reason (e.g., too many strings, too much near code or too much stack space), the program cannot be linked and will not run.

Throughout a program's design and coding, the developer should keep a lookout on this near space usage. One should take the extra time and effort from the beginning to minimize the use of logical space. If not, having to reduce logical space later is bound to come at a most inopportune time (e.g., close to the end of the project where deadlines are closing in fast). Do not put off logical memory savings today with the expectation of doing it later.

Near and far pointers can be mixed, as long as the result can be used in a far context. If an expression or variable is expecting a far pointer but receives a near pointer, a conversion is done at runtime from near pointer to far pointer. An analogy that concisely describes this is when float and int expressions are mixed. The ints are promoted to floats and the computation is done in floating point.

When near and far pointers are mixed, near pointers are promoted to far pointers such that they both point to the same physical memory address. Then the far pointer is used in the expression and the memory address is accessed using the far pointer. Once a pointer has been converted to far (or if it's far already), it cannot be converted back to a near pointer. Doing so will generate an error.

10.11.2 Using and Storing Strings

Normally, near code isn't the problem when running out of logical space because the SCRabbit compiler defaults to creating far code. What more often sneaks up on the developer is the use of strings which take up logical (near) space. The C standard has no provision to tell the compiler how to store a string. By definition, a string is *a pointer to array of char*. The key word here is *pointer*. A pointer is near; therefore the array is near and the string takes up near space. However, there are still alternatives that can be explored.

Strings have a nice feature the programmer may not realize at first—they are unnamed arrays. They can simply be used as an array of char. Some readers may think strings are ASCII only. In fact, initialization for a device, which is binary, can still be used in the form of a string. There are no limitations on what a string contains. With the fact presented that strings are just arrays, consider the following translation:

```
printf( "Hello world" );

const char helloWorld[] = "Hello world";
printf( helloWorld );
```

The two printfs are equivalent. Each one stores a string in near space (in fact, in the same segment in this case) and the pointer to the array is passed to printf. So far, this does not appear to accomplish too much, but the conversion to an array is the first step to illustrate our point—making strings far.

Softools—The Third Party Tool

The next part is easy:

```
const char far helloWorld[] = "Hello world";
printf( helloWorld );
```

This fragment stores `helloWorld` in far space (in segment FARCONST because far and const are both used[12]). This moves strings to far space lowering the use of near space. Menus, arrays of strings and other tables should be moved to far space as shown here.

This discussion would not be complete if we ended now. Because the strings have been made far doesn't mean our change is finished. The `printf` above requires a near pointer—as provided above it's a far array which is also a far poiner and it would produce a *Pointer conversion error* from the compiler. The compiler cannot convert a far pointer to a near pointer. This means the code that is converted as above needs a slight change to support far pointers. The correct example is as follows:

```
const char far helloWorld[] = "Hello world";
farprintf( helloWorld );
```

The WinIDE supports `farprintf` (and others flavors of `printf`) to allow this to work with a `printf`-compatible function. Of course, it's also very common for `printf` to take an argument other than the format string (the first argument) which would need to be a far pointer. Softools has solved this with a simple `printf` extension which can be seen in the next short example:

```
const char far helloWorld[] = "Hello world";
const char far format[] = "String: %ls; Number: %d";
farprintf( format, helloWorld, 4 );
```

Now, the format string is a far string. The `%ls` is a Softools extension like the standard `%s` format which specifies a far pointer as the `printf` argument. This example can be applied to uses other than `farprintf`. Moreover, `farstrcpy` and several other far pointer functions are available to complement the standard near pointer C library functions.

10.11.3 Dynamic Memory

Following the discussion on near and far pointers and memory mapping, we will briefly visit dynamic memory on the Rabbit—it is important to the program's use of physical and logical address spaces on Rabbit systems. Some developers think that dynamic memory and small embedded systems mix as well as oil and water. Indeed, 16 and 32-bit systems with hundreds of kilobytes of memory may handle dynamic memory perfectly well for extended (i.e., 24/7) execution. 8-bit systems are more constrained in both memory space and processing power.

---

[12] If const is not specified, the string is stored in segment FARDATA. There is no default support in the WinIDE for FARDATA. Initialized far data could be implemented in the same manner as initialized near data but it is not common to require this feature.

Chapter 10

This limited memory brings with it fragmentation problems in typical dynamic memory implementations. However, users who want to use dynamic memory can do so in the WinIDE environment. There are conditions where fragmentation is not an issue. One is when all of the allocations are the same sized blocks. However, the program must be designed to handle the condition of running out of memory.

There are three dynamic memory allocation methods, two similar and one which is compatible with Dynamic C's xmem.

- **malloc and free**: The WinIDE supports the standard malloc and free. malloc and free use near pointers for allocating and freeing dynamic memory. Traditionally, these pointers can be used with all of the standard C library functions that accept pointers—these are near pointers.

  malloc isn't ideal on the Rabbit because its memory pool comes out of the 56k of usable near space. Realistically, after the stack is allocated and other near space requirements are satisfied, having more than 40-42k of dynamic memory is unlikely.

  Some systems will find 30 to 36k to be more in the ball park. This may or may not be sufficient. If the allocations are small, this could well work out to be fine, and allocations small in size will reduce (but may not prevent) fragmentation. The heap is stored in segment BSS and is defined in cstart with the size determined by the assemble-time label HEAP_SIZE. In most cases, one would adjust HEAP_SIZE to use up the 56k of root space to maximize the heap's size.

- **farmalloc and farfree**: The WinIDE supports far versions of malloc and free. These allow for a much larger heap, but doing so requires using far pointers. Extended memory is used for the heap, which is good because its size is determined at runtime. This does mean that debugging in RAM will have a smaller far heap than when running in flash. In flash, the heap can jump up in size by the hundreds of kilobytes.

  The extended memory allocation, mentioned below, can coexist with a far heap. The far heap can either be a fixed size block of far memory allocated in C, or one allocated from extended memory.

  As with the far string pointer conversions mentioned above, the far pointers are not as compatible with most of the C runtime libraries. As discussed, far pointers are not a native type to the Rabbit processor; they take more code and processing time to load, to store, to read from and to write to than near pointers do. But they can be used to access much more memory than near pointers can. It's good to point out that one should avoid the temptation to use far data and far pointers indiscriminately—don't use them if near pointers can be used.

- **Extended memory**: The WinIDE supports extended memory. Its functions are compatible with Dynamic C's xmem functionality. Unlike the support provided by the two previously mentioned methods, extended memory does not support the unrestricted freeing of allocated memory. This means extended memory is suited only for one-time allocations that can not and will not be freed.

10.12 Real Time Operating Systems

Embedded systems development using the WinIDE for the Rabbit wouldn't be complete without a discussion of real time operating system (RTOS) support. An RTOS allows a program to multitask, running two or more threads[13] with the appearance that all threads are running concurrently. This often can significantly improve the management of processor utilization by allowing tasks waiting on resources or waiting on peripherals to consume a small fraction of processing time that might otherwise be used in polling the device or for waiting on an event.

As covered earlier in Chapter 8, multitasking comes in two modes: Preemptive and Cooperative. The former takes care of task switching automatically without a scheduler to relinquish the processor. Preemptive multitasking with priority-based scheduling yields quicker response times to events, shifting the processor to higher priority tasks allowing the critical functions to run as soon as possible. Priority based scheduling can, however, cause problems if a high priority task gets control of the processor and doesn't allow any other tasks to run. If not used carefully, priority based scheduling can allow priority inversion to occur. Basically, priority inversion occurs when a high priority task is blocked waiting on a resource that is locked by a low priority task, which in turn is preempted by a task with higher priority. The lower priority task is prevented from completing (for some period of time) and this prevents the first high priority task from running. This non-deterministic delay of the high priority task is not expected and can lead to timing or other errors in the program[14].

For example, we assume task 1, 2 and 3 have high, medium, and low priorities. Assume task 3, a low priority display task, has locked access to and is updating an LCD. The lock would be as short as possible—only long enough to write to the display—possibly in multiple accesses to reduce the maximum time the display is locked since high priority tasks may need to change the display. Task 1 needs to display an alert (in this example we hope it has already sounded the alarm). It waits what normally should be only very briefly for task 3 to unlock the display. But just as task 3 has locked the display, task 2 gets control because it needs to change something in the system. Task 3 is now taking a much longer time than usual. Task 1 is now waiting on task 2 (and 3). Task 1 essentially has a lower priority—that is like task 3, because it can't run until task 2 completes. Although the added delay could be small, it can detrimentally effect the timing of the system.

Cooperative multitasking puts the burden of task switching on the programmer and, at the same time, adding to the program's size. It does reduce the processor overhead to a slight degree.

Cooperative multitasking cannot support higher priority tasks for time critical sections of code. But it also prevents a task from being preempted at any possible point in the task (at the Rabbit instruction level). This can greatly simplify a program's use of resources and accesses to global memory objects.

---

[13] We refer to *threads* as *tasks* in Chapter 8.
[14] A high-profile example is the Mars Pathfinder. It had a problem caused by priority inversion, but fortunately it wasn't fatal to the mission. The bug was fixed on Earth and the system updated to prevent further priority inversion problems.

The WinIDE supports any RTOS written in standard C. This opens the possibilities up to several commercial, public domain, and open source RTOSes. Although the C part of the RTOS is portable, there are always a few functions that can be done only in Rabbit assembly code. This moves a complex task to the developer, who probably wishes to simply use the RTOS in the application and not work on the RTOS as part of the workload. Therefore, an off-the-shelf solution is usually a better choice from the development standpoint.

The WinIDE supports the popular Micro C OS II (µC/OS-II) developed by Micrium, Inc. This RTOS offers a lot of features in a portable OS that also has a couple of well-written comprehensive books by the author. The WinIDE also supports TurboTask, developed by Softools, for the Rabbit processor. Both support preemptive priority-based multitasking. TurboTask also supports round-robin multitasking in either priority based mode (tasks of the same priority are multitasked at a user-specified time interval) or in round-robin mode where all tasks get a processor time slice in which to run.

An additional feature to improve response time in round robin mode allows a task, which has been unsuspended by any TurboTask function called from an ISR, to run immediately upon exit of the ISR.

10.13 Ethernet and TCP/IP

Ethernet connectivity and TCP/IP are supported in the WinIDE development environment. Much of the functionality offered by Dynamic C for Ethernet and TCP/IP is supported in the Softools WinIDE environment with as much compatibility as possible. This is because most of the outstanding Dynamic C TCP/IP stack and drivers have been ported to WinIDE and are available in both binary library form and source code form.

The TCP/IP support includes significant functionality well above the functionality provided by just the UDP and TCP protocols alone. The CGI and SSI functionality provided by the TCP/IP library is comprehensive and powerful.

Although any ANSI C TCP/IP stack could theoretically be ported to the WinIDE, or one could be written from scratch if such an undertaking were warranted, it would be difficult to justify not using one that is already available. Especially when it is free of cost as well as royalty-free. An early release of the open source lwIP[15] TCP/IP stack was ported and is running under the Rabbit WinIDE environment, but lwIP even today isn't nearly as feature-rich as the Dynamic C stack.

10.14 WinIDE and the Book's Example Programs

Most of the examples that accompany this book show generally how to accomplish a task or how to use a feature of the Rabbit. The advice, approach and methods provided are fully applicable to the WinIDE environment. For example, the initialization shown in the LED examples apply to any development tool and language. In all cases, the C code provided on the CDROM will require changes to be made before it will run in the WinIDE. The installation of

---

[15] lwIP stands for "Lightweight TCP/IP Stack."

the WinIDE includes an LED example program much like the one shown in this book. It has the advantage of running on all Z-World development kit platforms, allowing the programmer to experiment with whatever hardware is at hand.

Should additional examples, errata, or information for the WinIDE which pertain to this book become available, they will be available on the WEB at http://rabbitbook.softools.com.

10.15 Conclusion

Tool choices are important for a developer to have. The creative gift that programmers have allows for numerous solutions for any given embedded project. Knowledge about the tool chains that are available for the work being performed influences many aspects of the development process.

Sometimes the tool choices create new possibilities or contribute to design ideas during project conception. Toolchains are different. Each one has strengths that may make some projects easier or faster to develop than other projects. It's not uncommon for a developer to use more than one toolchain for the project, even for the same processor, especially if each tool lends itself better to the task for which it was chosen. A book devoted to the Rabbit microprocessor would not be complete without a discussion of third party tool options.

APPENDIX A

Rabbit 3000A—Extending the Rabbit 3000's Architecture

In 2003, Rabbit introduced the Rabbit 3000A microprocessor. The Rabbit 3000A is similar in many respects to the Rabbit 3000 but there are notable improvements. An exhaustive enumeration of register details and timing examples can be found in the appendices B through D of the Rabbit 3000® Microprocessor User's Manual found on the accompanying CD. This appendix is intended as an introduction to the exciting new Rabbit 3000A features.

New Feature Overview

The following list gives an overview of the new Rabbit 3000A features.

- 15 new instructions
- 16-bit internal I/O address space
- Separate User and System modes
- Memory protection mechanisms
  - Selective write inhibits by address range
  - Stack protection
  - RAM segment relocation
- Secondary watchdog timer
- PWM improvements
- Quadrature decoder improvements
- Expanded low-power capability
- External I/O interface enhancements
- Integrated Schmitt trigger for 32 kHz oscillator input
- Alternate output port connections for many peripherals

The introduction of distinct User and System modes paved the way for new robust operating systems. System mode provides unbridled access to all of the system resources. Tasks running in System mode can explicitly grant privileges to User mode tasks. This can be used in a couple of interesting ways. More discussion will be found below in the section on User and System mode.

Several of the features on the above list relate directly to supporting the distinction between System and User mode. Others simply improve the functionality of existing peripherals. Some enhancements cover both areas.

Rabbit 3000A—Extending the Rabbit 3000's Architecture

For example, the Rabbit 3000A has a 16-bit internal I/O address space. The Rabbit 3000 only had an 8-bit internal address space. The Rabbit 3000A's enhanced I/O space allows new registers that both support separate User and System modes. Moreover, the enhanced I/O space has improved the peripheral interface.

The Rabbit 3000A is 100% backward compatible with the Rabbit 3000. Any system using a Rabbit 3000 can use a Rabbit 3000A without changing any code.

The Rabbit 3000A is designed to boot into System mode, thus giving the application complete access to the system's hardware resources. A Rabbit 3000 program would not know whether it is running on a Rabbit 3000 or a Rabbit 3000A in System mode.

The changes to the Rabbit 3000A's I/O peripherals, such as the extension of the Quadrature Decoder from 8-bits to 10-bits, will not affect a Rabbit 3000 program. The program must explicitly enable the new 10-bit extension by accessing a previously reserved bit in the Quadrature Decoder Control Register (QDCR: 0x0091).

In the following sections we will explore the exciting extensions that the Rabbit 3000A brings to the Rabbit 3000's architecture.

Additional Instructions

A whopping fifteen instructions have been added with the advent of the Rabbit 3000A. Fifteen may not seem like a big number, but in the world of 8-bit processors, fifteen is a lot of new instructions.

Two instructions have been added to help speed up the cryptographic math used in public-key calculations. Six block copy instructions have been added to make moving data to and from I/O address more efficient. Seven instructions have been added to support the new System/User modes.

The tables presented in this text use a format and nomenclature consistent with the Rabbit 2000 /3000 Microprocessor Instruction Reference Manual. Each instruction has an entry a table with the following headings.

Instruction	Bytes	Clk	A	I	S	Z	V	C	P	Operation

The **Instruction** column contains the instruction mnemonic and opcode format.

The **Bytes** column contains the number of bytes required to store the instruction in memory.

The **Clk** column indicates the number of machine cycles required for the instruction to execute.

The **A** column indicates what effect the ALTD prefix instruction has on the instruction. The following table shows the key for the "A" column.

Symbol	Description
F	ALTD selects alternate flags
R	ALTD selects the alternate destination register
SP	ALTD operation is a special case

The **I** column indicates what effect the IOI and IOE prefix instructions have on the instruction. The following table shows the key for the "I" column.

Appendix A

Symbol	Description
s	IOI and IOE affect source
d	IOI and IOE affect destination

The **S**, **Z**, **V**, and **C** columns correspond to the Sign, Zero, Overflow and Carry flags. These are found in the Rabbit's "Flags" register (sometimes called the Status or Status Flags register). The Overflow flag is sometimes referred to as the Logical/Overflow or **LV** flag in the Rabbit processor documentation. The following table shows the key for the symbols used in the flags column.

Symbol	Description
*	Flag affected
-	Flag unaffected
0	Flag is cleared
1	Flag is set
V	Arithmetic Overflow is stored
L	Logical Result is stored

The **P** column indicates whether the instruction is privileged or not. Privileged instructions do not allow an interrupt between it and the following instruction.

The following table summarizes the new instructions. The *Rabbit 2000/3000(A)® Microprocessor Instruction Reference Manual* has complete descriptions for all of the Rabbit instructions.

Instruction	Bytes	Clks	A	I	S	Z	V	C	P	Operation
UMA	2	8+8i	-	-	-	-	-	*	N	{ CY:DE':(HL) } = (IX) + [(IY) * DE + DE' + CY]; BC = BC −1; IX = IX + 1; IY = IY + 1; HL = HL +1; Repeat while BC != 0
UMS	2	8+8i	-	-	-	-	-	*	N	{ CY:DE':(HL) } = (IX) - [(IY) * DE + DE' + CY]; BC = BC −1; IX = IX + 1; IY = IY + 1; HL = HL +1; Repeat while BC != 0
LDDSR	2	6+7i	d	-	-	*	-	-	N	(DE) = (HL); BC = BC − 1; HL = HL − 1; Repeat while BC != 0
LDISR	2	6+7i	d	-	-	*	-	-	N	(DE) = (HL); BC = BC − 1; HL − HL + 1; Repeat while BC != 0
LSDR	2	6+7i	s	-	-	*	-	-	N	(DE) = (HL); BC = BC − 1; DE = DE − 1; HL = HL − 1; Repeat while BC != 0

Rabbit 3000A—Extending the Rabbit 3000's Architecture

Instruction	Bytes	Clks	A	I	S	Z	V	C	P	Operation
LSIR	2	6+7i			s	-	-	*	-	N	(DE) = (HL); BC = BC − 1; DE = DE + 1; HL = HL + 1; Repeat while BC != 0
LSDDR	2	6+7i			s	-	-	*	-	N	(DE) = (HL); BC = BC − 1; DE = DE − 1; Repeat while BC != 0
LSIDR	2	6+7i			s	-	-	*	-	N	(DE) = (HL); BC = BC − 1; DE = DE + 1; Repeat while BC != 0
SETUSR	2	4		-	-	-	-	-	Y	SU = { SU[5:0] , 0x01 }
PUSH SU	2	9		-	-	-	-	-	Y	(SP − 1) = SU; SP = SP − 1
POP SU	2	7		-	-	-	-	-	Y	SU = (SP); SP = SP + 1
SURES	2	4		-	-	-	-	-	Y	SU = {SU[1:0], SU[7:2]}
IDET	1	2	SP	-	-	-	-	-	N	Performs LD E, E but if (EDMF && SU[0]) then the System Violation interrupt flag is set; if ALTD appears before IDET always performs LD E', E
RDMODE	2	4		-	-	-	-	-	Y	CF = SU[0]
SYSCALL	2	10		-	-	-	-	-	N	SP=SP-2; PC={R,v} Where v = SYSCALL offset of 0x60

The UMA and UMS generate 24 bit results and are designed to optimize cryptographic math.

The new block move instructions (LDDSR, LDISR, LSDR, LSIR, LSDDR, LSIDR) fill a void in the Rabbit 3000 instruction set. Previously, the block move instructions could only write to an incrementing/decrementing destination address in I/O space. The LSDR and LSIR instructions allow I/O access for the source address. The LSIDR and LSDDR instructions maintain the same source address throughout the entire move, while the LDISR and LDDSR instruction maintain the same destination address. These allow both block-fills (to and from memory) and the option to move serialized data (from a parallel I/O port) into a memory block, or vice-versa.

With the advent of separate User and System modes, a method was needed to place the system in User mode. The SETUSR instruction places the processor into User mode.

Part of the separate User and System mode infrastructure is the SU register. As with other CPU registers, a programmer will want to be able to push and pop the SU register on and off the system stack. PUSH SU and POP SU were introduced to accommodate this.

The SU register is used as a one-byte stack that maintains a history of the System/User mode state of the processor. The SU register fills the roll of a stack in the same way that the IP register acts as a stack for interrupt priority (more detail in Chapter 7—Interrupts). The state of the processor is pushed into SU when one of the following conditions occur:

Appendix A

- SETUSR is executed (entering User mode)
- SYSCALL is executed (entering System mode)
- RST is executed (entering System mode)
- An interrupt occurs (entering System mode)

The SURES instruction was added to the Rabbit 3000A to allow the previous mode status to be popped off of the SU stack.

When executed in User mode, IDET causes a new System Violation Priority 3 interrupt to occur. When executed in System mode, IDET behaves just like a NOP. This will allow protection of System mode-only code from inadvertent execution while in User mode.

The RDMODE instruction was introduced to allow the detection of the current User/System mode status. The current mode will be placed in the carry flag (0 for System mode, 1 for User mode).

The SYSCALL instruction is a software interrupt that causes the CPU to be placed in System mode and begin executing at a location determined by the SYSCALL's 0x60 offset into the internal interrupt vector table. It is functionally equivalent to the RST instructions. The address of the internal interrupt vector table is derived from the IIR register in the usual way (see Chapter 7—Interrupts).

User Mode and System Mode

The introduction of separate System and User modes opens the way for the development of advanced multitasking kernels. The System mode has complete control over all of the CPUs resources. The User mode only has privileges that are granted through the manipulation of configuration registers by a System mode task.

If a User mode task happens to violate its privileges, say, by attempting to access memory for which the task is not authorized, the User mode task will be halted and the CPU will be switched to System mode and the kernel can handle the exception.

The Rabbit 3000A was carefully crafted so that all of the new features (except for the new instructions) must be explicitly enabled through control registers. The Rabbit 3000A boots up in System mode so that the all resources are available to the application code.

Rabbit 3000 code that doesn't manipulate reserved bits in configuration registers will run transparently on the Rabbit 3000A.

The Rabbit 3000A will not enter User mode unless the separate User/System mode feature is enabled and the SETUSR instruction is subsequently executed. To enable separate User and System modes, a configuration bit in the new Enable Dual Mode Register (EDMR: 0x0420) is set.

To further ensure that that a Rabbit 3000 program doesn't accidentally write to the EDMR, the new configuration register is given a 16-bit internal I/O address (0x0420). The Rabbit 3000A boots with the internal address bus in 8-bit mode to maintain compatibility with the Rabbit 3000. Setting the formerly reserved bit 7 in the MMU Instruction / Data Register

Rabbit 3000A—Extending the Rabbit 3000's Architecture

(MMIDR: 0x0010) enables 16-bit address decoding for internal I/O registers. This is required to access the EDMR to enable separate User and System modes.

As of this writing, there is no commercially available kernel that takes full advantage of the User/System mode feature. However, there is no reason that an application programmer can't use this new feature. Access to this, and all of the new features, is available through configuration registers—just like all of the Rabbit 3000 peripherals. Either C or assembly can be used to access these registers.

Memory Protection

The Rabbit 3000A has expanded the MMU's functionality to include the ability to selectively inhibit writes to physical memory. Write protection can be applied to either User mode programs or to both User and System mode code. Setting or clearing bit 0 in the Write Protect Control Register (WPCR: 0x0440) determines the scope of write protection..

Once the scope is determined, the specific write protect privileges must be assigned.

For the purposes of write protection, the 20-bit external address space is broken down into sixteen 64 KB blocks. Each block either has writes enabled or disabled. Two Write Protect Registers (Low and High), WPLR and WPHR, specify the write protection for each 64 KB memory block.

WPLR - Write Protect Low Register - Address = 0x0460
WPHR - Write Protect High Register - Address = 0x0461
WPSAR - Write Protect Segment A Register - Address = 0x0480
WPSHR - Write Protect Segment A High Register - Address = 0x0482
WPSLR - Write Protect Segment A Low Register - Address = 0x0481

Figure A.1 illustrates how the registers affect the memory protection scheme.

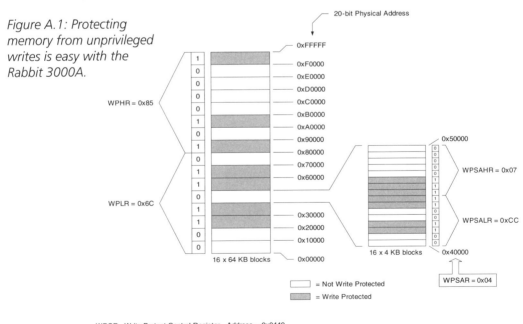

Figure A.1: Protecting memory from unprivileged writes is easy with the Rabbit 3000A.

443

Appendix A

In addition to the 64 KB granularity offered by the WPHR and WPLR registers, up to two of the 64 KB blocks can be selected for protection with 4 KB granularity. Two "segment" registers, Write Protect Segment <A,B> Register, WPSAR and WPSBR select the 64 KB block to be affected. In the example above, the WPSAR register has selected the 64 KB block starting at 0x40000.

In the example, the WPSALR and WPSAHR specify which 4 KB block is write protected. A 1 stored in these register's bits mean the corresponding 4 KB block is write protected.

The example does not show the "B" segment registers. WPSBR, WPSBLR and WPSBHR function identically to the "A" registers shown in Figure A.1.

If a program attempts an unauthorized write to a protected memory block, then a Priority 3 "Write-Protection" interrupt is generated.

Another very nice feature the Rabbit 3000A has added is the ability to detect stack overflows. If a violation occurs once this feature has placed constraints on the stack size, a Priority 3 "Stack-Violation" interrupt is generated.

Two registers are used to specify the stack upper and lower boundaries. The STKHLR and STKLLR are the Stack High Limit Register and Stack Low Limit Register. The boundaries are automatically given a 16-byte buffer that will allow a stack access to complete as the stack approaches a boundary without violating the boundary and corrupting adjacent memory.

Any access within the 16-byte buffer or beyond the boundaries will handled as a violation.

The stack protection feature, like the write protection feature, may be configured to apply to either the User mode or both the User and System modes. The Stack Limit Control Register, STKCR, determines to which operating modes the stack protection is applied.

The last memory protection feature is actually more of a memory mapping feature. Separate I & D space mode (see Chapter 2) provides greater memory ranges for code and data, but makes it more difficult to access code in RAM, which is needed for run-time modifications such as the interrupt vector table. To make this easier, Rabbit Semiconductor added the ability to map a small segment (1 KB, 2 KB, or 4 KB) of RAM into the code space when separate I & D space is enabled

This mapping is accomplished through the RAM Segment Register (RAMSR: 0x0448).

Secondary Watchdog

The Rabbit 3000A has added a secondary watchdog. When this watchdog times out, a priority 3 interrupt is generated. This differs from the primary watchdog, which will reset the entire CPU.

The secondary watchdog was added primarily to enhance the robustness of the periodic interrupt. However, this watchdog can be used in other ways as well. For example, the secondary watchdog can be used to generate a periodic priority 3 interrupt. This could be used as a high priority tick timer for a System mode preemptive multitasking kernel.

The secondary watchdog could also be used as a watchdog that User mode tasks must service to prevent a System mode kernel from assuming command.

By default the secondary watchdog is disabled. A program enables the secondary watchdog by writing a 0x5F to the Watchdog Timer Control Register (WDTCR: 0x0008).

Resetting (hitting) the secondary watchdog is accomplished by writing 0x5F to the WDTCR. This, in effect, restarts the secondary watchdog.

The secondary watchdog timer is a modulo n+1 counter and is clocked from the 32 kHz oscillator. The Secondary Watchdog Timer Register (SWDTR: 0x000C) holds the time constant "n." Values of n range over 0x00 to 0xFF and correspond to timeout intervals of 30.5 us to 7.8 ms.

PWM Enhancements

The Rabbit 3000A introduces a new PWM interrupt. The PWM peripheral can be setup to generate the interrupt at the end of every cycle, every 2nd cycle, every 4th cycle or every 8th cycle. The configuration registers PWL0R and PWL1R control the interrupt's behavior.

Rabbit Semiconductor has introduced a handy feature for controlling R/C servos. The PWM channels can be configured to suppress seven out of eight pulses. This allows a waveform to be created that will control commercially available R/C servos accurately.

R/C servos expect to see a 1 ms to 2 ms pulse that occurs every 20 ms. The pulse width corresponds to the angular displacement of the servo's shaft. Figure A.2 shows how the Rabbit 3000A's ability to suppress 7-of-8 PWM cycles can be used to produce such a pulse.

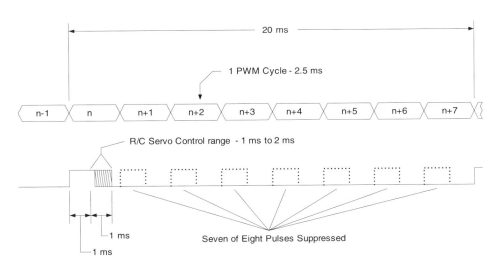

Figure A.2: The Rabbit 3000A has PWM channels capable of generating servo control waveforms.

In addition to being able to suppress 7-of-8 pulses, the PWM channels can also be programmed to suppress 3-of-4 or 1-of-2 cycles.

Appendix A

Quadrature Decoder Enhancements

The Rabbit 3000A has expanded the quadrature decoders from 8-bits to 10-bits. To maintain backward compatibility, upon reset the Rabbit 3000A defaults to an 8-bit mode for the quadrature decoders.

The programmer enables the extended counters by using the formerly reserved bit 5 in the Quadrature Decode Control Register (QDCR: 0x0091). Setting this bit to a 1 will enable the 10-bit mode.

The two-bit extension to each of the two quadrature decoder channels is available in the two new Quadrature Decode Count High Registers, QDC1HR and QDC2HR. The least two significant bits of these registers hold the two-bit extensions to the existing 8-bit counters.

Low Power Enhancements

The Rabbit 3000 provides the option of shortening read chip select strobes when running in a power saving reduced-speed mode. This allowed the Rabbit 3000 to save power by not enabling flash memories or SRAMs for the full period of the slower bus cycles. Since most programs fetch more instructions than they write data, the biggest power savings was realized by implementing the option for read cycles.

The Rabbit 3000A has extended this feature to write strobes. The Global Power Save Control Register (GPSCR: 0x000D) controls both of these features.

External I/O Interface Enhancements

Three I/O enhancements have been added to the Rabbit 3000A. The ability to invert the polarity of an I/O strobe is new. This is an extension of Rabbit's philosophy that adding I/O should ideally require zero glue logic.

The ability to shorten a read strobe by one full clock cycle allows the Rabbit to more efficiently communicate with faster I/O devices as well as provide data hold time for devices that require it. As with the shortened memory strobes discussed above, a shortened read cycle can also save power.

The Rabbit 3000A can direct a strobe to either the alternate I/O bus or the memory bus. This gives system designers added flexibility. As discussed in Chapter 5, moving I/O devices to the alternate I/O bus removes some of the capacitive burden from the high-speed memory bus.

The three new features are controlled with from the existing eight I/O Bank Control Registers (IB0CR..IB7CR). The lower two bits of the IBxCR registers were reserved and written as zeros in the Rabbit 3000. The Rabbit 3000A uses these two bits in each of the IBxCR registers to control the three new I/O enhancements.

Integrated Schmitt Trigger for 32 kHz Oscillator Input

The Rabbit 3000 and Rabbit 3000A have a very low power 32 kHz oscillator. To achieve the remarkable low power operation, the oscillator circuit required a very high-impedance input. One drawback to this arrangement is that noise on the slow changing input can cause false triggering of the oscillator circuit.

The solution to this dilemma has been to add a single Schmitt triggered gate to the oscillator. The CMOS device consumes essentially no power, but the hysteresis offered by the Schmitt triggered gate provides enough noise immunity that the 32 kHz oscillator would run reliably.

The Rabbit 3000A has added a Schmitt trigger to the 32 kHz oscillator input eliminating the need for the external gate.

Alternate output port connections for many peripherals

The Rabbit 3000A provides additional I/O multiplexing options. The following table indicates new options.

Pin Name	New Alternate function
PB6	/ASCS – Alternate slave port chip select input
PG3	APWM0 – Alternate PWM output, bit 0
PG7	APWM1 – Alternate PWM output, bit 1
PD5	APWM2 – Alternate PWM output, bit 2
PD7	APWM3 – Alternate PWM output, bit 3
PG5	ARXE – Alternate serial port E receive
PG4	ARCLKE – Alternate serial port E receive clock (HDLC)
PG1	ARXF – Alternate serial port F receive
PG0	ARCLKF – Alternate serial port F receive clock

About the Authors

Kamal Hyder

Kamal has been fascinated with embedded systems since his school days in India, when it would cost him a month's allowance to buy an issue of *Byte* magazine. After he discovered the Z80 processor and taught himself Z80 assembly language in his school days, he promised himself to live and work in Silicon Valley, where the Z80 was invented. It took him 12 years to achieve that goal and he still lives and works there.

Kamal started his career with an embedded microcontroller manufacturer. He then wrote CPU microcode for Tandem Computers for a number of years, and is now a Product Manager at Cisco Systems, working on next-generation switching platforms. Kamal's bachelor degree is in EE/CS and he has an MBA in finance/marketing.

Bob Perrin

Bob got his start in electronics at the age of nine when his mother gave him a "150-in-one Projects" kit from Radio Shack for Christmas. He grew up programming a Commodore PET. In 1990 Bob graduated with a BSEE from Washington State University. Since then Bob has been working as an engineer designing digital and analog electronics. He has published about twenty technical articles, most with *Circuit Cellar*.

Index

Symbols

#asm debug, 102
#asm root debug, 267
#class auto, 106
#class static, 106
#GLOBAL_INIT, 86, 203
#include, 209
#pragmas, 419
#use, 209
#web, 398
#ximport, 376, 397, 421
8051, 13
9-bit protocol, 42, 247
µC/OS-II, 106, 305, 310, 319

A

abort, 308, 311
accept, 350
accessing I/O ports, 185
active open, 361
active socket, 350
ADC, 321, 361
ADC errors, 144
ADC gain, 150
addressing modes, 417
address bus, 38
address resolution protocol, 335
advanced encryption standard, 337
ALTD prefix, 160, 174, 439
alternate I/O bus, 446
alternate register set, 40, 165
always_on, 307
analog to digital converter, 143
ANSI C, 67, 84
ANSI C compiler, 410
arithmetic operations, 167
assembly window, 103
assert(), 113
asynchronous communication, 245
asynchronous events, 226
asynchronous mode, 42

asynchronous tasks, 298
authentication, 376, 379, 381
auto variables, 71
auxiliary data bus, 126
auxiliary I/O bus, 53, 143
AVR, 20
AVR ATmega128, 33
A HREF tag, 402

B

bash, 321
battery-backed RAM, 65, 420
Berkeley sockets, 349
Berkeley socket API, 348
best-effort service, 336
bind, 350
BIOS code, 70
bipolar junction transistors, 130
BitRdPortE, 188
BitRdPortI, 188
BitWrPortE, 188
BitWrPortI, 188
Bit Set, Reset and Test, 170
blocking, 340
blocking functions, 355
block copy instructions, 174, 439
board-level controllers, 10
bootstrap, 43
breakdown voltages, 121
breakpoint, 66, 83, 102, 268, 427, 428
broadcast, 353
browser interface, 374
buffering, 352
buffer overruns, 99
bugs, 92
build, 413
byte order, 348

C

callback function, 376, 422
calls, 175
    subroutine, 176

Index

capacitive loading, 126
carrier current transmission, 196
carry flag, 161, 440
char promotion to int, 418
checkbox, 382, 403
check for valid battery-backed RAM, 420
circular buffer, 251
clamping, 136
client server paradigm, 348
clock doubler, 57
clock spreader, 57
CMOS latch-up, 271
cofunction, 305, 308, 311, 315
common gateway interface, 375
conditional jumps, 175
congestion, 352
connect, 350
connection-oriented data transfer, 336
connection establishment, 369
console application, 359
constant, 74, 85, 89, 98, 100, 117, 161, 264, 266, 283, 310, 419
consumer task, 300, 303
contact resistance, 136
context switching, 296, 300, 311
cooperative multitasking, 294, 297, 306, 435
coordinated protection, 120
CoPause, 307
CoResume, 307
core module, 12, 17, 65
costatement, 301, 305, 315, 364, 399
counter underflow, 254
CPLD, 253
CPU, 38
   Rabbit versus the Z180, 39
CPU context, 294
CRC, 43, 425
cryptographic math, 439
CTS flow control, 251
custom libraries, 209
C Wrappers, 180

D

DAQ channel, 144
datagram, 350, 369, 372
DATAORG, 71, 73, 74
data acquisition channel, 144
data bus, 38
data encryption standard, 337
data segment, 73
DCRTCP_DEBUG, 343
DCRTCP_STATS, 343
DCRTCP_VERBOSE, 343
deadlock condition, 299, 303, 314
debouncing, 205, 293

debug_on, 344
debugging, 66, 81, 92, 97, 426
   array bounds checking, 106
   custom run-time error handler, 111
   exit(N), 111
   run-time pointer checking, 106
   runtime errors, 106
   step into, 107
   step over, 107
debug functions, 105
debug kernel, 301
decrement instructions, 170
delay, 196
DelayMs, 305
DelaySec, 305
DelayTicks, 305
delta-sigma ($\Delta\Sigma$) converter, 143
demilitarized zone, 338
destination register, 160, 161, 162, 164
DHCP Client Table, 340
DHCP server, 340, 347
disabling interrupts, 233
disassembly, 67
displacement, 163
domain name system, 336
drivers, 26
duty cycle, 51, 315, 393
dynamic addressing, 340, 347, 355
dynamic C, 25, 66
dynamic host configuration protocol, 337
dynamic memory, 434
dynamic web page, 375
dynamic web server, 379

E

edge sensitive interrupts, 229
electro-static discharge, 122, 270
EMI, 12, 41, 54, 57, 270, 272
enabling interrupts, 233
enum, 399
error detection, 352
Ethereal, 408
Ethernet, 342, 387, 436
evaluate expression, 113
event timing, 45
exception handling, 367
exchange instructions, 165
extended memory, 434
extended memory segment, 71
external I/O registers, 173
external interrupts, 229

F

fallback IP addresses, 347
far, 421

452

Index

farfree, 434
farmalloc, 434
farprintf, 433
far pointer, 432
fast accumulator operations, 171
FIFO, 247, 251
file transfer protocol, 336
finite state automata, 19, 272, 289
firewall, 349, 355, 364
firsttime functions, 305, 315
flags register, 440
Flash converter, 143
flow control, 352
fly-back voltage, 136
flyback suppression, 389
format specifier, 402
format string, 433
form element, 402
FPGA, 253
fragmentation, 352
free, 434
freq_divider, 246
functions in XMEM, 72
function description headers, 209

G

galvanic isolation, 140
gas discharge tubes, 119, 271
general purpose registers, 40
generate a CRC at link time, 425
GetVectExtern3000, 232
GetVectIntern, 232
globally optimizing ANSI C, 416
global variables, 87
gnuplot, 321, 325
Gray code, 48
guaranteed delivery, 352
guard expression, 376, 381, 399

H

HDLC, 43, 244, 245
high-voltage transients, 389
high current driver, 130
HTML form, 401
http_handler, 377, 399
http_init, 377, 399
HTTP server, 375
HyperTerm, 247, 428
hypertext markup language, 336, 376
hypertext transfer protocol, 336
hysteresis, 326, 383

I

I/O, 41
ICE, 20
IDE, 66, 410
IDET, 442
ifconfig, 343
immediate address, 162
immediate addressing mode, 161
increment instructions, 170
index.html, 421
indexed addressing, 163, 416
indexed load and store, 162, 164
index register, 41, 163
index with displacement, 163
inductive flyback, 389
inductive load, 138
init_on, 307
inline assembly, 180, 423
input capture, 45, 229
INPUT tag, 402
instructions
    ADC, 168, 169
    ADD, 168, 169
    ALTD, 174
    AND, 168, 169
    BIT, 170
    BOOL, 169
    CALL, 72, 176, 231
    CCF, 178
    CPL, 171
    CP, 168
    DEC, 169, 171
    DJNZ j, 177
    EXX, 40, 166
    EX, 166
    INC, 169, 171
    IOE, 174
    IOI, 174
    IPRES, 178, 231, 234, 239
    JP, 175
    JR, 175
    LCALL, 72, 176
    LDD, 175
    LDDR, 175
    LDI, 175
    LDIR, 175
    LDP, 179
    LD, 161–165, 178
    LJP, 175
    LRET, 176
    LSDDR, 143
    LSIDR, 143
    MUL, 169
    NEG, 171
    NOP, 178
    OR, 168, 169
    POP, 167, 178, 234
    PUSH, 167, 178, 234

Index

RDMODE, 442
RES, 170
RET, 72, 176, 231
RETI, 176, 231, 235
RLA, 171
RLCA, 171
RLC, 172
RL (HL), 169, 172
RRA, 171
RRCA, 171
RRC, 172
RR, 169, 172
RST, 102, 105, 177, 231, 442
SBC, 168, 169
SCF, 178
SETUSR, 442
SET, 170
SLA, 172
SRA, 173
SRL, 173
SUB, 168
SYSCALL, 442
XOR, 168
instruction prefixes, 173
instruction set, 158
Intel386, 31
inter-process communication, 304, 312
inter-task communication, 300
internal I/O registers, 173
internal interrupts, 229
Internet connection firewall, 364
Internet control message protocol, 336
Internet protocol, 335
interrupts, 225, 301
interrupt latency, 234, 302, 304
interrupt levels, 234
interrupt priority, 227, 231, 234, 301, 423
interrupt service routine, 72, 176, 226, 236, 422, 423
interrupt sources, 228, 229
interrupt vector table, 165, 229, 231, 423, 442
IntervalMs, 306
IntervalSec, 306
IntervalTick, 306
IOE prefix, 98, 173
IOI prefix, 98, 173
IP address, 340
IP header, 352
IP register, 234
IrDA, 245

J

JavaScript, 402
Julian date, 148
jumps, 175
  conditional, 175
  unconditional, 175

K

kernel, 298
keywords
  root, 72, 91
  xmem, 72, 91

L

latency, 295
LCD, 10, 285, 435
LCD library, 67, 424
left shift, 172
level sensitive interrupts, 229
LIB.DIR, 209, 210, 345
library, 26, 67, 411, 423
Linux, 321
listen, 350
list box, 383
load and store to immediate address, 162
load immediate data, 161
locks, 295
logical address, 59, 69, 70, 415, 431
logical operations, 167
LonTalk, 7
loop-based delays, 197
loopback address, 350
LV flag, 161, 440
lwIP, 436

M

macro, 71, 98, 104, 106, 111, 113, 301, 341, 415
MAC address, 335, 426
make, 413
malloc, 434
map file, 89, 424
maskable interrupts, 233
match registers, 256
match value, 56
maximum transmission unit, 352
mean time between failure, 141
memory bank control registers, 430
memory dump, 67, 104
memory management unit, 178
memory protection, 443
memory spaces, 88
memory usage without separate I & D space, 69
memory usage with separate I & D space, 73
Micrium, Inc., 436
Micro C OS II (μC/OS-II), 436
MIME table, 398
Mini-ITX, 2
MITS Altair 8080, 157
MIU, 58
MMI, 6, 10

Index

MMU, 40, 58, 415, 427, 430
module header, 209
module key, 209
moisture sensor, 387
MOSFETs, 133
MOV, 271, 387, 389
MS_TIMER, 204, 276, 305
multicast, 353
multiplexed LED display, 211
multitasking, 288
multitasking kernel, 294
mutual wait, 303

N

Nassi Schneiderman, 290
NDEBUG, 113
near, 421
near pointer, 432
Netcat, 358, 366, 409
networking utilities, 406
network address translation, 339
network configuration, 340
network mask, 347
network time protocol, 286
new project wizard, 412
nodebug, 102
nodebug functions, 105
noise free resolution, 150, 156
nonblocking functions, 355
nonmaskable interrupts, 233
NULL-terminated, 100

O

opcode, 159
operands, 159
OTP, 20
out of variable data space, 146
overflow flag, 161, 440

P

packaged controllers, 10
parallel port, 213
parity, 42, 246, 251, 415
passing parameters, 189
passive socket, 350
PC, 272
PC104+, 1, 4
PCLK, 41, 245, 260
peak-to-peak noise, 150, 156
pending interrupts, 233
periodic timer, 423
peripheral clock, 254
physical address, 59, 69, 415, 430, 431
physical memory, 70, 71
PIC, 20

PICMG, 1, 10
ping, 336, 337, 346, 406
PLD, 19
point-to-point protocol, 337, 342
pointer, 41, 94, 97, 100, 106, 183, 188, 190, 193, 195, 307
pointer-based operations, 162
pointer conversion error, 433
polling, 226, 290
port A, 54
port B, 54
port F, 50, 52
port number, 349, 358, 361
POST method, 401
PPP over Ethernet, 342
preemptive multitasking, 294, 298, 310, 435
prefix instructions
    IOE, 161
    IOI, 161
printf(), 67, 104, 416, 433
    printf() and root code, 104
priority based scheduling, 435
priority inversion, 435
priority stack, 231
private address space, 344
privileged instructions, 440
producer task, 300, 303
program counter, 177
protection diodes, 122
prototyping board, 66
public-key calculations, 439
pulse width, 45
    pulse width modulation, 51
PWM frequency, 52

Q

quadrature decoder, 48, 446

R

RabbitWeb, 375
Rabbit versus the Z180, 417
Rabbit WinIDE, 411
race condition, 300, 303
radio button, 382
RAM trace log, 111
random number generator, 290
Raytheon RDS 500, 157
RCM3200, 65, 79
RCM3400, 116, 144, 387
RCM3700, 142
RdPortE, 188
RdPortI, 188
real-time clock, 221, 280
receive buffer, 229, 244
recvfrom, 350
registers

Index

BDCR, 63, 105
DATASEG, 59
EDMR, 442
EIR, 40, 232
GCSR, 245, 254, 256, 260, 281
GPSCR, 446
I0CR, 229
I1CR, 229
IBxCR, 446
ICCR, 45
ICCSR, 45
ICSxR, 45
ICTxR, 45
IIR, 40, 165, 232
IP, 40, 427
IX, 40
IY, 40, 422
MB0CR, 61
MB1CR, 61
MB2CR, 61
MB3CR, 61
MBxCR, 61
MMIDR, 60, 443
MTCR, 63
PBDDR, 186
PCFR, 186
PECR, 186
PEFR, 186
PGCR, 188
PGDCR, 188
PGDDR, 188
PGFR, 188
PWL0R, 445
PWL1R, 445
PWLxR, 52
PWM, 258, 393, 445
PWMxR, 52
QDC1HR, 446
QDC2HR, 446
QDCR, 439, 446
QDCSR, 50
RAMSR, 444
RTCCR, 281
SADR, 244
SASR, 244, 251
SBDR, 247
SBSR, 247
SEGSIZE, 59
shadow, 186
SP, 427
STACKSEG, 59
STKCR, 444
STKHLR, 444
STKLLR, 444
SU, 441

SWDTR, 445
SxAR, 247, 251
SxCR, 246
SxDR, 247, 251
SxLR, 247, 251
TACR, 245, 254, 259
TACSR, 55, 244, 254, 259
TAPR, 245, 254, 265
TAT1R through TAT7R, 254
TAT9R, 52
TATxR, 245
TBCR, 254, 256, 261, 282
TBCSR, 254, 259, 261, 263
TBL1R, 254, 261, 284
TBL2R, 254
TBLxR, 259
TBM1R, 254, 261, 284
TBM2R, 254
TBMxR, 259
TMRA1 through TMRA7, 254
WDTCR, 276, 445
WPCR, 443
WPHR, 443
WPLR, 443
WPSAHR, 444
WPSALR, 444
WPSAR, 444
WPSBHR, 444
WPSBLR, 444
WPSBR, 444
XPC, 40, 422, 427, 431
register addressing, 164
register to register moves, 164
register window, 67, 104
relays, 136
reliable data transfer, 336
reliable packet delivery, 352
remote boot, 245
retinal retention, 212
returns from subroutines, 176
return from interrupt, 176
Reverse ARP, 335
right shift, 172
root, 420, 422
root code, 70, 73, 104
root constants, 70
root data, 70, 73
root memory, 70
root segment, 71, 73
rotates, 171
round-robin, 228, 290, 299
round-robin multitasking, 436
round trip time, 352
RS-232, 7, 42, 124, 244
RS-485, 7, 9, 26, 42, 124, 244, 334

Index

RTC, 27, 290, 301, 322, 395, 405
RTOS, 288, 295, 435
RTS flow control, 251
run-time errors, 109, 329

S

scheduler, 298
Schottky diodes, 122
scofunc, 308
scofunctions, 309, 311
SDLC, 43, 244, 245
SEC_TIMER, 204, 276, 282, 305
secondary watchdog, 444
secure HTTP, 337
secure socket layer, 337, 405
segments, 59, 425
   base segment, 59, 70
   data segment, 59, 69, 73
   extended memory segment, 59, 69, 71
   root segment, 59, 69, 71, 73
   stack segment, 69, 71
   XPC segment, 59, 70
selection variable, 399
SELECT tag, 401
semaphores, 295, 299, 319
sendto, 350
sensor routine, 374
sensor task, 326, 328
separate instruction and data space, 60, 329, 444
sequencing, 352
serial communication, 244, 301
serial interrupts, 301
serial ports, 42, 245
   interrupt-driven, 249
   polled, 246
SetVectExtern3000, 232, 238
SetVectIntern, 232, 238
seven segment display, 212
shadow register, 186, 428
shared, 319
shifts, 171
sign flag, 161, 440
simple mail transfer protocol, 336
simple network management protocol, 337
simultaneous sampling, 126
single stepping, 67, 82, 107, 268, 427
slave port, 43
slice, 310, 315, 318
slicing, 305
SMODE0, 43
SMODE1, 43
snubber leakage, 138
snubber network, 137
sock_alive, 354
sock_awrite, 354
sock_bytesready, 354
sock_close, 354
sock_established, 354
sock_fastread, 354
sock_fastwrite, 354
sock_gets, 354
sock_init, 342, 353, 377
sock_puts, 354
sock_read, 355
sock_readable, 354
sock_writable, 354
sock_write, 355
sockets API, 349
socket connection, 351
Softools, Inc., 25, 410
Softools Rabbit WinIDE, 410
software interrupts, 231
solid-state relay, 138, 140, 387
spark gap suppressors, 119
SPI, 244, 245
sprinkler controller, 384
square-wave, 393
stack-violation interrupt, 444
stack manipulation instructions, 166
STACK segment, 71, 425
stack window, 67, 103
standalone assembly, 180
state machine, 331
static addressing, 340, 355
static IP address, 346
static variables, 106
static web page, 375
status register, 161, 440
stdio window, 67, 80, 347
step into, 82
step over, 82
stream, 350
streaming, 353
stream reader, 359
structure, 70, 163, 183, 281, 286, 293, 307, 309, 310, 323
stty, 147
stuck interrupt, 228
sub-timers, 254
subroutine calls, 176
successive approximation (SAR) converter, 143
suppression diode, 136
synchronous communication, 245
synchronous mode, 42
synchronous tasks, 298
system mode, 438
system violation interrupt, 442

T

table lookup, 194
target communication error, 107

457

Index

task, 288, 295
task priority, 299, 312
task setup, 303
task synchronization, 304
TCP/IP, 336, 425, 436
tcp_config.lib, 341
tcp_extopen, 354
tcp_listen, 354
tcp_open, 354
tcp_tick, 343
TCPCONFIG, 341
TCP client, 358, 361
TCP server, 355
TCP socket buffers, 343
TCP state machine, 355, 364
TCP utilities, 357
Telnet, 337, 358
temperature probe, 387
thermistor, 328, 379
threads, 288
three-way handshake, 350
tick, 52, 53, 226, 228, 253, 256, 298, 310
TICK_TIMER, 204, 276, 305
Timer A, 54
Timer A interrupt, 259
Timer A periods, 254
Timer B, 56
Timer B interrupt, 261
Timer B periods, 256
timer interrupts, 254
timer synchronization, 41
timestamp, 55
time slice, 298, 300
toggle breakpoint, 83
traceroute, 407
transient voltage, 120
transient voltage suppressor, 122
transmission control protocol, 336
transmit buffer, 229, 244
traps, 231
trivial file transfer protocol, 336
troubleshooting, 92
try/catch exception handling, 360, 367
TurboTask, 436

U

UARTs, 35
uCOS/II, 16
UDP, 336
  udp_recv, 354
  udp_recvfrom, 354
  udp_send, 354
udp_sendto, 354
UDP client, 372
UDP header, 353
UDP server, 369
UDP socket buffers, 343
unconditional jumps, 175
unterminated strings, 99
USE_RABBITWEB, 377, 397
user datagram protocol, 336
user interface, 395
user mode, 438

V

VERBOSE, 344, 364
Virtual Local Area Networks, 338
virtual watchdog, 274, 301
visual persistence, 212, 218
VLAN, 338

W

waitfordone, 308, 311
waitfor statements, 305, 308, 311
watchdog timer, 270, 423, 444
watch expressions, 67, 83, 102
watch window, 83, 427
web browser, 398
web server, 376, 398
wfd, 308
WinIDE, 411
WinIDE LIB files, 411
Winsock library, 360
write-protection interrupt, 444
WrPortE, 188
WrPortI, 188

X

X10 protocol, 196
xalloc(), 146
xgetint(), 146
xmem, 71, 286, 413, 434
xsetint(), 146

Y

yield, 308, 310, 311

Z

Z180, 37, 40, 42, 410, 417, 423
Z80, 37, 40, 410, 415, 431
zero crossing, 141
zero flag, 161, 440
ZHTML, 380, 398, 400
zhtml_handler, 400

GET YOUR HANDS ON A RABBIT.

Rabbit Semiconductor features a complete line of Development Kits that include a Microprocessor Core Module, hardware tools, and Dynamic C® development software.

Everything you need to get started!

We are offering specially priced Development Kits for readers of *Embedded Systems Design Using the Rabbit 3000 Microprocessor*.

Visit www.rabbitsemiconductor.com/books for more information.

Use promotional code 9067.

CIRCUIT CELLAR®

THE MAGAZINE FOR COMPUTER APPLICATIONS

And for our next trick...

No, we won't be pulling a rabbit out of a hat. But we'll certainly show you how design engineers across the globe are using microprocessors like the Rabbit 3000 in applications that look like magic.

If you're the type of person who needs to know all the tricks of the trade, then you're in luck. For more than 16 years, *Circuit Cellar* magazine has provided a forum where the most groundbreaking product designs in the embedded systems industry are published in detail. Our monthly magazine brings readers innovative fabrication and enterprise solutions direct from the field, complete with the schematics and code. With *Circuit Cellar* you'll get articles that entertain and educate with in-depth analysis and honest opinions.

If you haven't taken advantage of this information pipeline, you won't want to miss this special offer.

To help you further explore the design innovations of your peers, *Circuit Cellar* will include a full year of its online Electronic Edition publication, a $15 value, at no extra charge with a fully paid print subscription to the magazine. The Electronic Edition is a mirror version of *Circuit Cellar*'s print magazine in PDF format. The Electronic Edition provides the flexibility to, for example, download issues to your laptop or make your own archive CD-ROMs. With the Electronic Edition, you can search articles by keywords and access additional information by clicking on helpful hyperlinks. You can even download the issue one article at a time.

Lock in your bonus subscription today!

Visit www.circuitcellar.com/specials/rabbit.htm

SAVE 10% ON YOUR PURCHASE OF THESE POWERFUL DEVELOPMENT TOOLS FOR RABBIT!

Softools, Inc.
Rabbit WinIDE Development Tools

THIS COUPON GOOD FOR 10% OFF THE PURCHASE PRICE OF A LICENSE TO THE TOOLS

Fax: 860-236-4202

See www.softools.com/contact.htm for mailing address.

Credit Card No. _____ Expires: ___ ___ / ___ ___

Cardholder Name: _____

CARD STATEMENT ADDRESS:

Street: _____

City: _____ State: _____ Zip/Postal Code: _____

Phone: _____

Email: _____

Signature: _____

TERMS: Purchase must be made directly from Softools, Inc. • Offer valid on the purchase of one (1) single user license. • Offer valid only for one coupon per address or business. • Cannot be combined with other offers. • Offer expires December 15, 2005. • Completely filled out coupon must be mailed or faxed to Softools at the time of purchase.

VISIT WWW.SOFTOOLS.COM FOR MORE INFORMATION ON THESE AND OTHER DEVELOPMENT TOOLS

ELSEVIER SCIENCE CD-ROM LICENSE AGREEMENT

PLEASE READ THE FOLLOWING AGREEMENT CAREFULLY BEFORE USING THIS CD-ROM PRODUCT. THIS CD-ROM PRODUCT IS LICENSED UNDER THE TERMS CONTAINED IN THIS CD-ROM LICENSE AGREEMENT ("Agreement"). BY USING THIS CD-ROM PRODUCT, YOU, AN INDIVIDUAL OR ENTITY INCLUDING EMPLOYEES, AGENTS AND REPRESENTATIVES ("You" or "Your"), ACKNOWLEDGE THAT YOU HAVE READ THIS AGREEMENT, THAT YOU UNDERSTAND IT, AND THAT YOU AGREE TO BE BOUND BY THE TERMS AND CONDITIONS OF THIS AGREEMENT. ELSEVIER SCIENCE INC. ("Elsevier Science") EXPRESSLY DOES NOT AGREE TO LICENSE THIS CD-ROM PRODUCT TO YOU UNLESS YOU ASSENT TO THIS AGREEMENT. IF YOU DO NOT AGREE WITH ANY OF THE FOLLOWING TERMS, YOU MAY, WITHIN THIRTY (30) DAYS AFTER YOUR RECEIPT OF THIS CD-ROM PRODUCT RETURN THE UNUSED CD-ROM PRODUCT AND ALL ACCOMPANYING DOCUMENTATION TO ELSEVIER SCIENCE FOR A FULL REFUND.

DEFINITIONS

As used in this Agreement, these terms shall have the following meanings:

"Proprietary Material" means the valuable and proprietary information content of this CD-ROM Product including all indexes and graphic materials and software used to access, index, search and retrieve the information content from this CD-ROM Product developed or licensed by Elsevier Science and/or its affiliates, suppliers and licensors.

"CD-ROM Product" means the copy of the Proprietary Material and any other material delivered on CD-ROM and any other human-readable or machine-readable materials enclosed with this Agreement, including without limitation documentation relating to the same.

OWNERSHIP

This CD-ROM Product has been supplied by and is proprietary to Elsevier Science and/or its affiliates, suppliers and licensors. The copyright in the CD-ROM Product belongs to Elsevier Science and/or its affiliates, suppliers and licensors and is protected by the national and state copyright, trademark, trade secret and other intellectual property laws of the United States and international treaty provisions, including without limitation the Universal Copyright Convention and the Berne Copyright Convention. You have no ownership rights in this CD-ROM Product. Except as expressly set forth herein, no part of this CD-ROM Product, including without limitation the Proprietary Material, may be modified, copied or distributed in hardcopy or machine-readable form without prior written consent from Elsevier Science. All rights not expressly granted to You herein are expressly reserved. Any other use of this CD-ROM Product by any person or entity is strictly prohibited and a violation of this Agreement.

SCOPE OF RIGHTS LICENSED (PERMITTED USES)

Elsevier Science is granting to You a limited, non-exclusive, non-transferable license to use this CD-ROM Product in accordance with the terms of this Agreement. You may use or provide access to this CD-ROM Product on a single computer or terminal physically located at Your premises and in a secure network or move this CD-ROM Product to and use it on another single computer or terminal at the same location for personal use only, but under no circumstances may You use or provide access to any part or parts of this CD-ROM Product on more than one computer or terminal simultaneously.

You shall not (a) copy, download, or otherwise reproduce the CD-ROM Product in any medium, including, without limitation, online transmissions, local area networks, wide area networks, intranets, extranets and the Internet, or in any way, in whole or in part, except that You may print or download limited portions of the Proprietary Material that are the results of discrete searches; (b) alter, modify, or adapt the CD-ROM Product, including but not limited to decompiling, disassembling, reverse engineering, or creating derivative works, without the prior written approval of Elsevier Science; (c) sell, license or otherwise distribute to third parties the CD-ROM Product or any part or parts thereof; or (d) alter, remove, obscure or obstruct the display of any copyright, trademark or other proprietary notice on or in the CD-ROM Product or on any printout or download of portions of the Proprietary Materials.

RESTRICTIONS ON TRANSFER

This License is personal to You, and neither Your rights hereunder nor the tangible embodiments of this CD-ROM Product, including without limitation the Proprietary Material, may be sold, assigned, transferred or sub-licensed to any other person, including without limitation by operation of law, without the prior written consent of Elsevier Science. Any purported sale, assignment, transfer or sublicense without the prior written consent of Elsevier Science will be void and will automatically terminate the License granted hereunder.

TERM

This Agreement will remain in effect until terminated pursuant to the terms of this Agreement. You may terminate this Agreement at any time by removing from Your system and destroying the CD-ROM Product. Unauthorized copying of the CD-ROM Product, including without limitation, the Proprietary Material and documentation, or otherwise failing to comply with the terms and conditions of this Agreement shall result in automatic termination of this license and will make available to Elsevier Science legal remedies. Upon termination of this Agreement, the license granted herein will terminate and You must immediately destroy the CD-ROM Product and accompanying documentation. All provisions relating to proprietary rights shall survive termination of this Agreement.

LIMITED WARRANTY AND LIMITATION OF LIABILITY

NEITHER ELSEVIER SCIENCE NOR ITS LICENSORS REPRESENT OR WARRANT THAT THE INFORMATION CONTAINED IN THE PROPRIETARY MATERIALS IS COMPLETE OR FREE FROM ERROR, AND NEITHER ASSUMES, AND BOTH EXPRESSLY DISCLAIM, ANY LIABILITY TO ANY PERSON FOR ANY LOSS OR DAMAGE CAUSED BY ERRORS OR OMISSIONS IN THE PROPRIETARY MATERIAL, WHETHER SUCH ERRORS OR OMISSIONS RESULT FROM NEGLIGENCE, ACCIDENT, OR ANY OTHER CAUSE. IN ADDITION, NEITHER ELSEVIER SCIENCE NOR ITS LICENSORS MAKE ANY REPRESENTATIONS OR WARRANTIES, EITHER EXPRESS OR IMPLIED, REGARDING THE PERFORMANCE OF YOUR NETWORK OR COMPUTER SYSTEM WHEN USED IN CONJUNCTION WITH THE CD-ROM PRODUCT.

If this CD-ROM Product is defective, Elsevier Science will replace it at no charge if the defective CD-ROM Product is returned to Elsevier Science within sixty (60) days (or the greatest period allowable by applicable law) from the date of shipment.

Elsevier Science warrants that the software embodied in this CD-ROM Product will perform in substantial compliance with the documentation supplied in this CD-ROM Product. If You report significant defect in performance in writing to Elsevier Science, and Elsevier Science is not able to correct same within sixty (60) days after its receipt of Your notification, You may return this CD-ROM Product, including all copies and documentation, to Elsevier Science and Elsevier Science will refund Your money.

YOU UNDERSTAND THAT, EXCEPT FOR THE 60-DAY LIMITED WARRANTY RECITED ABOVE, ELSEVIER SCIENCE, ITS AFFILIATES, LICENSORS, SUPPLIERS AND AGENTS, MAKE NO WARRANTIES, EXPRESSED OR IMPLIED, WITH RESPECT TO THE CD-ROM PRODUCT, INCLUDING, WITHOUT LIMITATION THE PROPRIETARY MATERIAL, AN SPECIFICALLY DISCLAIM ANY WARRANTY OF MERCHANTABILITY OR FITNESS FOR A PARTICULAR PURPOSE.

If the information provided on this CD-ROM contains medical or health sciences information, it is intended for professional use within the medical field. Information about medical treatment or drug dosages is intended strictly for professional use, and because of rapid advances in the medical sciences, independent verification of diagnosis and drug dosages should be made.

IN NO EVENT WILL ELSEVIER SCIENCE, ITS AFFILIATES, LICENSORS, SUPPLIERS OR AGENTS, BE LIABLE TO YOU FOR ANY DAMAGES, INCLUDING, WITHOUT LIMITATION, ANY LOST PROFITS, LOST SAVINGS OR OTHER INCIDENTAL OR CONSEQUENTIAL DAMAGES, ARISING OUT OF YOUR USE OR INABILITY TO USE THE CD-ROM PRODUCT REGARDLESS OF WHETHER SUCH DAMAGES ARE FORESEEABLE OR WHETHER SUCH DAMAGES ARE DEEMED TO RESULT FROM THE FAILURE OR INADEQUACY OF ANY EXCLUSIVE OR OTHER REMEDY.

U.S. GOVERNMENT RESTRICTED RIGHTS

The CD-ROM Product and documentation are provided with restricted rights. Use, duplication or disclosure by the U.S. Government is subject to restrictions as set forth in subparagraphs (a) through (d) of the Commercial Computer Restricted Rights clause at FAR 52.22719 or in subparagraph (c)(1)(ii) of the Rights in Technical Data and Computer Software clause at DFARS 252.2277013, or at 252.2117015, as applicable. Contractor/Manufacturer is Elsevier Science Inc., 655 Avenue of the Americas, New York, NY 10010-5107 USA.

GOVERNING LAW

This Agreement shall be governed by the laws of the State of New York, USA. In any dispute arising out of this Agreement, you and Elsevier Science each consent to the exclusive personal jurisdiction and venue in the state and federal courts within New York County, New York, USA.